TEACHER'S GUIDE FOR

Chemistry

WILLIAM L. MASTERTON
University of Connecticut, Storrs, Connecticut

EMIL J. SLOWINSKI
Macalester College, St. Paul, Minnesota

EDWARD T. WALFORD
Cheyenne Mountain High School, Colorado Springs, Colorado

HOLT, RINEHART AND WINSTON, PUBLISHERS
New York • Toronto • London • Sydney

Printed in the United States of America
ISBN 0-03-056242-2
3 4 026 9 8 7 6 5 4

Contents

GUIDE TO THE LABORATORY

Experiment

Introduction to the Teacher's Guide

ORGANIZATION

The Teacher's Guide is divided into two parts, the Guide to the Text, which follows the Introduction, and the Guide to the Laboratory, which appears in the second part of the Guide. It is the author's intent that the Guide should help to smooth the way in what is a most demanding (and rewarding) profession—the teaching of high school chemistry. In turn, the authors would welcome your suggestions for additions or changes to the Guide.

ABOUT THE TEXT

The text, **Chemistry,** is designed for the mainstream course in high school chemistry. It is suitable for students who will not pursue the subject further as well as those who go on to college chemistry. The principles covered have been restricted to those which are basic to a beginning course. These principles are developed in considerable detail from an elementary point of view. More advanced ideas are treated in optional sections set off from the rest of the text. These may be covered at the discretion of the teacher, but are not built upon in subsequent chapters.

A major goal of the authors has been to write a text which is both readable and understandable by the students. We hope that the discussions leading to principles are thorough and logical. Numerous examples with solutions are provided for students to test their understanding of these principles. Each chapter is concluded with a summary, which reviews the major ideas, and a list of new terms. Each of these terms appears in the chapter and is specifically defined in a glossary in the appendix.

An extensive set of questions and problems is provided at the end of each chapter. The questions are an excellent source of material for classroom discussions. Each problem set begins with a subset of problems which test a specific listed principle (Simplest Formulas, Mass Relations, Molarity, etc.). Each of these problems is keyed to a numbered example of a similar problem within the chapter. In this way the student can refer to the example's solution for help. The answers to this subset are in the appendix. The second subset contains a mixture of problems, without answers, which relate to the objectives of the chapter. A final subset, marked with an asterisk, contains more advanced problems, some of which may relate to the optional sections. The answers to this subset are also included in the appendix.

The text can be considered to consist of seven units: Chapters 1–4, Chapters 5–8, Chapters 9–13, Chapters 14–16, Chapters 17–18, Chapters 19–23, and Chapters 24–27. Descriptive chemistry, both inorganic and organic, is emphasized throughout the book. Of the 27 chapters, 12 are primarily descriptive. These are distributed throughout the book and are designed to apply and illustrate chemical principles. While the material is essentially sequential, a certain degree of freedom is available. Particular chapters, or subjects within chapters, can be emphasized or not, depending upon the desires of the teacher (or students). No attempt is made to provide a rigid time schedule as most schools and classes present unique teaching situations.

Chapters 1–4 build the foundation for the course. After an introduction, Chapter 1 describes the important properties of matter and how they are measured. In Chapter 2 the student is introduced to the structure of matter and the mole. Chapter 3 begins a study of chemical reactions and how they are described using the language of chemistry. The energy changes associated with reactions are covered in Chapter 4. The concepts presented in this unit are developed at an elementary level, but are carried far enough to provide the necessary background for future work.

Chapters 5–8 describe and organize the chemistry of the more important elements and their compounds. The unit begins by introducing the physical and chemical behavior of gases (Chapters 5 and 6). In Chapter 7 the periodic table is developed from the properties of elements and used to explain and predict trends. The periodic table is then used to investigate the chemistry of the main group elements (Chapter 8).

Chapters 9–13 consider the structure of matter and the forces which hold it together. Chapter 9 develops the electronic structure of atoms and relates it to the periodic table. This leads to a discussion of chemical bonding (ionic and covalent) in Chapters 10 and 11. At this stage students have sufficient background to investigate liquids and solids (Chapter 12) and solutions (Chapter 13).

The unit of Chapters 14–16 is devoted to the chemistry of carbon compounds. Too often this material is squeezed into a single chapter and relegated to the rear of the text. In fact, organic chemistry is often the most interesting and relevant part of the course to the student. Chapter 14 covers hydrocarbons (and fuels) while the important organic oxygen compounds are introduced in Chapter 15. The material of these chapters is then used to introduce organic polymers (Chapter 16). Carbohydrates and proteins are discussed there as well as the synthetic polymers.

Chapters 17 and 18 are a small unit which serves to develop the ideas related to how and why reactions take place. Rate concepts are discussed in Chapter 17 and equilibrium concepts in Chapter 18. These concepts are vital to the understanding of the following unit on chemical reactions. While the mathematics associated with this material can be quite complex, the text restricts itself to a qualitative approach and to the simpler mathematical relationships.

Chapters 19–23 classify and investigate the different types of chemical reactions: acid-base (Chapter 19), precipitation (Chapter 20), and oxidation-reduction (Chapter 23). Chapters 21 and 22 show how these reactions are used in quantitative and qualitative analysis.

Chapters 24–27 cover some practical applications of the principles which have been previously presented. While some may consider this material to be "icing on the cake," we hope that most teachers and students will take the time to sample the icing. Electrochemistry (Chapter 24) and Metals and Their Ores (Chapter 25) expand the coverage of metals and relate to the important subject of mineral resources. The chemical aspects of environmental pollution and the "energy crisis" are investigated in the final chapters (Chapters 26 and 27).

ABOUT THE GUIDE TO THE TEXT

The guide to the text is divided into chapters which correspond to the text chapters. Each chapter is subdivided into the following six sections:

I	Basic Skills	IV	Suggested Activities
II	Chapter Development	V	Answers to Questions
III	Problem Areas	VI	Solutions to Problems

Teachers are advised to read through Sections I–IV carefully when using the text for the first time.

The *Basic Skills* section lists the things that students should be able to do by the completion of the chapter. (You may prefer to refer to them as "performance goals" or "behavioral objectives.") The section is subdivided into qualitative and quantitative skills. You can use it as a checklist of the major points to be covered. It also serves as a guide in the preparation of examination questions. Some teachers will want to give their students a copy of this section for their own use. (NOTE: Do not despair if all students do not master all skills.)

The *Chapter Development* section tries to explain and justify the particular approach taken in the chapter. The section is also used to point out how far a concept is developed and where related material will be covered again in later chapters.

The *Problem Areas* section attempts to warn teachers in advance where to expect teaching and learning problems. For new teachers, many of these problem areas are totally unexpected. In most cases, advice is also given in this section on how to solve, or minimize, the problem.

The *Suggested Activities* section gives detailed instructions on activities which relate to the chapter material. The activity may be a traditional teacher demonstration or may be a more student-involved activity. The list is not intended to be complete; many standard references on this subject are available. Teachers should also consider the Laboratory Manual as a source of possible activities (when the experiment is not done by the class). In all activities, the student should be involved as much as possible and the activity should be tested in advance.

The last two sections include *answers to questions* and *solutions to problems*. In most cases, detailed answers to the questions are given. Complete solutions, not just answers, are given for each problem. Additional explanation is given when a difficult-to-understand line of reasoning is involved.

CHAPTER 1

An Introduction to Chemistry

I. Basic Skills

A. Qualitative Students should be able to:

1. Distinguish among the classes of matter and give examples of each (Table 1.2).
2. Distinguish between physical and chemical properties and give examples of each.
3. Describe how differences in properties are used to separate pure substances in a mixture and to identify the substances.
4. Describe the techniques and units used in measuring length, volume, mass and temperature.
5. a. Understand the concept of density in terms of mass per unit volume.
   b. Describe the techniques of determining the density of a liquid or solid.
6. a. Learn the common prefixes of the metric system (kilo, centi, milli, and nano).
   b. Learn a metric-English conversion factor for length, volume, and mass (Table 1.4).
7. Explain the use of conversion factors in the unit analysis method when converting from one unit to a different unit.
8. a. Explain why all measurements have a precision error.
   b. Explain what is meant by a significant figure.
   c. Relate the sensitivity of an instrument to the number of significant figures obtained in a measurement (Table 1.5).

B. Quantitative Students should be able to:

1. Calculate one of the three quantities, density, mass, or volume, given the other two (Equation 1.1 and Example 1.1).
2. Convert temperatures between the Celsius and Fahrenheit scales (Equation 1.2 and Example 1.2).
3. a. Use the prefixes of the metric system to make conversions within the metric system.
   b. Apply unit analysis to convert measurements between the metric and English systems (Examples 1.3 and 1.4).
4. a. Give the number of significant figures in a measured quantity (Example 1.5).
   b. Determine the number of significant figures in a calculated value obtained by multiplication or division (Example 1.6).
   c. Determine the number of significant figures in a calculated value obtained by addition or subtraction (Example 1.7).

5. a. Express numbers in exponential notation or correct back to ordinary numbers (Appendix 1).
   b. Multiply and divide exponential numbers (Appendix 1).

II. Chapter Development

1. The development of science, and chemistry as a science, is briefly explained. What is sometimes referred to as the "scientific method" is explained as an experimental approach to answering questions of nature. You may wish to expand on the history of science and the processes of science by outside readings. The relevance of chemistry to society is introduced and should be emphasized throughout the course.
2. After the classification of matter, the student is led to the problem of separating and identifying the components of a mixture. This task should be related to the differences in properties of different substances. Only simple separation and identification methods are discussed at this point. Fractional distillation and colorimetric methods are discussed later. The point should be made that the separation and identification of pure substances is a major task of chemists. This theme is repeated throughout the text and is particularly important in quantitative analysis (Chapter 21) and qualitative analysis (Chapter 22).
3. The principles of chromatography are briefly discussed in the first optional section. The material in this section, and all optional sections which follow, is not essential to the continuity of the text. Optional sections may be assigned or not, at the discretion of each teacher, based on personal preferences and interests. They may also be used as extra material for the advanced class or student.
4. The choice of units used throughout the text is a compromise. The authors do not believe that it is practical at this time to convert completely to the *International System of Units (SI)*. The meter, cubic meter, and kilogram are not, as a rule, convenient laboratory units. Mass measurements are commonly made in grams and length measurements in centimeters or millimeters. In volume measurements the cubic centimeter (cm^3) is used in preference to the milliliter (ml). Yet, at the same time, the convenience of the liter has been retained. Energy units will be discussed in Chapter 4 and pressure units in Chapter 5. The use of the Kelvin temperature scale is also deferred until Chapter 5.
5. The determination of properties requires quantitative measurements. Stress that all measurements contain a precision error and show how the error is related to the correct use of significant figures. The discussion of accuracy error (experimental error) is covered in Experiment 1.
6. The unit analysis method is introduced and used in the conversions of units. We are beginning to see this method of problem solving taught in earlier science courses and even in some math courses. Do not assume previous knowledge, however. It is important to make a good beginning with this method in Chapter 1. Stress the proper use of units and their cancellation. Unit analysis is used throughout the text, in discussions and in the problem examples.

III. Problem Areas

1. Additional help is provided in the Appendix for students who have trouble using exponents.
2. While most students will have had previous instruction in the metric system, they will still need encouragement in its use. Have students calculate familiar quantities such as their height, weight, or distance from school. Point out the use of metric units in sports, automobiles, beverages and foods, etc.
3. In equations such as $D = m/V$, some students will have difficulty in solving for m or V. Cover the algebra carefully here as this type of problem will occur frequently.
4. Do not expect all students to be proficient in the unit analysis method by the end of Chapter 1. Maintain a steady push in this direction and try to achieve proficiency by the end of Chapter 3.
5. The use of significant figures is always a problem area. The text, Activities 3 and 4, and Experiment 1 all make a case for their use. The major stress on significant figures should be on their correct use in the collection and handling of experimental data in the laboratory.

IV. Suggested Activities

1. Elements, Compounds and Mixtures:
   Exhibit samples of the different classes of matter. Suggested examples are copper, lead, sulfur, iodine, bromine (sealed in flask), sugar, copper sulfate, water, copper sulfate solution, denatured alcohol, rocks, and dirt. Have students classify each sample.
2. Metric System of Measurement:
   Exhibit a meter stick or rule, beakers of various volumes to one liter, and objects of labeled mass to one kilogram. Relate the metric units to their approximate equivalents in the English system.
3. Precision and Significant Figures:
   Using a multiscale (four-sided) meter stick, measure the height of a student (or width of a desk) beginning with the least precise scale. Repeat the measurement with each scale of increasing precision. Point out that each measurement has an uncertainty of $\pm$ 1 unit in the last digit.
4. Calculators and Significant Figures:
   Have each student (or a minimum of five) measure the length, width, and thickness of a textbook. Use a metric rule marked in millimeters and estimate each measurement to 0.1 mm. Have each student calculate the book's volume using an 8-digit calculator. Record all answers on the board. Point out why the number of digits in an answer should be limited when using a calculator. Show how the correct number of digits can be obtained by using the significant figure rules.

5. Separation Methods:
   Demonstrate separation methods by separating the components of a simple mixture such as white sand and salt (a handy technique for the beach!). Add 10 cm^3 of water to a few grams of the mixture and shake to dissolve the salt. Filter the mixture to obtain the sand and a solution. Heat the solution and evaporate the water to obtain the salt. Explain that separations are based on differences in properties; salt is more soluble than sand (in water) and water has a lower boiling point than salt.

V. Answers to Questions

1. a. Lavoisier performed experiments which required quantitative measurements. The Greek philosophers reasoned from arbitrary theories and did not resort to experimentation.
   b. While the alchemists did perform experiments, they often failed to control conditions or to make quantitative measurements. As a result, their experiments were not always reproducible.
2. In doing this experiment, the student most closely resembles the alchemists.
3. Fill two ice cube trays, one with cold water and the other with hot water. Place both trays in similar locations in a freezer. Measure the time required for each liquid to freeze.
4. a. mixture (solution)
   b. pure (element)
   c. mixture
   d. pure (compound)
   (mixture, if iodized)
   e. pure (element)
   f. mixture (solution, if beverage)
   g. pure (compound)
5. Compounds: sugar, salt, baking soda, starch, water (if pure)
   Solutions: vinegar, corn oil, soft drinks, household ammonia, air
6. The answer to this question will vary depending on the background (and imagination) of the student. Some possible answers are:
   helium (balloons), carbon (graphite), neon (lights), copper (wire), mercury (thermometers)
7. a. Decompose water or make from its elements.
   b. Evaporate the water from the salt.
   c. Examine visually.
8. a. Add water to dissolve the sugar and then filter out the charcoal.
   b. Evaporate the water leaving the sugar behind as a solid.
   c. Separate by difference in density. When washed with water, the denser gold falls to the bottom first.
   d. Use a magnet and separate out the magnetic iron filings.
   e. If you can't tell by looking, try an odor test.
9. a. distillation b. recrystallization c. filtration
10. No. If the solubility of a substance does not vary with temperature, the substance will remain in solution regardless of the solution's temperature.

11. Colorless, odorless, tasteless, density of 1 g/cm^3, boiling point of 100°C, and freezing point of 0°C. Less likely answers might refer to water's solubility, viscosity, surface tension, conductivity, and specific heat.
12. a. same b. same c. different d. same
13. Physical properties:

    Density: 2.9 g/ℓ at 25°C, 1 atm

    Boiling point: −34°C

    Melting point: −101°C

    Color: greenish-yellow (gas)

    Chemical properties:

    Reacts with calcium to form a white solid.
    Reacts with hydrogen to form gaseous hydrogen chloride.
14. The element could be aluminum, chromium, copper, gold, iron, magnesium, nickel, or silver. It could be identified by measuring the element's density. If copper or gold, the metal would be colored.
15. a. Using a buret or pipet, deliver an exact volume of alcohol to a previously weighed container. Measure the mass of the container and alcohol.
    b. After measuring the mass of an aluminum sample, measure the volume of water displaced by the sample in a graduated cylinder.
16. Assuming a regular shape, measure the wood and obtain its volume by calculation. As salt is soluble in water, measure its volume displacement in a liquid in which the salt is insoluble.
17. In a refrigerator, heat flows from the food to the refrigerator. In a wood stove, heat flows from the stove to the room. In a sauna, heat flows from the air to the body.
18. a. 10^3 b. 10^{-3} c. 10^{-9} d. 10^{-2}
19. Milk by the liter would be the better buy as it is equivalent to 1.06 quarts.
20. The metric ton has the greater mass as it is equivalent to 2,200 pounds.
21. a. This number could be exact or contain an uncertainty depending upon its size and method of counting.
    b. uncertain c. uncertain d. exact e. uncertain at this time
22. Significant figures are all those which have been measured and hence are experimentally meaningful.

VI. Solutions to Problems

1. $$D = \frac{m}{V} = \frac{4.40\ \text{g}}{5.00\ \text{cm}^3} = 0.880\ \text{g/cm}^3$$

2. $$V = \frac{m}{D} = \frac{1.00\ \text{g}}{0.880\ \text{g/cm}^3} = 1.14\ \text{cm}^3$$

3. a. $°F = 1.8(°C) + 32$

$°F = 1.8(112) + 32 = 234$

b. $°C = \dfrac{°F - 32}{1.8} = \dfrac{68 - 32}{1.8} = 20$

4. $\text{Distance (km)} = 202 \text{ mi} \times \dfrac{1 \text{ km}}{0.621 \text{ mi}} = 325 \text{ km}$

5. Since: $1 \text{ cm} = 0.394 \text{ in}$

Then: $(1 \text{ cm})^3 = (0.394 \text{ in})^3 = 0.0612 \text{ in}^3$

$V = 25.2 \text{ cm}^3 \times \dfrac{0.0612 \text{ in}^3}{1 \text{ cm}^3} = 1.54 \text{ in}^3$

6. a. $D(g/\ell) = 0.799 \dfrac{g}{cm^3} \times \dfrac{10^3 \text{ cm}^3}{1 \ \ell} = 799 \text{ g}/\ell$

b. $D(lb/qt) = 799 \text{ g} \times \dfrac{1 \text{ kg}}{10^3 \text{ g}} \times \dfrac{2.20 \text{ lb}}{kg} \times \dfrac{1 \ \ell}{1.06 \text{ qt}} = 1.66 \text{ lb/qt}$

7. a. 4 b. 3 c. 2

8. $A = \text{length} \times \text{width}$

$A = 51 \text{ cm} \times 11.2 \text{ cm} = 5.7 \times 10^2 \text{ cm}^2$

9. $12.612 \text{ g} - 11.5 \text{ g} = 1.1 \text{ g}$

10. $D = \dfrac{m}{V} = \dfrac{27.5 \text{ g}}{15.1 \text{ cm}^3} = 1.82 \text{ g/cm}^3$

11. a. $m = D \times V = 13.5 \dfrac{g}{cm^3} \times 25.2 \text{ cm}^3 = 3.40 \times 10^2 \text{ g}$

b. $V = \dfrac{m}{D} = \dfrac{1.00 \text{ g}}{13.5 \text{ g/cm}^3} = 7.41 \times 10^{-2} \text{ cm}^3$

12. $°F = 1.8(°C) + 32$

$°F + 1.8 (-40) + 32 = -40$

13. $°F + 1.8(°C) + 32$

$°F = 1.8 (-20) + 32 = -4$

As the radiator froze at above $-20°F$, you can collect (hopefully).

14. a. $\text{mass (mg)} = 16.2\text{ g} \times \frac{10^3\text{ mg}}{1\text{ g}} = 1.62 \times 10^4\text{ mg}$

$\text{mass (lb)} = 16.2\text{ g} \times \frac{2.20\text{ lb}}{10^3\text{ g}} = 0.0356\text{ lb}$

b. $\text{volume } (\ell) = 248\text{ cm}^3 \times \frac{1\ \ell}{10^3\text{ cm}^3} = 0.248\ \ell$

$\text{volume (qt)} = 248\text{ cm}^3 \times \frac{1.06\text{ qt}}{10^3\text{ cm}^3} = 0.263\text{ qt}$

c. $\text{distance (ft)} = 1.52\text{ mi} \times \frac{5280\text{ ft}}{1\text{ mi}} = 8.03 \times 10^3\text{ ft}$

$\text{distance (km)} = 1.52\text{ mi} \times \frac{1\text{ km}}{0.621\text{ mi}} = 2.45\text{ km}$

15. Since: $1\text{ cm} = 0.394\text{ in}$

Then: $(1\text{ cm})^3 = (0.394\text{ in})^3 = 0.0612\text{ in}^3$

$$D(\text{g/cm}^3) = 0.031\frac{\text{lb}}{\text{in}^3} \times \frac{454\text{ g}}{\text{lb}} \times \frac{0.0612\text{ in}^3}{1\text{ cm}^3} = 0.86\text{ g/cm}^3$$

16. $\text{nerds} = 62.7\text{ curd} \times \frac{4\text{ troll}}{8.92\text{ curd}} \times \frac{1\text{ nerd}}{2.60\text{ troll}} = 10.8\text{ nerds}$

The use of ''nonsense'' units helps to create interest and demonstrates the utility of the unit-conversion method. Have your students develop other nonsense systems.

17. $\text{Distance (ft)} = 10\text{ m} \times \frac{39.4\text{ in}}{1\text{ m}} \times \frac{1\text{ ft}}{12\text{ in}} = 33\text{ ft}$

$\text{Extra distance} = 33\text{ ft} - 30\text{ ft} = 3\text{ ft}$

18. NOTE: The answer to this question will vary with the individual student. The solution below is based on a height of 5 ft 7 in (67 in) and a mass of 145 pounds.

$\text{Height (m)} = 67\text{ in} \times \frac{1\text{ m}}{39.4\text{ in}} = 1.7\text{ m}$

$\text{Mass (kg)} = 145\text{ lb} \times \frac{1\text{ kg}}{2.20\text{ lb}} = 65.9\text{ kg}$

19. a. 3 b. 4 c. 3 d. 2 e. 2
20. a. 4.7 cm^2 b. 5.2 g/cm^3 c. 3.0×10^1 cm^3
    d. 2.5 g/cm^3 e. 1.00×10^2 g
21. a. 19.9 g b. 8.0×10^1 cm^3 c. 5.9 m d. 9.09 m

    e. $12 \text{ in} - 2 \text{ cm} \times \frac{0.394 \text{ in}}{1 \text{ cm}} = 12 \text{ in} - 0.8 \text{ in} = 11 \text{ in}$

22. a. 9.99 g b. 0.77 g c. 2.60 g d. 9 g
23. First, find the volume of the oil film knowing its mass and density:

$$\text{volume} = \frac{\text{mass}}{\text{density}} = \frac{10^3 \text{ g}}{0.85 \text{ g/cm}^3} = 1200 \text{ cm}^3$$

Second, find the surface area of the film knowing its volume and thickness:

$$\text{area (cm}^2) = \frac{\text{volume}}{\text{thickness}} = \frac{1200 \text{ cm}^3}{1 \times 10^{-6} \text{ cm}} = 1 \times 10^9 \text{ cm}^2$$

Third, convert the surface area to units of square miles:
Since: 1 km = 0.621 mi
Then: $(1 \text{ km})^2 = (0.621 \text{ mi})^2 = 0.386 \text{ mi}^2$

$$\text{area (mi}^2) = 1 \times 10^9 \text{ cm}^2 \times \frac{1 \text{ m}^2}{(10^2 \text{ cm})^2} \times \frac{1 \text{ km}^2}{(10^3 \text{ m})^2} \times \frac{0.386 \text{ mi}^2}{1 \text{ km}^2}$$
$$= 4 \times 10^{-2} \text{ mi}^2$$

24. °F = 1.8(°C) + 32; Let °F = 2(°C)

$$2(°C) = 1.8(°C) + 32$$

$$0.2(°C) = 32; \; °C = \frac{32}{0.2} = 160$$

Check:

$$°F = 1.8\,(160) + 32 = 320$$

25. $°C = \frac{°F - 32}{1.8} = \frac{367 - 32}{1.8} = 186$

Lithium has a melting point of 186°C and density of 0.53 g/cm^3. (Reference: Table 7.1 of Chapter 7.)

CHAPTER 2

Atoms, Molecules, and Ions

I. Basic Skills

A. Qualitative   Students should be able to:

1. Describe the main features of modern atomic theory.
2. Explain the Laws of Conservation of Mass and Constant Composition and relate them to atomic theory.
3. List the subatomic particles and compare them in terms of mass, charge, and location in the atom (Table 2.1).
4. Define atomic number (Eqn 2.1) and mass number (Eqn 2.2).
5. Distinguish between atoms, molecules, and ions.
6. Describe the concept of atomic mass (and molecular mass) as a relative mass.
7. Locate the atomic symbol, atomic number, and atomic mass using the Table of Atomic Masses (inside back cover of text).
8. Describe what is meant by a mole and know the number of items in a mole.

B. Quantitative   Students should be able to:

1. a. Write the nuclear symbol of a particular isotope (Example 2.1a).
   b. Determine the particle composition of an isotope from its nuclear symbol (Example 2.1b).
2. Interpret a molecular formula in terms of atoms per molecule (Example 2.2).
3. Relate the charge of an ion to its proton and electron composition.
4. Use the principle of electrical neutrality to predict the formulas of ionic compounds (Example 2.3).
5. Determine the relative masses of any two atoms or molecules (Example 2.4).
6. Calculate the atomic mass of an element from the abundances of its isotopes (Example 2.5).
7. a. Calculate the formula mass (atomic, molecular or ionic) of a substance (Example 2.6).
   b. Calculate the molar mass of a substance (Example 2.7).
8. Make mole-gram conversions using unit analysis (Example 2.8).

II. Chapter Development

1. This chapter stresses the nature of the building units of matter. A brief historical development of the atomic theory is given. (The Law of Multiple Proportions is omitted.) A quick transition to modern atomic theory and the

Rutherford atom follows. The material presented is sufficient for the student until further expansion is presented later (Chapters 9–11).

2. The development of atomic masses was an essential step in making chemistry a quantitative science. Atomic masses and molecular masses should be presented as relative masses whose values are dependent on the mass of the standard ($^{12}_{6}C$) to which they are compared.
3. The concept of atomic mass unit (amu) is not used in this text. (Why introduce an unnecessary new mass unit?) The terms, atomic mass (AM) and molecular mass (MM) are used in preference to atomic weight and molecular weight when applied to individual atoms or molecules.
4. For molar amounts, the use of gram atomic weight (GAW) and gram molecular weight (GMW) has been avoided. The student is taught that the molar mass in grams is numerically equal to the formula mass of the substance. In examples and problems a molar mass is requested by asking for "the mass of one mole" of the substance.
5. The bulk properties of atomic, molecular, and ionic substances are discussed in Chapter 12, Liquids and Solids.

III. Problem Areas

1. Obtaining the correct charge on an ion presents some difficulty. Students frequently reverse the correct charge when electrons are gained or lost.
2. The concept of relative mass is clarified by using examples of familiar objects. Choose a "standard" student and have the others calculate their relative mass.
3. The mole is introduced as a necessary term used to express the amount of something. As chemists work with small items, a term is needed for the very large number of these items. The mole will be a new term to most students and you must work to gain its acceptance.
4. The size of Avogadro's number is difficult for anyone to grasp. Have students make calculations which point out its magnitude. Two examples are: 1. Calculate the depth of ping-pong balls (dia = 3.5 cm) when one mole of balls is used to cover the state of Texas (270,000 sq mi). 2. Calculate the time required to spend a mole of dollars if money is spent at the rate of one million dollars per second.
5. In making mole-gram conversions, the use of the unit-analysis method is continued. Student mastery of this type of conversion is essential to coming calculations in Chapter 3.

IV. Suggested Activities

1. Atoms, Molecules, and Ions:
   Use a ball-and-stick model kit to build the simple molecules discussed in Section 2.3. Distinguish between atoms and molecules. Show a model of an ionic compound such as NaCl. Contrast molecular and ionic substances

with regard to their type of structure and attractive forces. (Save details for Chapter 12.)

2. The Nature of Electrons: **Teacher Demonstration Only**
   Use a cathode ray tube to generate a beam of electrons. The electrons come from the cathode through a slit and strike a zinc sulfide screen which emits light. The intermittent nature of the light (flashes) indicates the particle nature of the electron beam. When the north pole of a magnet is brought up to the side of the tube the line of light is bent downward. The direction of the deflection indicates that the cathode rays are negatively charged.
3. The Mole Concept:
   Exhibit a mole of various substances labeled with their formula and mass. Examples are sucrose (342 g), NaCl (58.5 g), S_8 (256 g), I_2 (254 g), H_2O (18 g) and powdered aluminum (27 g). Point out that each volume contains one mole of the formula unit. Ask students for the number of atoms (or ions) in each sample. Have them calculate the relative masses of the particles.

V. Answers to Questions

1. Statements (a) and (d) are true. Statement (c) is not true because a very small percent of H atoms (^2H) have a mass about twice as great as ordinary H atoms. Statement (b) is obviously false.
2. The Law of Constant Composition states that the mass percent of the elements in a compound is constant. This is because the ratio of the different atoms in a compound is fixed and the atoms have a fixed mass. For example, the H/O atom ratio in water is always 2/1. As a result of this ratio, the mass composition of water is always 11.19% H and 88.81% O.
3. a. Thompson discovered the electron and showed that it was a particle common to all elements. He also measured the mass-to-charge ratio of the electron.
   b. Millikan determined the mass of the electron.
   c. Rutherford showed that atoms have a small, dense, positively charged nucleus.
4. a. electron b. neutron c. electron d. proton and neutron
5. The pairs of atoms in (a) and (d) are isotopes. Isotopes are atoms of the same element (atomic no.) which differ in mass (mass no.).
6. a. Ar b. B c. Sr d. Sn e. W
7. a. copper b. tin c. potassium d. phosphorus e. arsenic
8. The atomic mass is closer to the mass of X as isotope X is present in the larger amount.
9. A molecule is a particle of matter consisting of two or more atoms held together by chemical bonds. Examples of molecular substances are water, H_2O, and carbon dioxide, CO_2. The atom is the smallest unit of matter which retains the properties of a particular element.
10. A Cl^- ion has one more electron than a Cl atom. The Cl_2 molecule consists of two Cl atoms joined by a chemical bond.

11. In order of decreasing mass: Al > B > Be.
12. In order of decreasing mass: $C_2H_6 > C_2H_4 > CH_4 > C > H$.
13. Statements (a) and (c) are true. Statement (b) is false because 12.0 g is the mass of one mole of C, not one atom of C. Statement (d) is false because C has an atomic number of 6. Its mass number is 12.
14. Statements (a) and (c) are true. Statement (b) is false because different elements have a different atomic mass and, therefore, a different molar mass.

VI. Solutions to Problems

1.

	Nuclear Symbol	Number of Protons	Atomic Number	Number of Neutrons	Mass Number
a.	$^{12}_{6}C$	6	6	6	12
b.	$^{11}_{5}B$	5	5	6	11
c.	$^{25}_{12}Mg$	12	12	13	25
d.	$^{9}_{4}Be$	4	4	5	9

2.

	Molecular Formula	C atoms per molecule	H atoms per molecule	Total atoms per molecule
b.	C_2H_2	2	2	4
c.	C_2H_6	2	6	8
d.	C_4H_8	4	8	12

3. a. LiF b. Li_2S c. MgF_2 d. MgS

4. $\frac{\text{mass Zn}}{\text{mass Cl}} = \frac{65.38}{35.45} = 1.844$

5. X = 0.460 (58.70) = 27.0 X = Al (aluminum)

6. AM of X = $\frac{\%A}{100}$ (mass A) + $\frac{\%B}{100}$ (mass B)

AM = 0.188 (10.0) + 0.812 (11.0)

AM = 1.88 + 8.93 = 10.81 X = B (boron)

7. a. $MM(C_2H_2) = 2(12.01) + 2(1.01) = 26.04$

b. $MM(CO_2) = 12.01 + 2(16.00) = 44.01$

c. $MM(H_2S) = 2(1.01) + 32.06 = 34.08$

8. $\frac{\text{mass } H_2O}{\text{mass O}} = \frac{18.02}{16.00} = 1.126 \qquad \frac{\text{mass } H_2O}{\text{mass } H_2S} = \frac{18.02}{34.08} = 0.5288$

9. a. $MM(C_2H_2) = 2(12.01) + 2(1.01) = 26.04 \qquad 1 \text{ mol } C_2H_2 = 26.04 \text{ g}$

   b. $MM(CO_2) = 12.01 + 2(16.00) = 44.01 \qquad 1 \text{ mol } CO_2 = 44.01 \text{ g}$

   c. $MM(K_2S) = 2(39.10) + 32.06 = 110.26 \qquad 1 \text{ mol } K_2S = 110.26 \text{ g}$

   d. $MM(KMnO_4) = 39.10 + 54.94 + 4(16.00) = 158.04$
   $1 \text{ mol } KMnO_4 = 158.04 \text{ g}$

10. a. $\text{mass } C_2H_2 = 2.50 \text{ mol } C_2H_2 \times \frac{26.0 \text{ g } C_2H_2}{1 \text{ mol } C_2H_2} = 65.0 \text{ g}$

    b. $\text{mass } H_2S = 2.50 \text{ mol } H_2S \times \frac{34.1 \text{ g } H_2S}{1 \text{ mol } H_2S} = 85.2 \text{ g}$

    c. $\text{mass } KNO_3 = 2.50 \text{ mol } KNO_3 \times \frac{101.1 \text{ g } KNO_3}{1 \text{ mol } KNO_3} = 253 \text{ g}$

11. a. $\text{moles } N_2 = 1.00 \text{ g } N_2 \times \frac{1 \text{ mol } N_2}{28.0 \text{ g } N_2} = 3.57 \times 10^{-2} \text{ mol}$

    b. $\text{moles HF} = 1.00 \text{ g HF} \times \frac{1 \text{ mol HF}}{20.0 \text{ g HF}} = 5.00 \times 10^{-2} \text{ mol}$

    c. $\text{moles } COCl_2 = 1.00 \text{ g } COCl_2 \times \frac{1 \text{ mol } COCl_2}{99.0 \text{ g } COCl_2} = 1.01 \times 10^{-2} \text{ mol}$

12. NOTE: Answer depends on actual length of page. Solution below is for a measured length of 26.0 cm.

$$\text{Number of O atoms} = \frac{0.260 \text{ m}}{1.32 \times 10^{-10} \text{ m/atom}} = 1.97 \times 10^{9} \text{ atoms}$$

13. a. Li^+: 3p and 2e c. Mg^{2+}: 12p and 10e
    b. F^-: 9p and 10e d. S^{2-}: 16p and 18e

14. $\frac{\text{Avogadro's No.}}{\text{Walford's No.}} = \frac{6.02 \times 10^{23}}{1.00 \times 10^{23}} = 6.02 \text{ Walford's/mole}$

15.

Nuclear Symbol	Number of Protons	Number of Electrons	Number of Neutrons	Charge
b. $^{27}_{13}Al^{3+}$	13	10	14	+3
c. $^{24}_{12}Mg^{2+}$	12	10	12	+2
d. $^{36}_{17}Cl^{-}$	17	18	19	−1

16. a. $MM(C_6H_{12}O_6) = 6(12.01) + 12(1.01) + 6(16.00) = 180.18$

b. 1 mole $C_6H_{12}O_6 = 180.18$ g

c. 6.022×10^{23} molecules of anything/mol

d. $\text{atoms} = 6.022 \times 10^{23} \text{ molecules} \times \frac{24 \text{ atoms}}{\text{molecule}} = 1.445 \times 10^{25}$

17. a. TNT: $C_7H_5N_3O_6$

b. $MM(TNT) = 7(12.01) + 5(1.01) + 3(14.01) + 6(16.00) = 227.1$

c. $\text{mass TNT} = 1.20 \text{ mol TNT} \times \frac{227.1 \text{ g TNT}}{1 \text{ mol TNT}} = 273 \text{ g}$

18. a. $FM(CaCO_3) = 40.08 + 12.01 + 3(16.00) = 100.09$

b. $\text{mass } CaCO_3 = 0.239 \text{ mol } CaCO_3 \times \frac{100.09 \text{ g } CaCO_3}{1 \text{ mol } CaCO_3} = 23.9 \text{ g}$

c. $\text{moles } CaCO_3 = 24.5 \text{ g } CaCO_3 \times \frac{1 \text{ mol } CaCO_3}{100.09 \text{ g } CaCO_3} = 0.245 \text{ mol}$

19. a. $FM(KNO_3) = 39.10 + 14.01 + 3(16.00) = 101.11$

$\text{mass } KNO_3 = 1.42 \text{ } KNO_3 \times \frac{101.1 \text{ g } KNO_3}{1 \text{ mol } KNO_3} = 144 \text{ g}$

b. $MM(C_6H_6) = 6(12.01) + 6(1.01) = 78.12$

$\text{mass } C_6H_6 = 1.42 \text{ mol } C_6H_6 \times \frac{78.12 \text{ g } C_6H_6}{1 \text{ mol } C_6H_6} = 111 \text{ g}$

c. $FM(NH_4Cl) = 14.01 + 4(1.01) + 35.45 = 53.50$

$\text{mass } NH_4Cl = 1.42 \text{ mol } NH_4Cl \times \frac{53.50 \text{ g } NH_4Cl}{1 \text{ mol } NH_4Cl} = 76.0 \text{ g}$

20. a. $\text{moles } KNO_3 = 12.3 \text{ g } KNO_3 \times \frac{1 \text{ mol } KNO_3}{101.11 \text{ g } KNO_3} = 0.122 \text{ mol}$

b. $\text{moles } C_6H_6 = 12.3 \text{ g } C_6H_6 \times \frac{1 \text{ mol } C_6H_6}{78.12 \text{ g } C_6H_6} = 0.157 \text{ mol}$

c. $\text{moles } NH_4Cl = 12.3 \text{ g } NH_4Cl \times \frac{1 \text{ mol } NH_4Cl}{53.50 \text{ g } NH_4Cl} = 0.230 \text{ mol}$

21. $\text{atoms} = 1 \times 10^{-6}\ \text{g Cu} \times \dfrac{1\ \text{mol Cu}}{63.55\ \text{g Cu}} \times \dfrac{6.02 \times 10^{23}\ \text{atoms}}{1\ \text{mol Cu}}$

$= 9 \times 10^{15}$ atoms

22. $\text{AM of Cu} = \dfrac{\%\ ^{63}\text{Cu}}{100}(\text{mass}\ ^{63}\text{Cu}) + \dfrac{\%\ ^{65}\text{Cu}}{100}(\text{mass}\ ^{65}\text{Cu})$

Let X = % of ^{63}Cu and 100 − X = % of ^{65}Cu

$$\text{AM of Cu} = \frac{X}{100}(62.93) + \frac{100 - X}{100}(64.93) = 63.55$$

$$62.93\ X + 6493 - 64.93\ X = 6355$$

$$2.00\ X = 138;\ X = \frac{138}{2.00}$$

$$X = \%\ ^{63}\text{Cu} = 69.0 \qquad 100 - X = \%\ ^{65}\text{Cu} = 31.0$$

23. $\text{volume (ice cube)} = (\text{side})^3 = (2.50\ \text{cm})^3 = 15.6\ \text{cm}^3$

$$\text{volume (mole of cubes)} = 6.02 \times 10^{23}\ \text{cubes} \times \frac{15.6\ \text{cm}^3}{\text{cube}}$$
$$= 9.39 \times 10^{24}\ \text{cm}^3$$

$$\text{mass (mole of cubes)} = 9.39 \times 10^{24}\ \text{cm}^3 \times \frac{0.92\ \text{g}}{1\ \text{cm}^3} = 8.6 \times 10^{24}\ \text{g}$$

$$\frac{\text{mass of cubes}}{\text{mass of earth}} = \frac{8.6 \times 10^{24}\ \text{g}}{6 \times 10^{27}\ \text{g}} = 1 \times 10^{-3}\ \text{or}\ 0.1\%$$

24. $\text{mass proton} = \dfrac{1.0\ \text{g}}{\text{mol}} \times \dfrac{1\ \text{mol}}{6.02 \times 10^{23}\ \text{protons}} = 1.7 \times 10^{-24}\ \text{g}$

$$\text{radius proton} = \frac{\text{dia}}{2} = \frac{1 \times 10^{-12}\ \text{m}}{2} = 5 \times 10^{-13}\ \text{m or}\ 5 \times 10^{-11}\ \text{cm}$$

$$\text{volume proton} = \frac{4\pi r^3}{3} = \frac{4\pi}{3}(5 \times 10^{-11}\ \text{cm})^3 = 5 \times 10^{-31}\ \text{cm}^3$$

$$\text{density} = \frac{m}{V} = \frac{1.7 \times 10^{-24}\ \text{g}}{5 \times 10^{-31}\ \text{cm}^3} = 3 \times 10^{6}\ \text{g/cm}^3$$

CHAPTER 3

Chemical Formulas and Equations

I. Basic Skills

A. Qualitative Students should be able to:

1. a. Describe what is meant by a chemical formula.
   b. Distinguish between a simplest formula and a molecular formula.
2. Explain what is meant by the percentage composition of a compound.
3. Recognize the symbols for the physical states of substances.
4. Write the formulas and physical states of the elements (Table 3.2).
5. a. Interpret a chemical equation in terms of atoms and molecules.
   b. Interpret an equation in terms of moles.

B. Quantitative Students should be able to:

1. a. Calculate the percentage composition of a compound using experimental data (Equation 3.1 and Example 3.1).
   b. Calculate the mass of an element in a sample using the percentage composition (Equation 3.2 and Example 3.1).
2. a. Calculate the percentage composition from the formula of the compound (Examples 3.2 and 3.3).
   b. Calculate the simplest formula from the percentage composition (Examples 3.4 and 3.5).
3. Determine the molecular formula of a compound using its simplest formula and molecular mass (Example 3.6).
4. Write a balanced chemical equation given the formulas of all reactants and products (Example 3.7).
5. a. Obtain mole relations from balanced chemical equations (Examples 3.8, 3.9 and 3.10).
   b. Obtain mass relations from balanced chemical equations (Examples 3.9 and 3.10).
6. Use mole relations and mass relations in chemical calculations (Examples 3.8, 3.9 and 3.10).

II. Chapter Development

1. The language of chemistry is presented progressively from formulas through equations. Emphasis is placed on the experimental basis of a formula and its relation to percentage composition. Students are then led to describing what happens in a reaction by the use of an equation. Considerable discussion is given on writing and balancing equations.
2. Do not become over-involved with the ''chemistry'' of reactions at this stage.

It is enough to note that some substances are ionic and represented by simplest formulas, while others are molecular and represented by molecular formulas. The energy changes which accompany chemical reactions are covered in Chapter 4.

3. Equations are not an end in themselves but are needed for mole relations and mass relations. The mole relations from an equation are written as conversion factors and used in mole and mass problems. This material provides another extension of the unit-analysis method of problem solving.
4. The concept of a "limiting reactant" is not covered. Students find this very difficult at this stage and the authors believe it to be an unnecessary complication.

III. Problem Areas

1. In determining the simplest formula of a compound, XY, some students will round off the moles of X and Y to whole numbers before finding the mole ratio. This frequently gives an incorrect ratio. Caution students to only round off after the mole ratio is obtained, and only if the calculated ratio approximates an integer. If not, then a multiplier should be used to obtain the simplest ratio. You can increase interest in simplest formula calculations by switching to "nonsense" formulas such as HIOAg (a golden oldie) and C_3PO and Ar_2Dy_2 of Star Wars fame.
2. Many students will have difficulty in writing balanced chemical equations. Emphasize that this is a basic skill which is needed from this chapter on. Point out clearly that it is atoms which are conserved, not molecules. This is in agreement with Dalton's atomic theory and the Law of Conservation of Mass.
3. The use of incorrect formulas and states for the elements is a common error in writing equations. A little drill with Table 3.2 will be time well spent. Sulfur's formula is given as S, instead of S_8, in the interest of simplicity.
4. Mole-mole and mole-mass problems will challenge many students. Point out the practical need for information of this type in industry and laboratories. Use the unit-analysis method in problem solutions and discourage the use of the proportion method.
5. The use of electronic calculators has greatly simplified the actual arithmetic which is involved in chemical calculations. Grab the opportunity in this chapter to reinforce the correct use of significant figures.

IV. Suggested Activities

1. Balancing Equations Using Models:
   Use molecular models to demonstrate the rearrangement of atoms in chemical reactions. Show how atoms (and mass) are conserved when changing reactant models into product models. Point out the mole relations between reactants and products. Use simple reactions such as:

$H_2 + Cl_2 \rightarrow 2\ HCl$, $2H_2 + O_2 \rightarrow 2H_2O$, and $H_2 + O_2 \rightarrow H_2O_2$.

2. Writing the Equation for a Demonstration Reaction:
   Do a series of reaction demonstrations. As far as possible have students identify the reactants and products by name and formula. Finally, have the students write a balanced equation for each reaction. Example reactions are: heat powdered zinc and sulfur on a heat resistant pad, burn natural gas (CH_4), burn a candle ($C_{25}H_{52}$), decompose H_2O_2. Point out that an equation is a quantitative statement which describes the chemical changes which take place in a reaction.

V. Answers to Questions

1. $\%\ O = 100.0 - 71.5 = 28.5$
2. Since there is one atom of carbon in one molecule of CO_2, divide the atomic mass of carbon by the molecular mass of CO_2 and multiply the ratio by 100.
3. (1) Determine the mass of each element in 100 grams of the compound.
   (2) Convert the mass of each element to moles.
   (3) Determine the simplest mole ratio between the elements and set it equal to the atom ratio in the formula.
4. No. The same simplest formula will be obtained regardless of the sample size. If a 200 g sample is used, twice as many moles of C and H will be obtained but the mole ratio, C:H, remains constant.
5. A mole of atoms contains a fixed number (Avogadro's number) of atoms regardless of the element involved. If both the numerator and denominator of a mole ratio are multiplied by this number to obtain an atom ratio, the numerical value is not changed.
6. $\frac{2.50 \text{ mol O}}{\text{mol V}} = \frac{5.00 \text{ mol O}}{2 \text{ mol V}}$ Simplest formula is V_2O_5
7. $\frac{4 \text{ mol O}}{3 \text{ mol Fe}} = \frac{1\frac{1}{3} \text{ mol O}}{\text{mol Fe}}$
8. Formulas (a), (b) and (e) must be molecular formulas. Their simplest formulas would be CH, C_2H_5, and NH_2. Formulas (c) and (d) are simplest formulas but could also be molecular formulas.
9. Two molecules of SO_2 react with one molecule of O_2 to give two molecules of SO_3.
   Two moles of SO_2 react with one mole of O_2 to give two moles of SO_3.
10. Statements (a) and (d) are generally true. Statement (b) is false because the number of molecules of reactants and products (as noted by the coefficients) need not be equal in order to have equal numbers of each type of atom. It follows from this that statement (c) is also false in that the number of moles need not be equal.

11. c. Hg(l) d. $H_2(g)$ e. Mg(s) f. $N_2(g)$ g. Xe(g)
    h. Cu(s) i. $F_2(g)$ j. $Br_2(l)$
12. a. He, Ne, Ar, Kr, Xe, or Rn b. Hg c. I_2 d. Br_2
13. a. H_2 and O_2 are diatomic molecules.
    b. The H atoms are not balanced.
    c. The Fe atoms are not balanced.
    d. Phosphorus consists of P_4 molecules.
    e. H_2 is a gas at room temperature and pressure.

VI. Solutions to Problems

1. $$\text{mass N} = 2.56 \text{ g NH}_3 \times \frac{82.3 \text{ g N}}{100 \text{ g NH}_3} = 2.11 \text{ g}$$

$$\text{mass H} = 2.56 \text{ g NH}_3 \times \frac{17.7 \text{ g H}}{100 \text{ g NH}_3} = 0.453 \text{ g}$$

2. $$\% \text{ Mg} = \frac{1.123 \text{ g Mg}}{1.862 \text{ g compound}} \times 100 = 60.31$$

$$\% \text{ O} = 100.00 - 60.31 = 39.69$$

3. a. $$\text{MM } (C_3H_8) = 3(12.01) + 8(1.008) = 44.09$$

$$\% \text{ C} = \frac{36.03 \text{ g C}}{44.09 \text{ g C}_3\text{H}_8} \times 100 = 81.71$$

$$\% \text{ H} = 100.00 - 81.71 = 18.29$$

b. $$\text{MM } (C_2H_6O) = 2(12.01) + 6(1.008) + 16.00 = 46.07$$

$$\% \text{ C} = \frac{24.02 \text{ g C}}{46.07 \text{ g C}_2\text{H}_6\text{O}} \times 100 = 52.14$$

$$\% \text{ O} = \frac{16.00 \text{ g O}}{46.07 \text{ g C}_2\text{H}_6\text{O}} \times 100 = 34.73$$

$$\% \text{ H} = 100.0 - 52.14 - 34.73 = 13.13$$

c. $$\text{MM}(UF_6) = 238.0 + 6(19.00) = 352.0$$

$$\% \text{ U} = \frac{238.0 \text{ g U}}{352.0 \text{ g UF}_6} \times 100 = 67.61$$

$$\% \text{ F} = 100.00 - 67.62 = 32.39$$

4. $$\text{FM } (Fe_2O_3) = 2(55.85) + 3(16.00) = 159.7$$

$$\text{mass Fe} = 1.000 \times 10^3 \text{ g } Fe_2O_3 \times \frac{111.7 \text{ g Fe}}{159.7 \text{ g } Fe_2O_3} = 699.4 \text{ g}$$

5. $$\text{moles Cu} = 47.3 \text{ g Cu} \times \frac{1 \text{ mol Cu}}{63.5 \text{ g Cu}} = 0.745$$

$$\text{moles Cl} = 52.7 \text{ g Cl} \times \frac{1 \text{ mol Cl}}{35.5 \text{ g Cl}} = 1.48$$

$$\frac{\text{mole Cl}}{\text{mole Cu}} = \frac{1.48 \text{ mol Cl}}{0.745 \text{ mol Cu}} = 1.99 = 2/1$$

Simplest formula is $CuCl_2$

6. $$\text{moles C} = 52.14 \text{ g C} \times \frac{1 \text{ mol C}}{12.01 \text{ g C}} = 4.341$$

$$\text{moles H} = 13.13 \text{ g H} \times \frac{1 \text{ mol H}}{1.008 \text{ g H}} = 13.03$$

$$\text{moles O} = 34.73 \text{ g O} \times \frac{1 \text{ mol O}}{16.00 \text{ g O}} = 2.171$$

$$\frac{\text{moles C}}{\text{mole O}} = \frac{4.341}{2.171} = 2.000 \qquad \frac{\text{moles H}}{\text{mole O}} = \frac{13.03}{2.171} = 6.002$$

Simplest formula is C_2H_6O

7. $FM(C_2H_5) = 2(12.0) + 5(1.0) = 29.0$

$$\frac{MM}{FM} = \frac{58}{29.0} = 2.0$$

Molecular formula is C_4H_{10}

8. a. $C_2H_4(g) + 3\ O_2(g) \rightarrow 2\ CO_2(g) + 2\ H_2O(l)$

   b. $C_2H_4(g) + 2\ O_2(g) \rightarrow 2\ CO(g) + 2\ H_2O(l)$

   c. $2\ Fe(s) + 3\ Cl_2(g) \rightarrow 2\ FeCl_3(s)$

   d. $6\ Li(s) + N_2(g) \rightarrow 2\ Li_3N(s)$

9. $N_2(g) + O_2(g) \rightarrow 2\ NO(g)$

   a. $$\text{moles NO} = 2.26 \text{ mol } N_2 \times \frac{2 \text{ mol NO}}{1 \text{ mol } N_2} = 4.52$$

b. $0.128 \text{ mol } O_2 \triangleq 0.128 \text{ mol } N_2$

c. $\text{moles } O_2 = 2.34 \times 10^6 \text{ mol NO} \times \dfrac{1 \text{ mol } O_2}{2 \text{ mol NO}} = 1.17 \times 10^6$

10. a. $\text{mass NO} = 8.19 \text{ mol } N_2 \times \dfrac{2 \text{ mol NO}}{1 \text{ mol } N_2} \times \dfrac{30.0 \text{ g NO}}{1 \text{ mol NO}} = 491 \text{ g}$

b. $\text{moles NO} = 8.19 \text{ g } N_2 \times \dfrac{1 \text{ mol } N_2}{28.0 \text{ g } N_2} \times \dfrac{2 \text{ mol NO}}{1 \text{ mol } N_2} = 0.585$

11. a. $\text{mass NO} = 1.00 \text{ g } O_2 \times \dfrac{1 \text{ mol } O_2}{32.0 \text{ g } O_2} \times \dfrac{2 \text{ mol NO}}{1 \text{ mol } O_2} \times \dfrac{30.0 \text{ g NO}}{1 \text{ mol NO}}$

$= 1.88 \text{ g}$

b. $\text{mass } O_2 = 1.00 \text{ g NO} \times \dfrac{1 \text{ mol NO}}{30.0 \text{ g NO}} \times \dfrac{1 \text{ mol } O_2}{2 \text{ mol NO}} \times \dfrac{32.0 \text{ g } O_2}{1 \text{ mol } O_2}$

$= 0.533 \text{ g}$

12. a. $\% \text{ O} = 100.00 - 5.93 = 94.07$

b. $\text{mass O} = 6.46 \text{ g compound} \times \dfrac{94.07 \text{ g O}}{100 \text{ g compound}} = 6.08 \text{ g}$

c. $\text{mass compound} = 1.00 \text{ g H} \times \dfrac{100 \text{ g compound}}{5.93 \text{ g H}} = 16.9 \text{ g}$

13. $\text{FM}(KClO_3) = 39.10 + 35.45 + 3(16.00) = 122.55$

a. $\% \text{ K} = \dfrac{39.10 \text{ g K}}{122.55 \text{ g } KClO_3} \times 100 = 31.91$

$\% \text{ Cl} = \dfrac{35.45 \text{ g Cl}}{122.55 \text{ g } KClO_3} \times 100 = 28.93$

$\% \text{ O} = 100.00 - 31.91 - 28.93 = 39.16$

b. $\text{mass } O_2 = 1.00 \text{ g } KClO_3 \times \dfrac{39.16 \text{ g O}}{100 \text{ g } KClO_3} = 0.392 \text{ g}$

c. $\text{mass } KClO_3 = 1.00 \text{ g } O_2 \times \dfrac{100 \text{ g } KClO_3}{39.16 \text{ g O}} = 2.55 \text{ g}$

14. $\text{moles B} = 31.0 \text{ g B} \times \dfrac{1 \text{ mol B}}{10.8 \text{ g B}} = 2.87$

$$\text{moles O} = 69.0 \text{ g O} \times \frac{1 \text{ mol O}}{16.0 \text{ g O}} = 4.31$$

$$\frac{\text{moles O}}{\text{mole B}} = \frac{4.31}{2.87} = 1.50 \text{ or } 3/2$$

Simplest formula is B_2O_3

15. mass Sn = 15.499 g − 15.254 g = 0.245 g

mass oxide = 15.565 g − 15.254 g = 0.311 g

mass O = 0.311 g − 0.245 g = 0.066 g

a. $\% \text{ Sn} = \frac{0.245 \text{ g Sn}}{0.311 \text{ g oxide}} \times 100 = 78.8$

$\% \text{ O} = 100.0 - 78.8 = 21.2$

b. $\text{moles Sn} = 78.8 \text{ g Sn} \times \frac{1 \text{ mol Sn}}{119 \text{ g Sn}} = 0.662$

$$\text{moles O} = 21.2 \text{ g O} \times \frac{1 \text{ mol O}}{16.0 \text{ g O}} = 1.32$$

$$\frac{\text{moles O}}{\text{mole Sn}} = \frac{1.32}{0.662} = 1.99 = 2{:}1$$

Simplest formula is SnO_2

16. The molecular mass must be a whole-number multiple of the formula mass. Possible molecular masses for CH_2 are 14, 28, and 42.

The formulas would be (a) CH_2, (c) C_2H_4, and (e) C_3H_6.

17. a. $Fe_2O_3(s) + 3\ H_2(g) \rightarrow 2\ Fe(s) + 3\ H_2O(l)$

b. $2\ K(s) + Br_2(l) \rightarrow 2\ KBr(s)$

c. $2\ C_2H_2(g) + 5\ O_2(g) \rightarrow 4\ CO_2(g) + 2\ H_2O(l)$

18. To have the same number of atoms of each element on both sides of the equation:
a. $NxHy = NH_3$ b. $NxOy = NO_2$ c. $CxHy = C_5H_{12}$

19. a. $\text{mass H} = 120 \text{ g } H_2O \times \frac{11.2 \text{ g H}}{100 \text{ g } H_2O} = 13.4 \text{ g}$

b. $\text{mass O} = 6.24 \text{ g H} \times \frac{88.8 \text{ g O}}{11.2 \text{ g H}} = 49.5 \text{ g}$

c. $\text{mass } H_2O = 10.0 \text{ g O} \times \frac{100 \text{ g } H_2O}{88.8 \text{ g O}} = 11.3 \text{ g}$

20. $2.0 \text{ g X} + 8.0 \text{ g Y} = 10.0 \text{ g Z}$

a. $\% \text{ X} = \frac{2.0 \text{ g X}}{10.0 \text{ g Z}} \times 100 = 20 \qquad \% \text{ Y} = 100 - 20 = 80$

b. $\text{mass X} = 6.0 \text{ g Z} \times \frac{2.0 \text{ g X}}{10.0 \text{ g Z}} = 1.2 \text{ g}$

c. $\text{mass Z} = 1.2 \text{ g Y} \times \frac{10.0 \text{ g Z}}{8.0 \text{ g Y}} = 1.5 \text{ g}$

d. The atomic masses of X and Y are needed to obtain the simplest formula of Z.

21. $\text{mass O} = 2.45 \text{ g oxide} - 0.98 \text{ g S} = 1.47 \text{ g}$

$\text{moles O} = 1.47 \text{ g O} \times \frac{1 \text{ mol O}}{16.0 \text{ g O}} = 0.0919$

$\text{moles S} = 0.98 \text{ g S} \times \frac{1 \text{ mol S}}{32 \text{ g S}} = 0.031$

$\frac{\text{moles O}}{\text{mole S}} = \frac{0.0919}{0.031} = 3.0$

Simplest formula is SO_3. As the formula mass of SO_3 is 80, the molecular formula is also SO_3.

22. a. $\% \text{ Ca} = \frac{1.296 \text{ g Ca}}{1.813 \text{ g oxide}} \times 100 = 71.48$

$\% \text{ O} = 100.00 - 71.48 = 28.52$

b. $\text{moles Ca} = 71.48 \text{ g Ca} \times \frac{1 \text{ mol Ca}}{40.08 \text{ g Ca}} = 1.783$

$\text{moles O} = 28.52 \text{ g O} \times \frac{1 \text{ mol O}}{16.00 \text{ g O}} = 1.783$

Simplest formula is CaO

c. $2 \text{ Ca(s)} + O_2\text{(g)} \rightarrow 2 \text{ CaO (s)}$

23. $2 H_2\text{(g)} + O_2\text{(g)} \rightarrow 2 H_2O\text{(l)}$

a. $\text{moles } H_2 = 1.69 \text{ mol } O_2 \times \dfrac{2 \text{ mol } H_2}{1 \text{ mol } O_2} = 3.38$

b. $0.918 \text{ mol } H_2 \simeq 0.918 \text{ mol } H_2O$

c. $\text{moles } H_2O = 1.62 \text{ g } H_2 \times \dfrac{1 \text{ mol } H_2}{2.016 \text{ g } H_2} \times \dfrac{2 \text{ mol } H_2O}{2 \text{ mol } H_2} = 0.804$

d. $\text{mass } O_2 = 12.2 \text{ g } H_2 \times \dfrac{1 \text{ mol } H_2}{2.016 \text{ g } H_2} \times \dfrac{1 \text{ mol } O_2}{2 \text{ mol } H_2} \times \dfrac{32.00 \text{ g } O_2}{1 \text{ mol } O_2} = 96.8 \text{ g}$

24. $N_2(g) + 3\ H_2(g) \rightarrow 2\ NH_3(g)$

$0.50 \text{ mol } N_2 + \underline{1.5} \text{ mol } H_2 \rightarrow \underline{1.0} \text{ mol } NH_3$

$\underline{0.640} \text{ mol } N_2 + \underline{1.92} \text{ mol } H_2 \rightarrow 1.28 \text{ mol } NH_3$

$\underline{17.9} \text{ g } N_2 + \underline{3.87} \text{ g } H_2 \rightarrow 1.28 \text{ mol } NH_3$

$\underline{74.1} \text{ g } N_2 + 16.0 \text{ g } H_2 \rightarrow \underline{90.1} \text{ g } NH_3$

25. a. $2\ Na(s) + Cl_2(g) \rightarrow 2\ NaCl(s)$

b. $\text{mass NaCl} = 3.00 \text{ mol Na} \times \dfrac{2 \text{ mol NaCl}}{2 \text{ mol Na}} \times \dfrac{58.45 \text{ g NaCl}}{1 \text{ mol NaCl}} = 175 \text{ g}$

26. a. $2\ Al(s) + 6\ HCl(g) \rightarrow 2\ AlCl_3(s) + 3\ H_2(g)$

b. $\text{mass Al} = 5.00 \text{ mol } H_2 \times \dfrac{2 \text{ mol Al}}{3 \text{ mol } H_2} \times \dfrac{27.0 \text{ g Al}}{1 \text{ mol Al}} = 90.0 \text{ g}$

27. a. $2\ KClO_3(s) \rightarrow 2\ KCl(s) + 3\ O_2(g)$

b. $FM(KClO_3) = 39.10 + 35.45 + 3(16.00) = 122.55$

$\text{mass } O_2 = 1.00 \text{ g } KClO_3 \times \dfrac{1 \text{ mol } KClO_3}{122.55 \text{ g } KClO_3} \times \dfrac{3 \text{ mol } O_2}{2 \text{ mol } KClO_3} \times \dfrac{32.0 \text{ g } O_2}{1 \text{ mol } O_2}$
$= 0.392 \text{ g}$

$\text{mass KCl} = 1.00 \text{ g } KClO_3 \times \dfrac{74.55 \text{ g KCl}}{122.55 \text{ g } KClO_3} = 0.608 \text{ g}$

28. $\%\ O = 100.0 - 15.3 - 37.1 - 5.3 = 42.3$

$\text{moles Cr} = 15.3 \text{ g Cr} \times \dfrac{1 \text{ mol Cr}}{52.0 \text{ g Cr}} = 0.294$

$\text{moles N} = 37.1 \text{ g N} \times \dfrac{1 \text{ mol N}}{14.0 \text{ g N}} = 2.65$

$$\text{moles H} = 5.33 \text{ g H} \times \frac{1 \text{ mol H}}{1.01 \text{ g H}} = 5.28$$

$$\text{moles O} = 42.3 \text{ g O} \times \frac{1 \text{ mol O}}{16.0 \text{ g O}} = 2.64$$

$$\frac{\text{moles O}}{\text{mole Cr}} = \frac{2.64}{0.294} = 8.98 = 9:1$$

$$\frac{\text{moles H}}{\text{mole Cr}} = \frac{5.28}{0.294} = 18.0$$

$$\frac{\text{mole N}}{\text{mole Cr}} = \frac{2.65}{0.294} = 9.01 = 9:1$$

Simplest formula is $CrN_9O_9H_{18}$

29. First, determine the moles of C in the sample from the mass of CO_2 produced and the moles of H from the mass of H_2O produced:

$$\text{mass C} = 1.998 \text{ g } CO_2 \times \frac{12.01 \text{ g C}}{44.01 \text{ g } CO_2} = 0.5452 \text{ g}$$

$$\text{moles C} = 0.5452 \text{ g C} \times \frac{1 \text{ mol C}}{12.01 \text{ g C}} = 0.04540$$

$$\text{mass H} = 0.818 \text{ g } H_2O \times \frac{2.016 \text{ g H}}{18.02 \text{ g } H_2O} = 0.0915 \text{ g}$$

$$\text{moles H} = 0.0915 \text{ g H} \times \frac{1 \text{ mol H}}{1.008 \text{ g H}} = 0.0908$$

The mass of oxygen in the sample must be determined by difference.

$$\text{mass O} = 1.000 \text{ g sample} - 0.5452 \text{ g C} - 0.0915 \text{ g H} = 0.3633 \text{ g}$$

$$\text{moles O} = 0.3633 \text{ g O} \times \frac{1 \text{ mol O}}{16.00 \text{ g O}} = 0.02271$$

Find mole ratios by comparing all elements to the element present in the smallest amount:

$$\frac{\text{moles C}}{\text{mole O}} = \frac{0.04540}{0.02271} = 1.999 = 2:1$$

$$\frac{\text{moles H}}{\text{mole O}} = \frac{0.0908}{0.02271} = 4.00 = 4:1$$

Simplest formula is C_2H_4O

30. $2\ K(s) + Br_2(l) \rightarrow 2\ KBr(s)$

$$\text{moles K} = 4.0 \text{ g K} \times \frac{1 \text{ mol K}}{39.1 \text{ g K}} = 0.10$$

$$\text{moles } Br_2 = 10.0 \text{ g } Br_2 \times \frac{1 \text{ mol } Br_2}{159.8 \text{ g } Br_2} = 0.0626$$

Calculate the ratio in which the reactants are present:

$$\frac{\text{moles K}}{\text{mole } Br_2} = \frac{0.10}{0.0626} = 1.6$$

As the reaction ratio of K/Br_2 is 2:1 as given in the equation, insufficient K is present to react with all of the Br_2. In this case the maximum amount of KBr is determined by K, the reactant which is in short supply.

$$\text{mass KBr} = 0.10 \text{ mol K} \times \frac{2 \text{ mol KBr}}{2 \text{ mol K}} \times \frac{119 \text{ g KBr}}{1 \text{ mol KBr}} = 12 \text{ g}$$

31. $6\ CO_2(g) + 6\ H_2O(l) \rightarrow C_6H_{12}O_6(s) + 6\ O_2(g)$

First, convert to grams of CO_2:

$$25 \times 10^9 \text{ ton } CO_2 \times \frac{10^6 \text{ g } CO_2}{1 \text{ ton } CO_2} = 25 \times 10^{15} \text{ g } CO_2$$

From the equation:

$$6(44.0 \text{ g})\ CO_2 \simeq 1(180 \text{ g})\ C_6H_{12}O_6$$

$$264 \text{ g } CO_2 \simeq 180 \text{ g } C_6H_{12}O_6$$

Hence: $\text{mass } C_6H_{12}O_6 = 25 \times 10^{15} \text{ g } CO_2 \times \frac{180 \text{ g } C_6H_{12}O_6}{264 \text{ g } CO_2}$

$$= 1.7 \times 10^{16} \text{ g}$$

Also, from the equation:

$$264 \text{ g } CO_2 \simeq 6(32.0 \text{ g})\ O_2 \simeq 192 \text{ g } O_2$$

Hence: $\text{mass } O_2 = 25 \times 10^{15} \text{ g } CO_2 \times \frac{192 \text{ g } O_2}{264 \text{ g } CO_2} = 1.8 \times 10^{16} \text{ g}$

CHAPTER 4

Energy Changes

I. Basic Skills

A. Qualitative Students should be able to:

1. a. Describe the different forms of energy.
   b. Explain what is meant by potential energy and kinetic energy.
2. State the Law of Conservation of Energy and explain its application to physical and chemical changes.
3. Define the energy unit, the calorie.
4. a. Explain what is meant by endothermic and exothermic processes.
   b. Classify a change of state as endothermic or exothermic (Table 4.1).
5. Describe what is meant by ΔH (Equation 4.3) and how the proper sign is assigned to its value (Equations 4.4 and 4.5).
6. Define heat of fusion (ΔH_{fus}) and heat of vaporization (ΔH_{vap}).
7. Describe what is meant by heat of formation and heat of combustion.

B. Quantitative Students should be able to:

1. Make conversions between different energy units (Equations 4.1 and 4.2 and Example 4.1).
2. Write the thermochemical equation for a physical or chemical change (Example 4.3).
3. Draw the heat content diagram for a physical or chemical change showing the ΔH (Example 4.3).
4. Use the Heats of Formation Table (Table 4.2) and the Heats of Combustion Table (Table 4.3).
5. a. Determine the ΔH for any amount of reactant or product using the thermochemical equation (Example 4.5).
   b. Determine the amount of a substance required to produce a particular ΔH (Example 4.6).
6. Calculate the ΔH of a process by adding equations; Hess' Law (Equation 4.13 and Example 4.7).

II. Chapter Development

1. The different forms of energy (heat, light, etc.) are introduced and related to everyday events. Energy is then classified as potential or kinetic. A boulder-on-the-mountain analogy is used to explain chemical potential energy.
2. The primary energy units used are the calorie and the kilocalorie. The joule and the kilowatt hour are related to the calorie through conversion factors. Because

of its use in calorimetry and in dietary subjects the calorie is more meaningful to students.

3. The authors have chosen to use the term "heat content" instead of "enthalpy" for the symbol H. Accordingly, ΔH is read as "the change in heat content."
4. The ΔH for changes in state is presented prior to that of chemical changes because of its simplicity. The effect of attractive forces on heats of fusion and heats of vaporization is discussed in Chapter 12.
5. Heats of formation (Table 4.2) are used as a source of thermochemical equations for calculations and for Hess' Law problems. The ΔH_f values are not used directly to determine the ΔH of a reaction.
6. Calorimetry is presented in an optional section as an experimental technique used to determine the ΔH of a process. The meaning of specific heat follows from the definition of the calorie. Calorimetry and specific heat are best discussed in the context of related laboratory experiments.
7. Additional material on fossil fuels and the energy crisis is provided in Sections 1–3 of Chapter 27, Energy Resources. You may want to review this material for background information. The subject of nuclear energy is also covered in Chapter 27.

III. Problem Areas

1. The terms "endothermic" and "exothermic" are often intermixed. The same problem exists for the signs on ΔH. Use heat content diagrams to help solve these problems. If necessary, return to the boulder-mountain analogy.
2. The correct use of thermochemical equations is another stumbling block. Stress that the ΔH value given in a thermochemical equation is related to the molar amounts in the equation. Demonstrate this by changing the coefficients in an equation and adjusting the ΔH value.
3. The unit analysis method is extended in this chapter to handle calculations involving energy. An energy conversion factor is used to make mole-energy conversions. This provides additional practice for those students who are still striving to learn the method.

IV. Suggested Activities

1. Energy Changes in Chemical Reactions: **Teacher Demonstration Only**
   Demonstrate chemical reactions in which the energy change is obvious. The examples given below are all eye-opening "spectaculars." Stress the energy effects and avoid having to give reaction equations. Draw generalized heat content diagrams.
   a. For an exothermic reaction, cover the bottom of a paper cup with finely powdered $KMnO_4$ to a depth of 2–3 mm. (If necessary, grind the $KMnO_4$ in small amounts with a clean, dry mortar and pestle.) Place the cup on

a heat resistant pad and add 5–10 drops of glycerol with a dropper. A purple fire will occur in a few seconds. The reaction is believed to be:

$$14\ KMnO_4(s) + 4\ C_3H_5(OH)_3(l) \rightarrow$$

$$7\ K_2CO_3(s) + 7\ Mn_2O_3(s) + 5\ CO_2(g) + 16\ H_2O(g)$$

b. For an endothermic reaction, place 32 g of $Ba(OH)_2 \cdot 8\ H_2O$ into a 250-ml flask and 16 g of NH_4SCN into a second 250-ml flask. Wet the outside of the NH_4SCN flask with water and add the barium hydroxide crystals. Shake the flask vigorously until the mass begins to liquify. Measure the liquid's temperature and note the ice formed on the outside of the flask. (Spontaneous endothermic chemical reactions are rare.) The reaction is:

$$Ba(OH)_2 \cdot 8\ H_2O(s) + 2\ NH_4SCN(s) \rightarrow$$

$$Ba(SCN)_2(aq) + 2\ NH_3(g) + 10\ H_2O(l)$$

2. Conversion of Chemical Energy to Light Energy: **Teacher Demonstration Only** Demonstrate the conversion of chemical energy to light energy using a photochemical reaction. Darken the room and simultaneously pour 50 ml of solutions A and B over a column of ice cubes in a tall cylinder. Solution A is 0.5 g of luminol and 0.3 g NaOH in 1 liter of water. Solution B is 10 ml of 3% H_2O_2 and 1.25 g $K_3Fe(CN)_6$ in 1 liter of water.

V. Answers to Questions

1. a. house furnace (gas or oil), water heater (gas), fireplace (wood)
   b. auto engine (gasoline), chemical rockets, gun (gun powder)
   c. auto battery, dry cells, electrical power plant

2. a. decrease b. increase c. no change d. decrease
3. Energy is neither created nor destroyed in ordinary physical and chemical changes.
4. a. kilocalorie b. calorie c. kilowatt hour
5. a. exothermic (combustion rx) b. endothermic (decomposition rx)
   c. endothermic (decomposition rx) d. exothermic (formation rx)
6. a. negative b. positive c. positive d. negative
7. a. negative b. positive c. less than
8. a. exothermic
   b. $\Delta H = +288$ kcal
   c. The heat content of 2 mol MgO is 288 kcal less than that of 2 mol Mg + 1 mol O_2.
9. a. endothermic b. positive c. above
10. a. exothermic b. negative
11. a. $C_6H_6(l) \rightarrow C_6H_6(g)$; $\Delta H = +7.36$ kcal

    b. $H_2O(l) \rightarrow H_2O(s)$; $\Delta H = -1.44$ kcal

    c. $H_2O(g) \rightarrow H_2O(l)$; $\Delta H = -9.72$ kcal
12. a. positive b. positive c. usually negative d. negative
13. a. $2\ Ag(s) + 1/2\ O_2(g) \rightarrow Ag_2O(s)$; $\Delta H = -7.3$ kcal

    b. $C(s) + O_2(g) \rightarrow CO_2(g)$; $\Delta H = -94.1$ kcal

    c. $C_3H_8(g) + 5\ O_2(g) \rightarrow 3\ CO_2(g) + 4\ H_2O(l)$; $\Delta H = -531$ kcal
14. Forward and reverse reactions are opposite processes. In order for energy to be conserved, the energy changes must have the same magnitude but opposite signs. Otherwise, energy could be created by running the reaction first in one direction and then the other.
15. When the equations for two reactions are additive, their ΔH values are additive.
16. a. $\Delta H = +57.8$ kcal, b. $\Delta H = +68.3$ kcal c. $\Delta H = +26.4$ kcal

VI. Solutions to Problems

1. a. 219 cal = 0.219 kcal

   b. $$E(\text{kcal}) = 6.283\ \text{kJ} \times \frac{1\ \text{kcal}}{4.184\ \text{kJ}} = 1.502\ \text{kcal}$$

   c. $$E(\text{kcal}) = 4.19\ \text{kwh} \times \frac{860\ \text{kcal}}{1\ \text{kwh}} = 3.60 \times 10^3\ \text{kcal}$$

2. a. $$\text{Cost} = 1\ \text{kcal} \times \frac{1\ \text{kwh}}{860\ \text{kcal}} \times \frac{3.5¢}{\text{kwh}} = 4.1 \times 10^{-3}¢$$

   b. $$\text{Cost} = 58.2\ \text{kcal} \times \frac{4.1 \times 10^{-3}¢}{1\ \text{kcal}} = 0.24¢$$

3.

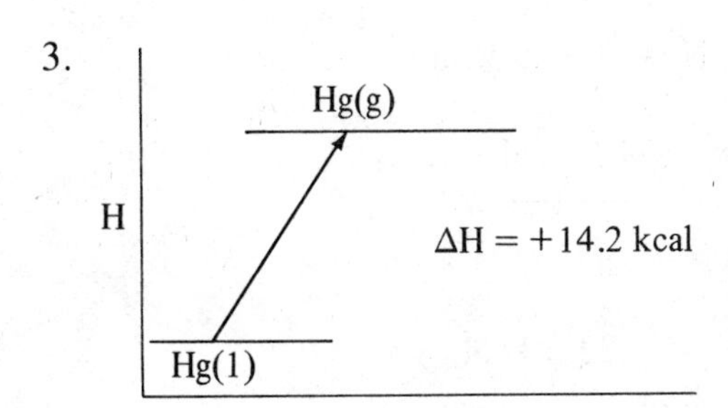

4.

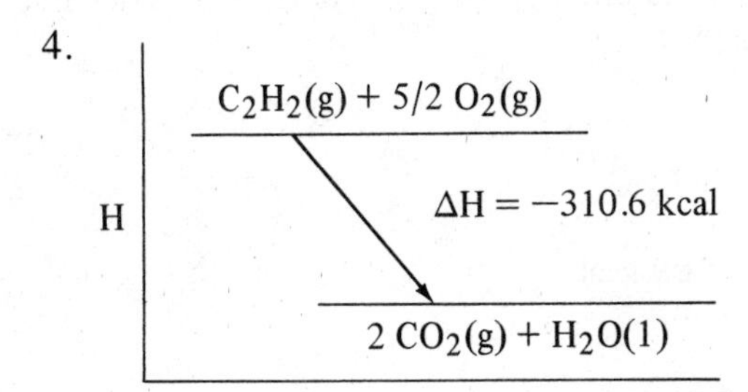

5. $CuO(s) + H_2(g) \rightarrow Cu(s) + H_2O(l)$; $\Delta H = -31.2$ kcal

a. $\Delta H = 2 \text{ mol CuO} \times \dfrac{-31.2 \text{ kcal}}{1 \text{ mol CuO}} = -62.4 \text{ kcal}$

b. $\Delta H = 1.00 \text{ g CuO} \times \dfrac{1 \text{ mol CuO}}{79.5 \text{ g CuO}} \times \dfrac{-31.2 \text{ kcal}}{1 \text{ mol CuO}} = -0.392 \text{ kcal}$

c. $\Delta H = 1.00 \text{ g Cu} \times \dfrac{1 \text{ mol Cu}}{63.5 \text{ g Cu}} \times \dfrac{-31.2 \text{ kcal}}{1 \text{ mol Cu}} = -0.491 \text{ kcal}$

6. a. $\Delta H = 0.210 \text{ mol } Br_2 \times \dfrac{7.16 \text{ kcal}}{1 \text{ mol } Br_2} = 1.50 \text{ kcal}$

b. $\Delta H = 15.2 \text{ g } Br_2 \times \dfrac{1 \text{ mol } Br_2}{160 \text{ g } Br_2} \times \dfrac{7.16 \text{ kcal}}{1 \text{ mol } Br_2} = 0.680 \text{ kcal}$

7. $CaO(s) + SO_3(g) \rightarrow CaSO_4(s)$; $\Delta H = -96.0$ kcal

$\text{mass } CaSO_4 = 1.00 \text{ kcal} \times \dfrac{1 \text{ mol } CaSO_4}{96.0 \text{ kcal}} \times \dfrac{136 \text{ g } CaSO_4}{1 \text{ mol } CaSO_4} = 1.42 \text{ g}$

8. $H_2O(s) \rightarrow H_2O(l)$; $\Delta H = +1.44$ kcal

$H_2O(l) \rightarrow H_2O(g)$; $\Delta H = +9.72$ kcal

$H_2O(s) \rightarrow H_2O(g)$

$\Delta H = +1.44 \text{ kcal} + 9.72 \text{ kcal} = +11.16 \text{ kcal}$

9. $S(s) + 3/2\ O_2(g) \rightarrow SO_3(g)$; $\Delta H = -94.5$ kcal

   $SO_2(g) \rightarrow S(s) + O_2(g)$; $\Delta H = +71.0$ kcal

   ---

   $SO_2(g) + 1/2\ O_2(g) \rightarrow SO_3(g)$

   $\Delta H = -94.5 \text{ kcal} + 71.0 \text{ kcal} = -23.5 \text{ kcal}$

   NOTE: In order to make the above equations additive, the second equation must be reversed from the way that it appears in Table 4.2. This makes it necessary to change the sign of ΔH.

10. a.

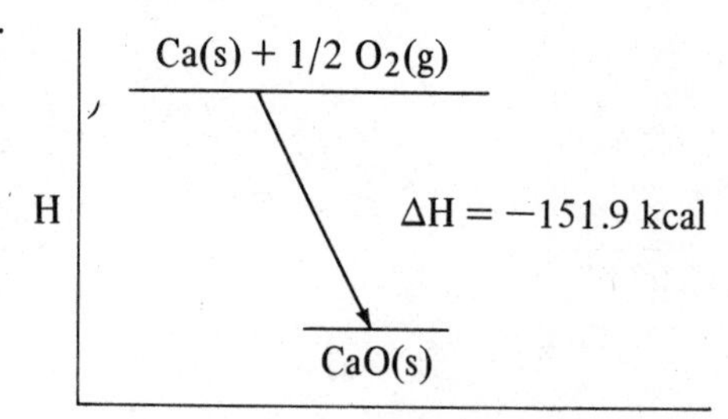

b.

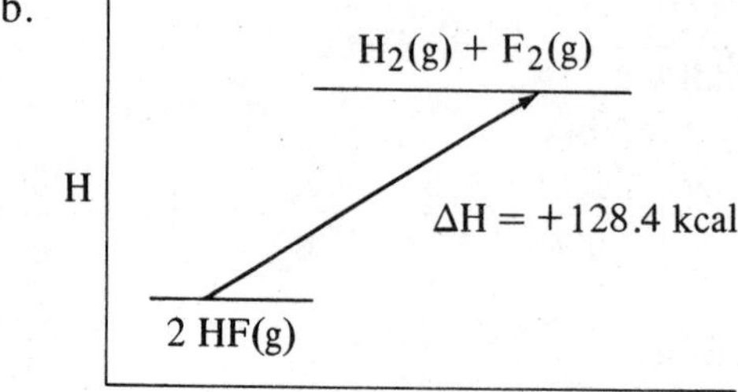

11. a. $\Delta H = +5.0$ kcal b. $\Delta H = -2.0$ kcal

    c. $\Delta H = -3.0$ kcal d. $\Delta H = -5.0$ kcal

12. a. $\Delta H = +4.8$ kcal

    b. $\Delta H = 1.00 \text{ g Ca} \times \dfrac{1 \text{ mol Ca}}{40.1 \text{ g Ca}} \times \dfrac{-151.9 \text{ kcal}}{1 \text{ mol Ca}} = -3.79 \text{ kcal}$

    c. $\Delta H = 1.00 \text{ g CO} \times \dfrac{1 \text{ mol CO}}{28.0 \text{ g CO}} \times \dfrac{-26.4 \text{ kcal}}{1 \text{ mol CO}} = -0.943 \text{ kcal}$

13. $\Delta H = 40.0 \text{ g glucose} \times \dfrac{1 \text{ mol glucose}}{180 \text{ g glucose}} \times \dfrac{-673 \text{ cal}}{1 \text{ mol glucose}}$

    $= -1.50 \times 10^2 \text{ kcal}$

14. $\Delta H = 1.00 \times 10^3 \text{ g U} \times \dfrac{1 \text{ mol U}}{238 \text{ g U}} \times \dfrac{-505 \text{ kcal}}{1 \text{ mol U}} = -2.12 \times 10^3 \text{ kcal}$

15. $C(gr) + O_2(g) \rightarrow CO_2(g); \Delta H = -94.05$ kcal

$CO_2(g) \rightarrow C(dia) + O_2(g); \Delta H = +94.50$ kcal

---

$C(gr) \rightarrow C(dia)$

$\Delta H = 94.50 \text{ kcal} - 94.05 \text{ kcal} = +0.45 \text{ kcal}$

16. $2\ Al(s) + 3/2\ O_2(g) \rightarrow Al_2O_3(s); \Delta H = -400$ kcal

$Fe_2O_3(s) \rightarrow 2\ Fe(s) + 3/2\ O_2(g); \Delta H = +200$ kcal

---

$2\ Al(s) + Fe_2\ O_3(s) \rightarrow 2\ Fe(s) + Al_2O_3(s)$

$\Delta H = -400 \text{ kcal} + 200 \text{ kcal} = -200 \text{ kcal}$

17. $N_2H_4(g) + O_2(g) \rightarrow N_2(g) + 2\ H_2O(l); \Delta H = -150$ kcal

$$\text{mass } N_2H_4 = 1.0 \times 10^6 \text{ kcal} \times \frac{32.0 \text{ g } N_2H_4}{150 \text{ kcal}} \times \frac{1 \text{ kg } N_2H_4}{10^3 \text{ g } N_2H_4} = 210 \text{ kg}$$

18. $$\text{heat absorbed by } H_2O = 1.26 \times 10^3 \text{ g} \times \frac{1.00 \text{ cal}}{\text{g} \cdot {}^\circ\text{C}} \times 5.8^\circ\text{C} = 7.31 \times 10^3 \text{ cal}$$

$\Delta H_{Rx} = -(\text{heat absorbed}) = -7.31 \times 10^3$ cal or -7.31 kcal

19. $$\text{SpHt} = \frac{\text{heat absorbed}}{\text{mass} \times \Delta T}$$

$$\text{SpHt} = \frac{214 \text{ cal}}{40.0 \text{ g} \times 50.0^\circ\text{C}} = 0.107 \text{ cal/g} \cdot {}^\circ\text{C}$$

20. $$C_2H_2\text{: } \frac{310.6 \text{ kcal/mol}}{26.04 \text{ g/mol}} = 11.93 \text{ kcal/g}$$

$$C_2H_6\text{: } \frac{372.8 \text{ kcal/mol}}{30.07 \text{ g/mol}} = 12.40 \text{ kcal/g}$$

21. $$\text{Cost} = 1 \text{ kcal} \times \frac{1 \text{ mol } CH_4}{212.8 \text{ kcal}} \times \frac{0.52¢}{1 \text{ mol } CH_4} = 2.4 \times 10^{-3}¢$$

$$\text{Cost} = 1 \text{ kwh} \times \frac{860 \text{ kcal}}{1 \text{ kwh}} \times \frac{2.4 \times 10^{-3}¢}{1 \text{ kcal}} = 2.1¢$$

$$\frac{\text{Cost (electricity)}}{\text{Cost (methane)}} = \frac{3.5¢/\text{kwh}}{2.1¢/\text{kwh}} = 1.7$$

At these rates, it is less expensive to use natural gas.

CHAPTER 5

The Physical Behavior of Gases

I. Basic Skills

A. Qualitative Students should be able to:

1. Explain what is meant by an absolute temperature scale.
2. Describe how a barometer works.
3. Explain what is meant by partial pressure.
4. Describe how a gas differs from a solid or liquid with regard to distance between particles.
5. Describe the characteristics of molecular motion in gases.
6. Describe how translational kinetic energy, molecular mass, and speed are related (Equation 5.7).
7. Describe the basic postulates of the kinetic molecular theory of gases.
8. Explain qualitatively the effect of pressure and temperature changes on the volume of a fixed sample of gas (Equations 5.9 and 5.11).
9. State the molar volume of gases at standard conditions.
10. Explain the Law of Combining Volumes.

B. Quantitative Students should be able to:

1. Convert between:
   a. liters and cubic centimeters (Equation 5.1).
   b. the Celsius and Kelvin temperature scales (Equation 5.2 and Example 5.1).
   c. pressure units of atm, mm Hg, lb/in^2, and kPa (Equations 5.3 and 5.4 and Example 5.2).
2. Relate the total pressure of a gas mixture to the partial pressures of the gases present (Equation 5.5 and Example 5.3).
3. Relate final and initial volumes and pressures for a fixed sample of gas at constant temperature (Equation 5.10 and Example 5.4).
4. Relate final and initial volumes and temperatures for a fixed sample of gas at constant pressure (Equation 5.12 and Example 5.5).
5. Relate final and initial volumes, pressures, and temperatures for a fixed sample of gas (Equation 5.14 and Example 5.6).
6. Relate the mass of a gas sample to its volume at standard conditions (Equation 5.16 and Example 5.8).
7. Relate the molecular masses of two gases to their densities (Equation 5.17 and Example 5.9).
8. Apply the Law of Combining Volumes to relate the volumes of different gases in a reaction (Example 5.10).
9. Relate the volume of a gas in a reaction to the amount (moles or grams) of a reactant or product (Example 5.11).

II. Chapter Development

1. The pressure units used in this chapter, and future chapters, are primarily atmospheres or millimeters of mercury. These units are most suitable for students at this stage. Kilopascals and pounds per square inch are introduced for comparison and completeness.
2. The algebraic approach to the solution of gas law problems is used throughout the chapter. The student must rearrange equations to solve for the unknown variable. Students will need this skill in any future science courses including college chemistry. From this respect, the use of the factor approach is less desirable. We do suggest, however, that students use the factor approach where applicable to check that their answers are physically reasonable.
3. Dalton's Law is covered only to the point that students can relate total to partial pressure. The relationship between partial pressure and mole fraction is not covered.
4. The treatment of kinetic molecular theory is entirely qualitative. A discussion of average speeds is given in an optional section. The effect of temperature on speed (and energy) will be discussed further in Chapter 17 with regard to reaction rates.
5. The relationship between speed and molecular mass (Graham's Law) is not covered.
6. The experimental determination of molecular masses is discussed only in terms of relative densities at the same temperature and pressure or the density of an individual gas at STP. We do not expect most students to be able to calculate the molecular mass of a gas from density data at conditions other than STP.
7. The ideal gas law has been presented in an optional section near the end of the chapter. To us, it does not seem essential to cover this law in a student's first exposure to the behavior of gases. None of the given problems requires its use. If you choose to expand your coverage into the ideal gas law, you will need to supply additional problems.
8. The last section of the chapter, dealing with the volumes of gases involved in reactions, is reviewed extensively in Chapter 6, The Chemical Behavior of Gases.

III. Problem Areas

1. Some students have difficulty understanding the principle behind the barometer. Show a mercury barometer and collect daily readings for a period to show the change in atmospheric pressure. Students in high altitude locations may wonder why the pressure given in TV weather reports differs from that obtained in the laboratory.
2. In spite of all warnings, students persist in using °C rather than K for the temperature in gas law calculations. Point out that when using Charles' Law, gas

volumes are not zero at 0°C or do not become negative at temperatures lower than 0°C.

3. Expect students to have difficulty in the algebraic manipulations of the gas laws. It is worth spending some time here to review the algebra again. The Review of Mathematics in the Appendix provides additional help on this subject.
4. The relationships between P, V, and T can be clarified for many students by emphasizing a graphical approach. Point out the distinction between direct relationships (V vs T and P vs T) and an inverse relationship (P vs V). Relationships of these types will occur frequently in the course. A review of graphing principles is presented in the Appendix.
5. Students need to be repeatedly cautioned that the molar volume value of 22.4 ℓ is only applicable to gases at standard conditions. The majority of problems given in this area involve gases at STP. Students should be able to convert the molar volume to other conditions if the need arises.

IV. Suggested Activities

1. Effect of Atmospheric Pressure: **Teacher Demonstration Only**
   Demonstrate the existence of atmospheric pressure by collapsing a large metal can. Add a small amount of water (about 50 cm^3) to a 1 gallon size metal can. (Used cans may be available or new ones can be obtained from paint supply stores.) Heat the water to boiling, allow it to boil briefly, and then cap the can quickly. During the boiling, the steam forces most of the air from the can. As the can cools, the can will gradually collapse due to the reduced pressure inside the can caused by the condensation of the steam. The can will collapse more rapidly if a small amount of cold water is poured on its top. If the can is not ruptured, a windy student can blow the can back to near its original shape.
2. Kinetic Molecular Theory: **Teacher Demonstration Only**
   a. Use a molecular vibration tube (Sargent-Welch or equivalent) to show the random motion of gas particles. The sealed tube contains colored glass chips and liquid mercury at reduced pressure. Heat the tube gently to vaporize the the mercury. The invisible mercury particles collide with the glass chips. The violent random motion of the visible particles is evidence of the motion of the gas particles.
   b. Exhibit a stoppered flask containing a small amount of bromine vapor and point out that some gases are colored (absorb visible light). CAUTION: Do not unstopper the bromine flask. Pass around a bottle of household ammonia solution and have students waft test the vapors. Point out that some gases have a detectable odor.
   c. Briefly unstopper side-by-side flasks containing concentrated ammonia and concentrated hydrochloric acid. The invisible vapors will mix and produce a visible ammonium chloride smoke. Have the students explain the meaning

of their observations.

3. Boyle's Law:
   a. Demonstrate the compressibility of gases and Boyle's Law by compressing a gas in a syringe. The syringe should be 10 cm^3 or larger and mounted so that weights can be applied to the plunger. (An ideal kit including a 30 ml plastic syringe and two wooden stands is available from suppliers of the *IPS* program.) When filled with air, the syringe can be compressed by hand to show the compressibility of gases. This should be related to the large distance between gas particles. Pressure-volume data can be collected by gradually increasing the mass applied to the plunger and noting the new volume. Books are commonly used with the *IPS* kit for applying pressure. Have the students plot the data to see the inverse relationship of P and V. NOTE: The first reading is usually in error due to friction in the system. High pressures may cause leakage so that these readings should also be discarded.
   b. Most students will need to be reminded that the pressure on the plunger is not zero when no mass is applied. Because the atmospheric pressure is neglected, the data will not fit Boyle's Law, PV = k. If desired, the equivalent effect of the atmosphere can be predetermined by the teacher. This value is then added to the "applied" pressure to obtain the total pressure. For example, with the *IPS* kit the atmospheric pressure is equivalent to 1–2 books.

V. Answers to Questions

1. a. Since K = °C + 273, the temperature in K is always 273° greater than that in °C.
   b. Since one atmosphere of pressure supports a column of mercury 760 mm in height, the pressure in mm of Hg is always greater than that in atmospheres.
2. The barometer would read low. The height of the mercury column reflects the difference in pressure outside and inside the tube. If there were air over the mercury instead of a vacuum, the pressure difference would be smaller.
3. The partial pressure of a gas in a mixture is the pressure the gas would exert if it occupied the container "by itself" at the same temperature. The partial pressure of a gas is always less than the total pressure.
4. a. Dalton's Law: The total pressure of a gas mixture is the sum of the partial pressures of the individual gases.

$$P_{total} = P_1 + P_2 + \cdots$$

   b. Boyle's Law: The volume of a gas sample is inversely proportional to its pressure, at constant temperature.

$$P = k\left(\frac{1}{V}\right) \quad \text{or} \quad PV = k$$

   c. Charles' Law: The volume of a gas sample at constant pressure is directly proportional to its absolute temperature (K).

$$V/T = k \quad \text{or} \quad V = kT$$

5. The basic postulates of the kinetic molecular theory are listed on page 112. Have students explain the meaning of such terms as "translational kinetic energy," "elastic" collisions, and "average" speed.
6. a. The low density of gases, their high compressibility, and the ease with which they diffuse into each other suggest that the molecules are far apart.
   b. Brownian motion is probably the simplest evidence. Simply quoting average molecular speeds is not an acceptable answer.
   c. The fact that the pressure of a gas does not decrease with time indicates that collisions are elastic.
7. a. Molecules in the gaseous state are farther apart. As there is less mass per unit volume, the density is lower.
   b. Since molecules are far apart in a gas, compression simply reduces the large amount of empty space. There is very little empty space in a liquid.
   c. Since gas molecules are far apart, they move past one another readily. This is more difficult in liquids where the closeness of molecules creates a traffic jam.
   d. Since molecules have mass and are in motion, they exert a force on the walls of their container. Pressure is the measure of this force over a unit area.

8. a. Divide both sides by V_2: $P_2 = P_1 \times \frac{V_1}{V_2}$

   b. Divide both sides by P_2: $V_2 = V_1 \times \frac{P_1}{P_2}$

   c. Divide both sides by V_1: $P_1 = P_2 \times \frac{V_2}{V_1}$

   d. Divide both sides by P_1: $V_1 = V_2 \times \frac{P_2}{P_1}$

9. a. $V_2 = V_1 \times \frac{P_1}{P_2} \times \frac{T_2}{T_1}$ b. $P_2 = P_1 \times \frac{V_1}{V_2} \times \frac{T_2}{T_1}$

   c. $T_2 = T_1 \times \frac{V_2}{V_1} \times \frac{P_2}{P_1}$ d. $V_1 = V_2 \times \frac{P_2}{P_1} \times \frac{T_1}{T_2}$

   e. $P_1 = P_2 \times \frac{V_2}{V_1} \times \frac{T_1}{T_2}$ f. $T_1 = T_2 \times \frac{V_1}{V_2} \times \frac{P_1}{P_2}$

10. A direct proportionality requires that $y = ax$, where "a" is a constant. Answers (a) and (d) are of this type.

An inverse proportionality requires that $y = a/x$ or $yx = a$. Answers (b) and (e) are of this type. Answers (c) and (f) are in neither category.

11. a. P increases as more gas is added (constant T and V).
    b. P decreases as volume is increased (constant T and n).
    c. P increases as temperature is increased (constant V and n).
12. a. Yes. Both pressure terms must be in like units.
    b. No. V is directly proportional to T in K, but not in °C.
13. a. When the volume is reduced, molecules strike the walls more frequently.
    b. When the temperature is increased, the average molecular speed increases. The molecules collide with walls more frequently and with greater force.
    c. A molecule's speed and energy increases with increasing temperature. In order to hold pressure constant, volume must be increased.
14. a. The N_2 and N_2O molar volumes are the same. All gases have the same molar volume at the same T and P.
    b. Molar volume is smaller at 0°C. (Charles' Law)
    c. Molar volume is larger at 1 atm. (Boyle's Law)
15. Answers (a) and (b) have the same volume, 22.4 ℓ. Answer (c) has a volume of 11.2 ℓ because it represents a half mole of H_2. Answer (d) has a volume of only 18 cm^3 as H_2O is a liquid at STP. The order is: (d) < (c) < (a) = (b)
16. The density of a gas is directly proportional to its molecular mass. The molecular masses are: N_2 (28.0), O_2 (32.0), Ar (39.9), H_2O (18.0). Hence the order is: $H_2O < N_2 < O_2 <$ Ar
17. a. True: Law of Combining Volumes
    b. False: 1 mole of C_3H_8 would produce 3 moles of CO_2, which is 3(22.4 ℓ) or 67.2 ℓ.
    c. False: Water is a liquid at the reaction conditions, not a gas.

VI. Solutions to Problems

1. a. $T(K) = t(°C) + 273 = 80 + 273 = 353\ K$

   b. First convert to °C:

   $°F = 1.8(°C) + 32$

   $98.6°F = 1.8(°C) + 32;\ 1.8(°C) = 66.6$

   $t(°C) = 66.6/1.8 = 37°C$

   $T(K) = 37 + 273 = 310\ K$

2. a. $$P(atm) = 742\ mm\ Hg \times \frac{1\ atm}{760\ mm\ Hg} = 0.976\ atm$$

   b. $$P(psi) = 742\ mm\ Hg \times \frac{14.7\ lb/in^2}{760\ mm\ Hg} = 14.4\ lb/in^2$$

c. $P(\text{kPa}) = 742 \text{ mm Hg} \times \frac{101.3 \text{ kPa}}{760 \text{ mm Hg}} = 98.9 \text{ kPa}$

3. a. From Table 5.1: $P_{H_2O} = 19$ mm Hg at 21°C

$$P_{O_2} = P_{total} - P_{H_2O} = 739 \text{ mm Hg} - 19 \text{ mm Hg} = 720 \text{ mm Hg}$$

b. From Table 5.1: $P_{H_2O} = 15$ mm Hg at 18°C

$$P_{total} = P_{H_2} + P_{H_2O} = 720 \text{ mm Hg} + 15 \text{ mm Hg} = 735 \text{ mm Hg}$$

4. In Boyle's Law, both pressures must be expressed in the same units:

$$V_2 = V_1 \times \frac{P_1}{P_2}; \quad P_2 = 730 \text{ mm Hg} \times \frac{1 \text{ atm}}{760 \text{ mm Hg}} = 0.961 \text{ atm}$$

$$V_2 = 50.0\ \ell \times \frac{120 \text{ atm}}{0.961 \text{ atm}} = 6.24 \times 10^3\ \ell$$

5. In Charles' Law, temperature must be expressed in K.

$$V_2 = V_1 \times \frac{T_2}{T_1}; \quad T_1 = 20 + 273 = 293 \text{ K}; \quad T_2 = -20 + 273 = 253 \text{ K}$$

$$V_2 = 2.10 \times 10^5 \text{ m}^3 \times \frac{253 \text{ K}}{293 \text{ K}} = 1.81 \times 10^5 \text{ m}^3$$

6. $V_2 = V_1 \times \frac{P_1}{P_2} \times \frac{T_2}{T_1}$

$V_1 = 3.82 \times 10^3 \text{ m}^3$; $P_1 = 740$ mm Hg; $T_1 = 273 + 22 = 295$ K

$P_2 = 219$ mm Hg; $T_2 = 273 - 15 = 258$ K

$$V_2 = 3.82 \times 10^3 \text{ m}^3 \times \frac{740 \text{ mm Hg}}{219 \text{ mm Hg}} \times \frac{258 \text{ K}}{295 \text{ K}} = 1.13 \times 10^4 \text{ m}^3$$

Note that the decrease in P more than compensates for the decrease in T.

7. $P_2 = P_1 \times \frac{V_1}{V_2} \times \frac{T_2}{T_1}$

$V_1 = 106\ \ell$; $P_1 = 42.7 \text{ lb/in}^2$; $T_1 = 288$ K

$V_2 = 110\ \ell$; $T_2 = 273 + 45 = 318$ K

$$P_2 = 42.7 \text{ lb/in}^2 \times \frac{106\ \ell}{110\ \ell} \times \frac{318 \text{ K}}{288 \text{ K}} = 45.4 \text{ lb/in}^2$$

8. a. $V = 0.500 \text{ mol } O_2 \times \frac{22.4\ \ell}{1 \text{ mol } O_2} = 11.2\ \ell$

b. As the MM of O_2 is 32.0, 32.0 g $O_2 \simeq 22.4\ \ell$

$$V = 1.00 \text{ g } O_2 \times \frac{22.4\ \ell}{32.0 \text{ g } O_2} = 0.700\ \ell$$

c. As the MM of N_2 is 28.0, 28.0 g $N_2 \simeq 22.4\ \ell$:

$$V = 1.00 \text{ g } N_2 \times \frac{22.4\ \ell}{28.0 \text{ g N}} = 0.800\ \ell$$

9. $\frac{(MM)_2}{(MM)_1} = \frac{d_2}{d_1}$

$(MM)_1$ of N_2 = 28.0; $d_2/d_1 = 2.72$

$$(MM)_2 = (MM)_1 \times \frac{d_2}{d_1} = 28.0 \times 2.72 = 76.2$$

10. $N_2(g) + 3\ H_2(g) \rightarrow 2\ NH_3(g)$

$$V_{H_2} = 11.0\ \ell\ N_2 \times \frac{3\ \ell\ H_2}{1\ \ell\ N_2} = 33.0\ \ell$$

$$V_{NH_3} = 11.0\ \ell\ N_2 \times \frac{2\ \ell\ NH_3}{1\ \ell\ N_2} = 22.0\ \ell$$

11. $NH_4Cl(s) \rightarrow NH_3(g) + HCl(g)$

a. $V_{NH_3} = 1.42 \text{ mol } NH_4Cl \times \frac{22.4\ \ell\ NH_3}{1 \text{ mol } NH_4Cl} = 31.8\ \ell$

b. As the FM of NH_4Cl is 53.5, 53.5 g $NH_4Cl \simeq 22.4\ \ell\ NH_3$.

$$V_{NH_3} = 16.0 \text{ g } NH_4Cl \times \frac{22.4\ \ell\ NH_3}{53.5 \text{ g } NH_4Cl} = 6.70\ \ell$$

12. a. mass $O_2 = 0.912 \text{ mol } O_2 \times \frac{32.0 \text{ g } O_2}{1 \text{ mol } O_2} = 29.2$ g

b. volume $O_2 = 0.912 \text{ mol } O_2 \times \frac{22.4\ \ell}{1 \text{ mol } O_2} = 20.4\ \ell$

c. $P(\text{atm}) = 102 \text{ kPa} \times \frac{1 \text{ atm}}{101.3 \text{ kPa}} = 1.01$ atm

d. $P(\text{mm Hg}) = 102 \text{ kPa} \times \frac{760 \text{ mm Hg}}{101.3 \text{ kPa}} = 765$ mm Hg

13. a. $P_{N_2} = P_{total} - P_{H_2O}$

From Table 5.1: $P_{H_2O} = 18$ mm Hg at 20°C

$P_{N_2} = 759 \text{ mm Hg} - 18 \text{ mm Hg} = 741 \text{ mm Hg}$

b. $V_2 = V_1 \times \frac{P_1}{P_2}$ P_1 = pressure of dry N_2

$$V_2 = 244 \text{ cm}^3 \times \frac{741 \text{ mm Hg}}{760 \text{ mm Hg}} = 238 \text{ cm}^3$$

14. $P_2 = P_1 \times \frac{V_1}{V_2}$ $V_2 = 1/8\ (V_1)$ or $V_1 = 8\ V_2$

$$P_2 = 1.00 \text{ atm} \times 8 \frac{V_2}{V_2} = 8.00 \text{ atm}$$

15. The relationship between P and T, while not specifically covered in the text, can be derived from the combined gas relation when the volume remains constant.

$$\frac{P_2V_2}{T_2} = \frac{P_1V_1}{T_1} \quad \text{and} \quad V_1 = V_2$$

Therefore: $P_2/T_2 = P_1/T_1$

$T_1 = 273 + 22 = 295$ K; $T_2 = 273 + 100 = 373$ K

$$P_2 = P_1 \times \frac{T_2}{T_1} = 740 \text{ mm Hg} \times \frac{373 \text{ K}}{295 \text{ K}} = 936 \text{ mm Hg}$$

16. $\frac{V_1}{T_1} = \frac{V_2}{T_2}$ $T_1 = 273 + 25 = 298$ K

$$T_2 = T_1 \times \frac{V_2}{V_1} = 298 \text{ K} \times \frac{192 \text{ cm}^3}{162 \text{ cm}^3} = 353 \text{ K}$$

$t(°C) = 353 - 273 = 80°C$

17. Point out to students that it is the air pressure in the tire which is directly proportional to T. For this reason, the gauge pressure must be converted to air pressure:

$P_{air} = P_{gauge} + P_{atm}$

$P_{air} = 28.2 \text{ lb/in}^2 + 14.7 \text{ lb/in}^2 = 42.9 \text{ lb/in}^2$

$T_1 = 273 + 15 = 288$ K; $T_2 = 273 + 45 = 318$ K

$$P_2 = P_1 \times \frac{T_2}{T_1} = 42.9\ \text{lb/in}^2 \times \frac{318\ \text{K}}{288\ \text{K}} = 47.4\ \text{lb/in}^2$$

P_{gauge} (final) $= 47.4\ \text{lb/in}^2 - 14.7\ \text{lb/in}^2 = 32.7\ \text{lb/in}^2$

NOTE: If worked directly with gauge pressure, an incorrect answer of 31.1 lb/in^2 will be obtained.

18. a. $\frac{V_1}{T_1} = \frac{V_2}{T_2}$; $T_1 = 273 + 25 = 298$ K; $T_2 = 273$ K

$$V_2 = V_1 \times \frac{T_2}{T_1} = 1.41\ \ell \times \frac{273\ \text{K}}{298\ \text{K}} = 1.29\ \ell$$

b. 1 mole gas $\hat{=}$ 22.4 ℓ at STP

$$\text{mass} = 22.4\ \ell \times \frac{1.60\ \text{g}}{1.29\ \ell} = 27.8\ \text{g}; \quad \text{molecular mass} = 27.8$$

19. a. MM SO_2 = 64.1 Therefore, 64.1 g $SO_2 \hat{=} 22.4\ \ell$

$$\text{density} = \frac{64.1\ \text{g}}{22.4\ \ell} = 2.86\ \text{g}/\ell$$

b. $$V\ (25°\text{C}) = V\ (0°\text{C}) \times \frac{298\ \text{K}}{273\ \text{K}} = 22.4\ \ell \times \frac{298\ \text{K}}{273\ \text{K}} = 24.5\ \ell$$

$$\text{density} = \frac{64.1\ \text{g}}{24.5\ \ell} = 2.62\ \text{g}/\ell$$

20. MM UF_6 = 238 + 6(19) = 352; 352 g $UF_6 \hat{=} 22.4\ \ell$

$$\text{density} = \frac{352\ \text{g}}{22.4\ \ell} = 15.7\ \text{g}/\ell$$

21. $$\frac{\text{density } NO_2}{\text{density NO}} = \frac{\text{MM } NO_2}{\text{MM NO}}$$

$$\text{density } NO_2 = \text{density NO} \times \frac{\text{MM } NO_2}{\text{MM NO}} = 1.23\ \text{g}/\ell \times \frac{46.0}{30.0} = 1.89\ \text{g}/\ell$$

22. $NH_4Cl(s) \rightarrow NH_3(g) + HCl(g)$

a. $$\text{mass } NH_4Cl = 1.00\ \ell\ NH_3 \times \frac{53.5\ \text{g}\ NH_4Cl}{22.4\ \ell\ NH_3} = 2.39\ \text{g}$$

b. As 1 ℓ $NH_3 \hat{=} 1\ \ell$ HCl, the same mass of NH_4Cl, 2.39 g, is required.

23. a. mass $= 22.4\ \ell \times \dfrac{0.670\ \text{g}}{0.500\ \ell} = 30.0\ \text{g}$

b. Molecular mass of unknown gas is 30.0.
MM $N_2 = 28.0$; $O_2 = 32.0$; $NO = 30.0$; $CO = 28.0$
Hence, the unknown gas could be NO.

24. $Ca(OH)_2(s) \rightarrow CaO(s) + H_2O(g)$

$$\text{moles } H_2O = 5.00\ \text{g } Ca(OH)_2 \times \frac{1\ \text{mol } Ca(OH)_2}{74.1\ \text{g } Ca(OH)_2} \times \frac{1\ \text{mol } H_2O}{1\ \text{mol } Ca(OH)_2}$$
$$= 0.0675\ \text{mol}$$

$$\frac{V}{\text{mol}}(110°C) = 22.4\ \ell \times \frac{383\ \text{K}}{273\ \text{K}} = 31.4\ \frac{\ell}{\text{mol}}$$

$$V = 0.0675\ \text{mol} \times \frac{31.4\ \ell}{\text{mol}} = 2.12\ \ell$$

25. According to the Law of Combining Volumes, the volume ratios of gases equals their mole ratios. Therefore:

$$1\ C_xH_y(g) + \underline{\ \ ?\ \ }\ O_2(g) \rightarrow 2\ CO_2(g) + 2\ H_2O(g)$$

where x is the number of C atoms and y is the number of H atoms per molecule. Since all these atoms come from the CO_2 and H_2O, $x = 2$ and $y = 4$. The complete balanced equation is:

$$C_2H_4(g) + 3\ O_2(g) \rightarrow 2\ CO_2(g) + 2\ H_2O(g)$$

26. Volume CO $= \dfrac{0.40}{100}(2.0 \times 10^5\ \ell) = 8.0 \times 10^2\ \ell$

$$\text{mass CO} = 8.0 \times 10^2\ \ell \times \frac{28.0\ \text{g CO}}{22.4\ \ell} = 1.0 \times 10^3\ \text{g } (1.0\ \text{kg})$$

27. a. P is directly proportional to T. At $T = 0$, $P = 0$.

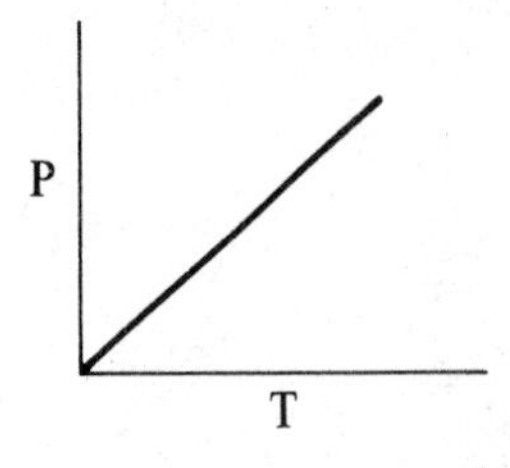

b. V is directly proportional to T in K, or $V = kT$.

As $T = t(°C) + 273$, $V = k(t + 273)$ or $V = kt + 273\ k$
This equation gives a straight line with a y-intercept of 273 k.

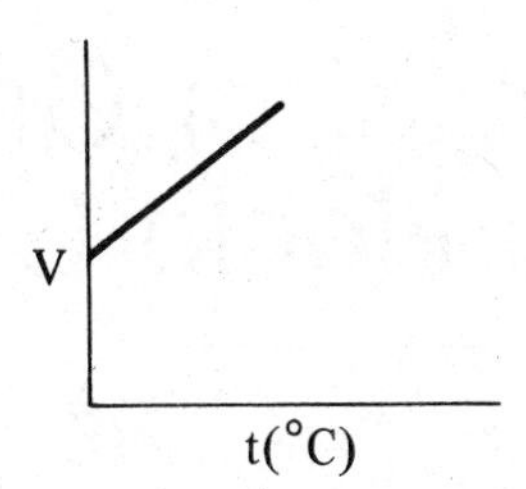

c. As $\frac{PV}{T} = k$, then $PV = kT$. This shows that the product PV is directly proportional to T. At $T = 0$, $PV = 0$.

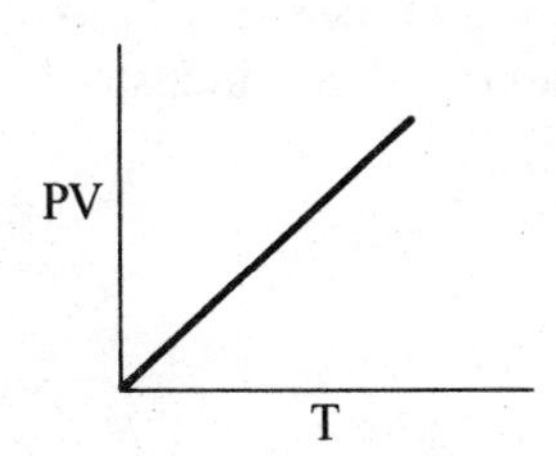

28. The density of the gas is given as 1.619 g/ℓ at 50°C and 720 mm Hg. The volume of one liter must first be corrected to standard conditions.

$$V_2 = V_1 \times \frac{P_1}{P_2} \times \frac{T_2}{T_1}$$

$V_1 = 1.00\ \ell$; $P_1 = 720$ mm Hg; $T_1 = 273 + 50 = 323$ K

$P_2 = 760$ mm Hg; $T_2 = 273$ K

$$V_2 = 1.00\ \ell \times \frac{720\text{ mm Hg}}{760\text{ mm Hg}} \times \frac{273\text{ K}}{323\text{ K}} = 0.801\ \ell$$

As 22.4 ℓ is the molar volume of any gas, the mass of this volume is equal to the molecular mass of the gas.

$$MM = \frac{22.4\ \ell}{\text{mol}} \times \frac{1.619\text{ g}}{0.801\ \ell} = 45.3\text{ g/mol}$$

CHAPTER 6

The Chemical Behavior of Gases: H_2, O_2, N_2, and the Noble Gases

I. Basic Skills

A. Qualitative Students should be able to:

1. Describe the major properties, occurrence, and uses of H_2, O_2, and N_2.
2. Describe a laboratory preparation method and an industrial preparation method for H_2, O_2, and N_2.
3. Describe characteristic chemical reactions of H_2, O_2, and N_2 with metals and nonmetals.
4. Explain what is meant by nitrogen fixation.
5. Describe the unique properties of the noble gases.

B. Quantitative

Chapter 6 reinforces quantitative skills presented previously in Chapters 3–5. No new calculation skills are presented.

II. Chapter Development

1. The physical behavior of gases, presented in Chapter 5, is mostly independent of the identity of the gas. An attempt is made in this chapter to present the distinctive chemical properties of the elemental gases. The Group 7 gases, F_2 and Cl_2, are discussed in Chapter 8.
2. The descriptive material of this chapter provides an opportunity to review many principles presented previously. In this respect, it offers the slower students a chance to catch up. On the other hand, as no new principles are presented, the chapter may be omitted with no effect on subsequent material.
3. Additional material on the chemistry of gaseous compounds is included in Chapter 26. It emphasizes the environmental problems caused by carbon monoxide, the sulfur oxides, and the nitrogen oxides.
4. Acids are introduced as a method of producing hydrogen gas. The use of acids and bases also occurs in the laboratory prior to detailed discussion in the text. It is sufficient for students at this stage to view acids as water solutions which contain the H^+ ion. Bases, in turn, are water solutions which contain the OH^- ion and react with acids to form water.
5. The symbol (aq), for water solutions, was introduced (but not used) in Chapter 3. The reactions of this chapter which involve acidic solutions are the first which require the use of the (aq) symbol.

III. Problem Areas

No new problems are expected. By this time, students should have attained an acceptable skill level in using unit analysis in calculations. In addition, students should be able to write correctly balanced chemical equations.

IV. Suggested Activities

1. Electrolysis of Water: **Teacher Demonstration Only**
   Demonstrate the production of H_2 and O_2 gases and the Law of Combining Volumes by electrolyzing water. Use a standard electrolysis apparatus, if available, adding a small amount of sulfuric acid to increase the rate. The endothermic reaction is:

$$2H_2O_{(l)} \rightarrow 2H_{2(g)} + O_{2(g)}; \quad \Delta H = + 136.6 \text{ kcal}$$

2. Composition of Air: **Teacher Demonstration Only**
   Determine the ratio of N_2/O_2 in air by removing the O_2. Add 0.5 g of pyrogallol to a 50-cm^3 eudiometer tube. Add sufficient 1M NaOH (amount determined in advance) to leave a volume equivalent to 50 cm^3 of air. Stopper the tube immediately and invert a few times to mix the contents. Place the stoppered end under the water level in a 1-liter beaker and remove the stopper. Water should rise in the tube to about one fifth of its height. Adjust the height of the tube until the water levels are equal and read the volume of the remaining N_2 gas. Calculate the N_2/O_2 volume ratio. If students believe in Avogadro's Law, then the results can be extended to give the N_2/O_2 molecule ratio.

V. Answers to Questions

1. a. O_2, N_2 and Ar b. H_2, O_2 and N_2 c. He and Ar
2. Laboratory preparations should be simple, fast, safe, and give a good yield of a pure product. Industrial preparations must produce a product at the least possible cost by using inexpensive materials and energy. The electrolysis of water (Equation 6.1) requires a large amount of expensive electrical energy. The reaction of zinc with acid (Equation 6.2) to make H_2 requires the use of expensive reactants. The reaction of methane with steam (Equation 6.3) is not satisfactory for the laboratory because the H_2 produced is mixed with CO and

the equipment required is complex.

3. a. $Zn(s) + 2\ H^+(aq) \rightarrow Zn^{2+}(aq) + H_2(g)$

   $2\ H_2O(l) \rightarrow 2\ H_2(g) + O_2(g)$

   b. $2\ HgO(s) \rightarrow 2\ Hg(l) + O_2(g)$

   $2\ KClO_3(s) \rightarrow 2\ KCl(s) + 3\ O_2(g)$

   $2\ H_2O_2(aq) \rightarrow 2\ H_2O(l) + O_2(g)$

   c. $NH_4NO_2(aq) \rightarrow 2\ H_2O(l) + N_2(g)$

4. $2\ Al(s) + 6\ H^+(aq) \rightarrow 2\ Al^{3+}(aq) + 3\ H_2(g)$

5. $H_2(g) + Br_2(l) \rightarrow 2\ HBr(g)$

6. a. $H_2(g) + Cl_2(g) \rightarrow 2\ HCl(g)$

   b. $2\ H_2(g) + O_2(g) \rightarrow 2\ H_2O(l)$

   c. $3\ H_2(g) + N_2(g) \rightarrow 2\ NH_3(g)$

7. $SnO_2(s) + 2\ H_2(g) \rightarrow Sn(s) + 2\ H_2O(l)$

8. Hydrogen is mostly used in making ammonia and solid fats. Lesser amounts are used in balloons and for extracting metals from oxides.
9. a. The molecular mass of N_2 (28) is less than the average molecular mass of air (29).
   b. The molecular mass of CO_2 (44) is greater than the average molecular mass of air (29).
10. CO_2, H_2O, Al_2O_3 and SiO_2
11. The boiling point of N_2 is lower than that of O_2.
12. In using HgO, the mercury vapor produced is toxic. In using $KClO_3$, molten $KClO_3$ may react explosively with any combustible substance.

13. a. $2\ KClO_3(s) \rightarrow 2\ KCl(s) + 3\ O_2(g)$

    b. $2\ H_2O(l) \rightarrow 2\ H_2(g) + O_2(g)$

    c. $2\ H_2O_2(aq) \rightarrow 2\ H_2O(l) + O_2(g)$

14. a. $4\ Li(s) + O_2(g) \rightarrow 2\ Li_2O(s)$

    b. $4\ Al(s) + 3\ O_2(g) \rightarrow 2\ Al_2O_3(s)$

    c. $2\ C(s) + O_2(g) \rightarrow 2\ CO(g)$

    d. $N_2(g) + O_2(g) \rightarrow 2\ NO(g)$

15. a. $2\ CO(g) + O_2(g) \rightarrow 2\ CO_2(g)$

    b. $2\ NO(g) + O_2(g) \rightarrow 2\ NO_2(g)$

    c. $2\ SO_2(g) + O_2(g) \rightarrow 2\ SO_3(g)$

16. a. $C_6H_{12}O_6(s) + 6\ O_2(g) \rightarrow 6\ CO_2(g) + 6\ H_2O(l)$

    b. $CH_4(g) + 2\ O_2(g) \rightarrow CO_2(g) + 2\ H_2O(l)$

c. $C_5H_{12}(l) + 8\ O_2(g) \rightarrow 5\ CO_2(g) + 6\ H_2O(l)$

17. Liquid O_2 will react violently with many combustible substances, whereas N_2 is relatively stable even at higher temperatures. Argon, a noble gas, is even less reactive than N_2 and would, therefore, be safer to work with than liquid oxygen.
18. a. N^{3-} b. NH_4^+ c. NH_3
19. a. NH_4Cl b. $(NH_4)_2Cr_2O_7$ c. NH_4NO_3 d. $(NH_4)_2SO_4$
20. a. $6\ Li(s) + N_2(g) \rightarrow 2\ Li_3N(s)$

    b. $3\ Ca(s) + N_2(g) \rightarrow Ca_3N_2(s)$

    c. $3\ Mg(s) + N_2(g) \rightarrow Mg_3N_2(s)$
21. a. $N_2(g) + O_2(g) \rightarrow 2\ NO(g)$

    b. $N_2(g) + 3\ H_2(g) \rightarrow 2\ NH_3(g)$

    c. $N_2(g) + 3\ Cl_2(g) \rightarrow 2\ NCl_3(g)$
22. Helium is used in balloons and as a cooling agent at low temperatures. Argon is used to give an inert atmosphere in welding and in incandescent lamps to reduce evaporation of the tungsten filament.
23. a. Rn b. Ar c. He d. Xe
24. a. O_2, N_2 and Ar b. H_2 and O_2 c. H_2, O_2 and N_2
    d. H_2, O_2 and N_2

VI. Solutions to Problems

1. a. $MM(C_6H_{10}O_5) = 162.14$

$$\%\ C = \frac{72.06}{162.14} \times 100 = 44.44 \qquad \%\ H = \frac{10.08}{162.14} = 6.217$$

$$\%\ O = 100.00 - 44.44 - 6.22 = 49.34$$

b. $FM(NH_4NO_3) = 80.05$

$$\%\ N = \frac{28.01}{80.05} \times 100 = 35.00 \qquad \%\ H = \frac{4.032}{80.05} \times 100 = 5.037$$

$$\%\ O = \frac{48.00}{80.05} \times 100 = 59.96$$

c. $MM(CON_2H_4) = 60.05$

$$\%\ C = \frac{12.01}{60.05} \times 100 = 20.00 \qquad \%\ O = \frac{16.00}{60.05} \times 100 = 26.64$$

$$\%\ H = \frac{4.032}{60.05} \times 100 = 6.714 \qquad \%\ N = \frac{28.01}{60.05} \times 100 = 46.64$$

2. $2\ KClO_3(s) \rightarrow 2\ KCl(s) + 3\ O_2(g)$

a. $\text{moles } O_2 = 10.0 \text{ g } KClO_3 \times \dfrac{1 \text{ mol } KClO_3}{122.5 \text{ g } KClO_3} \times \dfrac{3 \text{ mol } O_2}{2 \text{ mol } KClO_3} = 0.122$

b. $\text{mass } O_2 = 0.122 \text{ mol } O_2 \times \dfrac{32.0 \text{ g } O_2}{\text{mol } O_2} = 3.90 \text{ g}$

3. $6\ Li(s) + N_2(g) \rightarrow 2\ Li_3N(s)$

a. $\text{mass Li} = 1.00 \text{ g } Li_3N \times \dfrac{1 \text{ mol } Li_3N}{34.83 \text{ g } Li_3N} \times \dfrac{6 \text{ mol Li}}{2 \text{ mol } Li_3N} \times \dfrac{6.94 \text{ g Li}}{1 \text{ mol Li}}$
$= 0.598 \text{ g}$

b. $\text{moles } N_2 = 1.00 \text{ g } Li_3N \times \dfrac{1 \text{ mol } Li_3N}{34.83 \text{ g } Li_3N} \times \dfrac{1 \text{ mol } N_2}{2 \text{ mol } Li_3N}$
$= 1.44 \times 10^{-2}$

4. $4\ Li(s) + O_2(g) \rightarrow 2\ Li_2O(s); \quad \Delta H = -284 \text{ kcal}$

$\Delta H = 1.00 \text{ g Li} \times \dfrac{1 \text{ mol Li}}{6.94 \text{ g Li}} \times \dfrac{-284 \text{ kcal}}{4 \text{ mol Li}} = 10.2 \text{ kcal}$

5. $Zn(s) + 2H^+(aq) \rightarrow Zn^{2+}(aq) + H_2(g)$

a. $\text{moles } H_2 = 20.0 \text{ g Zn} \times \dfrac{1 \text{ mol Zn}}{65.4 \text{ g Zn}} \times \dfrac{1 \text{ mol } H_2}{1 \text{ mol Zn}} = 0.306$

b. $\text{volume } H_2 = 0.306 \text{ mol } H_2 \times \dfrac{22.4\ \ell}{1 \text{ mol } H_2} = 6.85\ \ell$

6. $(NH_4)_2Cr_2O_7(s) \rightarrow N_2(s) + 4\ H_2O(l) + Cr_2O_3(s)$

a. $\text{volume } N_2 = 1.00 \text{ g ADC} \times \dfrac{1 \text{ mol ADC}}{252 \text{ g ADC}} \times \dfrac{1 \text{ mol } N_2}{1 \text{ mol ADC}} \times \dfrac{22.4\ \ell}{1 \text{ mol } N_2}$
$= 8.89 \times 10^{-2}\ \ell$

b. $V_1/T_1 = V_2/T_2$

$V_2 = 8.89 \times 10^{-2}\ \ell \times \dfrac{298 \text{ K}}{273 \text{ K}} = 9.70 \times 10^{-2}\ \ell\ N_2$

7. $D = \dfrac{4.00 \text{ g He}}{1 \text{ mol He}} \times \dfrac{1 \text{ mol}}{22.4\ \ell} = 0.179 \text{ g}/\ell$

From Table 6.4: $D = 0.1785 \text{ g}/\ell$

8. $C_8H_{18}(l) + 12.5\ O_2(g) \rightarrow 8\ CO_2(g) + 9\ H_2O(l);\ \Delta H = -1308\ kcal$

$$\text{mass } C_8H_{18} = 1.00\ kcal \times \frac{1\ mol\ C_8H_{18}}{1308\ kcal} \times \frac{114\ g\ C_8H_{18}}{1\ mol\ C_8H_{18}} = 8.72 \times 10^{-2}\ g$$

9. $$\text{mass } KClO_3 = 1.00\ \ell\ O_2 \times \frac{1\ mol\ O_2}{22.4\ \ell\ O_2} \times \frac{2\ mol\ KClO_3}{3\ mol\ O_2} \times \frac{122.5\ g\ KClO_3}{1\ mol\ KClO_3}$$
$$= 3.65\ g$$

10. $CH_4(g) + H_2O(g) \rightarrow CO(g) + 3\ H_2(g)$

a. $20.0\ \ell\ CH_4 \simeq 20.0\ \ell\ H_2O$

b. volume of products $= 20.0\ \ell\ CO + 60.0\ \ell\ H_2 = 80.0\ \ell$

11. a. $$\Delta H = 1\ mol\ H_2 \times \frac{49.3\ kcal}{3\ mol\ H_2} = 16.4\ kcal$$

b. $$\text{moles } H_2 = 1.00\ kcal \times \frac{3\ mol\ H_2}{49.3\ kcal} = 6.09 \times 10^{-2}$$

12. $WO_3(s) + 3\ H_2(g) \rightarrow W(s) + 3\ H_2O(g)$

a. $$\text{volume } H_2 = 1.25\ g\ WO_3 \times \frac{1\ mol\ WO_3}{232\ g\ WO_3} \times \frac{3\ mol\ H_2}{1\ mol\ WO_3} \times \frac{22.4\ \ell}{1\ mol\ H_2}$$
$$= 0.362\ \ell$$

b. $$\text{mass } W = 1.25\ g\ WO_3 \times \frac{1\ mol\ WO_3}{232\ g\ WO_3} \times \frac{1\ mol\ W}{1\ mol\ WO_3} \times \frac{184\ g\ W}{1\ mol\ W}$$
$$= 0.991\ g$$

13. a. $2\ HgO(s) \rightarrow 2\ Hg(l) + O_2(g)$

$$\text{moles } O_2 = 100\ g\ HgO \times \frac{1\ mol\ HgO}{216.6\ g\ HgO} \times \frac{1\ mol\ O_2}{2\ mol\ HgO} = 0.231$$

$$\text{volume } O_2 = 0.231\ mol \times \frac{22.4\ \ell}{1\ mol} = 5.17\ \ell$$

b. $2\ KClO_3(s) \rightarrow 2\ KCl(s) + 3\ O_2(g)$

$$\text{moles } O_2 = 100\ g\ KClO_3 \times \frac{1\ mol\ KClO_3}{122.5\ g\ KClO_3} \times \frac{3\ mol\ O_2}{2\ mol\ KClO_3} = 1.22$$

$$\text{volume } O_2 = 1.22\ mol \times \frac{22.4\ \ell}{1\ mol} = 27.3\ \ell$$

c. $2\ H_2O_2(l) \rightarrow 2\ H_2O(l) + O_2(g)$

$$\text{moles } O_2 = 100 \text{ g } H_2O_2 \times \frac{1 \text{ mol } H_2O_2}{34.0 \text{ g } H_2O_2} \times \frac{1 \text{ mol } O_2}{2 \text{ mol } H_2O_2} = 1.47$$

$$\text{volume } O_2 = 1.47 \text{ mol} \times \frac{22.4\ \ell}{1 \text{ mol}} = 32.9\ \ell$$

14. a. For Li_2O: $\frac{0.50 \text{ mol } O_2}{2 \text{ mol Li}} = \frac{0.25 \text{ mol } O_2}{1 \text{ mol Li}}$

For MgO: $\frac{0.50 \text{ mol } O_2}{1 \text{ mol Mg}}$

For Al_2O_3: $\frac{1.5 \text{ mol } O_2}{2 \text{ mol Al}} = \frac{0.75 \text{ mol } O_2}{1 \text{ mol Al}}$

Since volume is directly proportional to number of moles, the volumes of oxygen required are in the ratio 0.25:0.50:0.75 or 1:2:3. Aluminum requires the largest volume of oxygen per mole.

b. For Li_2O: $\frac{0.500 \text{ mol } O_2}{2(6.94) \text{ g Li}} = 0.0360 \text{ mol } O_2/\text{g Li}$

For MgO: $\frac{0.500 \text{ mol } O_2}{24.0 \text{ g Mg}} = 0.0208 \text{ mol } O_2/\text{g Mg}$

For Al_2O_3: $\frac{1.50 \text{ mol } O_2}{2(27.0) \text{ g Al}} = 0.0278 \text{ mol } O_2/\text{g Al}$

Hence, lithium requires the greatest number of moles, or volume, of oxygen per gram.

15. a. $\text{mass N} = 1.00 \text{ g } NH_3 \times \frac{14.01 \text{ g N}}{17.03 \text{ g } NH_3} = 0.823 \text{ g}$

b. $\text{mass N} = 1.00 \text{ g } HNO_3 \times \frac{14.01 \text{ g N}}{63.02 \text{ g } HNO_3} = 0.222 \text{ g}$

c. $\text{mass N} = 1.00 \text{ g } NH_4NO_2 \times \frac{28.02 \text{ g N}}{64.05 \text{ g } NH_4NO_2} = 0.437 \text{ g}$

16. $\text{mass } NH_4NO_3 = 100 \text{ g fz} \times \frac{11 \text{ g N}}{100 \text{ g fz}} \times \frac{80 \text{ g } NH_4NO_3}{28 \text{ g N}} = 31 \text{ g}$

17. $2\ CO(g) + O_2(g) \rightarrow 2\ CO_2(g)$; $\Delta H = -135 \text{ kcal}$

a. $\Delta H = 16.4 \text{ g CO} \times \frac{1 \text{ mol CO}}{28.0 \text{ g CO}} \times \frac{-135 \text{ kcal}}{2 \text{ mol CO}} = -39.5 \text{ kcal}$

b. $\text{mass } CO_2 = 16.4 \text{ g CO} \times \frac{1 \text{ mol CO}}{28.0 \text{ g CO}} \times \frac{2 \text{ mol } CO_2}{2 \text{ mol CO}} \times \frac{44.0 \text{ g } CO_2}{1 \text{ mol } CO_2}$

$= 25.8 \text{ g}$

c. $\text{volume } O_2 = 16.4 \text{ g CO} \times \frac{1 \text{ mol CO}}{28.0 \text{ g CO}} \times \frac{1 \text{ mol } O_2}{2 \text{ mol CO}} \times \frac{22.4\ \ell}{1 \text{ mol } O_2}$

$= 6.56\ \ell$

18. $FM(NH_4)_3PO_4 = 149$ Therefore: 149 g AP $\simeq$ 42 g N

$\text{mass AP} = 50 \times 10^3 \text{ g fz} \times \frac{9.4 \text{ g N}}{100 \text{ g fz}} \times \frac{149 \text{ g AP}}{42 \text{ g N}} = 1.7 \times 10^4 \text{ g}$

19. a. $\text{mass Ar} = 1 \times 10^3 \text{ g air} \times \frac{1.29 \text{ g Ar}}{100 \text{ g air}} = 12.9 \text{ g}$

b. moles air without O_2 = 100.00 − 20.95 = 79.05

$\text{mole \% } N_2 = \frac{78.08 \text{ mol } N_2}{79.05 \text{ mole ``air''}} \times 100 = 98.77$

c. MM(air) = 0.7808 (28.01) + 0.2095 (32.00) + 0.00934 (39.95) + 0.00031 (44.01)

MM(air) = 21.87 + 6.704 + 0.373 + 0.014 = 28.96

20. a. $D(NO) = \frac{30.0 \text{ g NO}}{22.4\ \ell} = 1.34 \text{ g}/\ell$

$D(NO_2) = \frac{46.0 \text{ g } NO_2}{22.4\ \ell} = 2.05 \text{ g}/\ell$

b. To find the density when not at STP, first find the molar volume at the non-standard conditions:

$\text{volume} = \frac{22.4\ \ell}{\text{mol}} \times \frac{760 \text{ mm Hg}}{740 \text{ mm Hg}} \times \frac{298 \text{ K}}{273 \text{ K}} = 25.1\ \ell/\text{mol}$

$\text{density NO} = \frac{30.0 \text{ g NO}}{25.1\ \ell} = 1.20 \text{ g}/\ell$

$\text{density } NO_2 = \frac{46.0 \text{ g } NO_2}{25.1\ \ell} = 1.83 \text{ g}/\ell$

21. a. H_2 is combustible.
b. O_2 supports combustion and is reactive.
c. He is expensive.
d. CF_2Cl_2 destroys the ozone layer in the upper atmosphere.

CHAPTER 7

The Periodic Table

I. Basic Skills

A. Qualitative Students should be able to:

1. Explain what is meant by the Periodic Law and give examples of periodic properties.
2. a. Describe the organization of the modern periodic table.
   b. Explain what is meant by periods and groups.
   c. Locate the main-group elements, the transition elements, and the inner transition elements on the periodic table.
3. a. Describe the major properties of metals.
   b. Describe how the transition metals differ from the main-group metals.
4. Distinguish between a metal, a metalloid, and a nonmetal on the periodic table.
5. Describe what is meant by an element's:
   a. ionization energy
   b. electronegativity
   c. atomic radius (Figure 7.10)

B. Quantitative Students should be able to:

1. Predict the physical properties of an element given those of its neighbors in the periodic table (Example 7.1).
2. Predict the formulas of compounds knowing those of similar compounds of the same group (Example 7.2).
3. Describe the general trends of the following properties as a function of the element's atomic number:
   a. metallic character (Figure 7.7 and Example 7.3)
   b. ionization energy (Figure 7.8 and Example 7.4)
   c. electronegativity (Figure 7.9 and Example 7.5)
   d. atomic radius (Figure 7.11 and Example 7.6)

II. Chapter Development

1. The main emphasis of the chapter is on the periodic table and its use as a predictive device. The periodic law is introduced by showing the periodicity of ionization energies and that of the formulas of elemental chlorides. This leads to the organization of the modern periodic table. For simplicity, the A and B symbols for main-groups and sub-groups are not used. The theoretical basis for the periodic table, electronic structure, will be discussed in Chapter 9.

2. Distinction is made between metals, metalloids, and nonmetals but their chemistry is left for later. Considerable descriptive material on Groups 1, 2, 6, and 7 is presented in Chapter 8. Additional information on selected metals is presented in Chapter 25, Metals and Their Ores.
3. Ionization energy is presented as an important property of metals. The ability of nonmetals to gain electrons is discussed using the concept of electronegativity (Pauling's scale). The concept of electron affinity is not introduced. Knowledge of ionization energy and electronegativity trends will be needed later in Chapters 10 and 11 when chemical bonding is covered.

III. Problem Areas

1. Predictions of physical properties based on the periodic table are not always reliable. Neither a horizontal nor a vertical comparison of neighbors would predict mercury to be a liquid at room temperature! While Mendeleev was successful, students should realize that predictions involve an element of risk. While the prediction of formulas is more reliable, and is a useful device, many exceptions to the expected formulas exist.
2. The use of trends based on the periodic table has some difficulties, also. Students are not expected to understand the zigs and zags in the ionization energy table (Figure 7.1); the general trend is sufficient. The decrease in atomic size within a period is a surprise to most. This is best explained as due to the increased nuclear charge causing an increased attractive force.

IV. Suggested Activities

1. Exhibit of Elements: **Teacher Demonstration Only**
   Display a varied group of elements, including metals, nonmetals, solids, liquids, and gases. Suggested elements are C, N_2 and O_2 (air), Na, Mg, Al, P(red), S, Cl_2 Ca, Fe, Cu, Zn, Br_2, I_2, Hg, and Pb. Label each element with its formula, melting point, boiling point, and density. Have students identify the group and period of each element and classify each as a metal or nonmetal. Discuss which of the elements occurs naturally in the uncombined state.
2. The Periodic Law:
   Have students graph a set of physical property data such as that given in Table 7.1 (density, melting point, and boiling point). Plot atomic number on the x-axis and the related property on the y-axis. Discuss whether or not the property is a periodic property. Compare the graph to that given in Figure 7.1 for ionization energy. (Atomic radii are graphed in Experiment 13.)
3. Evolution of the Periodic Table:
   Display different forms of the periodic table for comparison. There are short forms (like Mendeleev's), long forms (inside back cover) and extended forms

which do not pull out the inner transition elements. If you look around you can also find circular forms, spiral forms, and figure-8 forms. The student should not feel that there is only one kind of periodic table.

V. Answers to Questions

1. Graph (b) represents a type of periodic function.
2. Ionization energy, (a), and electronegativity, (c), are periodic functions of atomic number. Atomic mass, (b), increases steadily so it is not.
3. Ionization energy is the energy required to remove an electron from an atom. Its value is positive because removing an electron from a positive nucleus absorbs energy.
4. a. 2 b. 8 c. 18 d. 32
5. There are ten elements in each transition series. The first series has atomic numbers of 21–30.
6. There are fourteen lanthanides and fourteen actinides.
7. The element that completes the 7th period would have an atomic number of 118. It would be a member of Group 8.
8. Atomic numbers 105 and 110 would be transition elements. Atomic number 115 would be a main-group element.
9. The transition metals which students may have seen are: chromium, Cr (plate on autos); iron, Fe; nickel, Ni (coins); copper, Cu (coins and wire); zinc, Zn (galvanized metal); silver, Ag (jewelry); gold, Au (jewelry); mercury, Hg (thermometers).
10. The percentage of metals will increase as new elements are discovered; these elements are transition metals.
11. a. true b. false c. true
12. a. false b. true c. false (upper left to lower right) d. true
13. a. decreases b. increases c. right
14. Transition metals are less reactive and have higher melting points than metals in the main groups. They also differ by forming many colored compounds with nonmetals and by forming alloys with other metals.
15. Metals are good conductors of electricity and heat, in contrast to nonmetals. Metals have luster and are ductile and malleable. Some metals give up electrons when heated or exposed to light.
16. a. increases b. decreases c. decreases d. increases
17. a. decreases b. increases c. increases d. decreases

VI. Solutions to Problems

1. $\text{Density} = \frac{1.55 + 3.51}{2} = 2.53 \text{ g/cm}^3$ (Actual: 2.60 g/cm^3)

$\text{Melting Pt.} = \frac{845 + 725}{2} = 785°\text{C}$ (Actual: 770°C)

$$\text{Boiling Pt.} = \frac{1420 + 1640}{2} = 1530°\text{C (Actual: 1380°C)}$$

$$\text{Ionization E.} = \frac{142 + 122}{2} = 132 \text{ kcal/mol (Actual 133 kcal/mol)}$$

2. $RaCl_2$, $RaCO_3$, $Ra_3(PO_4)_2$
3. Metallic character: Ge < Ga < In
4. Ionization energy: Ge > Ga > In
5. Electronegativity: Ge > Ga > In (Actual: Ga = In)
6. Atomic radius: In > Ga > Ge

7.

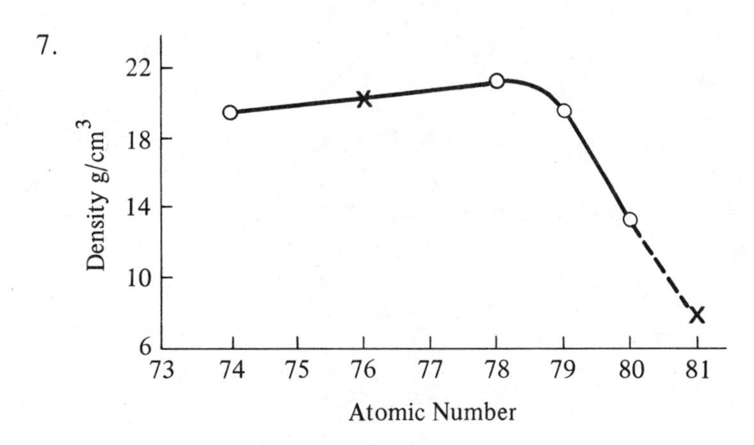

Predicted density: Os(76), 20 g/cm^3, Tl(81), 8 g/cm^3
Actual density: Os, 22.6 g/cm^3, Tl, 11.8 g/cm^3

8. The formulas of the carbon oxides are CO and CO_2.
   a. CS (does not exist) and CS_2 b. PbO and PbO_2

9.

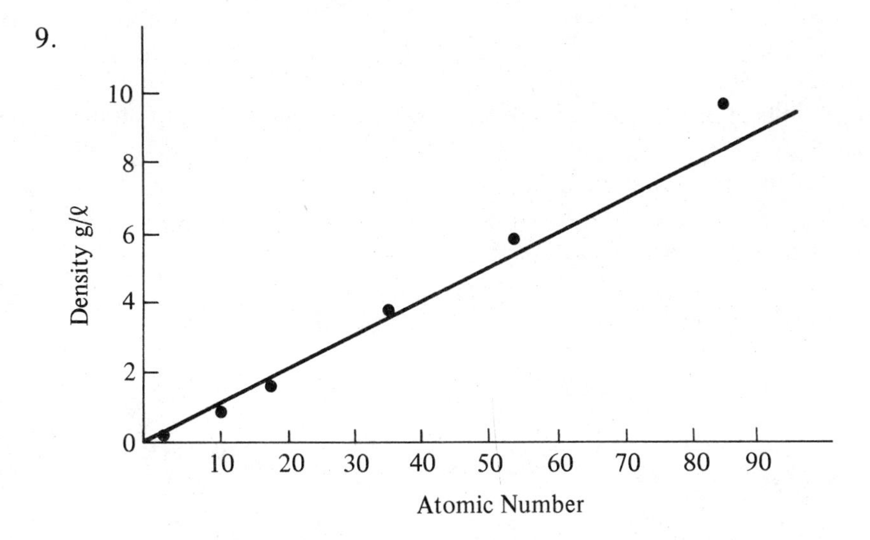

The heavier noble gases (Kr, Xe, and Rn) lie above the line and the lighter gases

(He, Ne, and Ar) lie below the line. (This graph indicates that the density increases faster than the atomic number.)

10. a. Ionization E: K $<$ Mg $<$ S

    b. Electronegativity: K $<$ Mg $<$ S

    c. Atomic radius: S $<$ Mg $<$ K

11. Ge and Sb, because they lie along the diagonal line from upper left to lower right.

12. a. Na_2CrO_4 and $Na_2Cr_2O_7$

    b. Na_2WO_4 and $Na_2W_2O_7$

13. The relationship, specific heat = 6/atomic mass, is an inverse relationship as shown below:

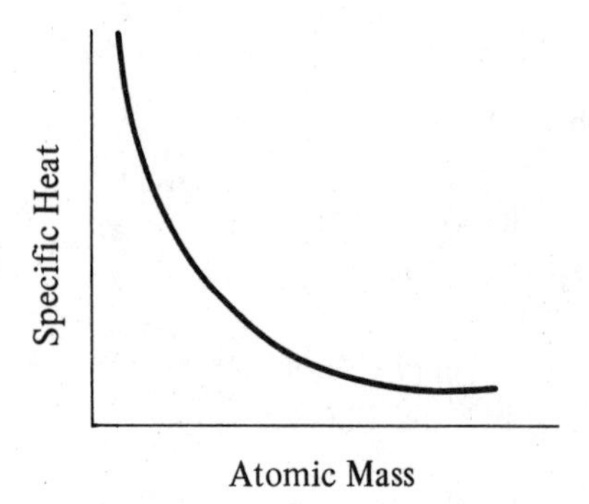

Specific heat is not a periodic function of either atomic mass or atomic number.

14. a. 28 inner transition elements

    b. 31 transition elements

    c. 44 main group elements

15.

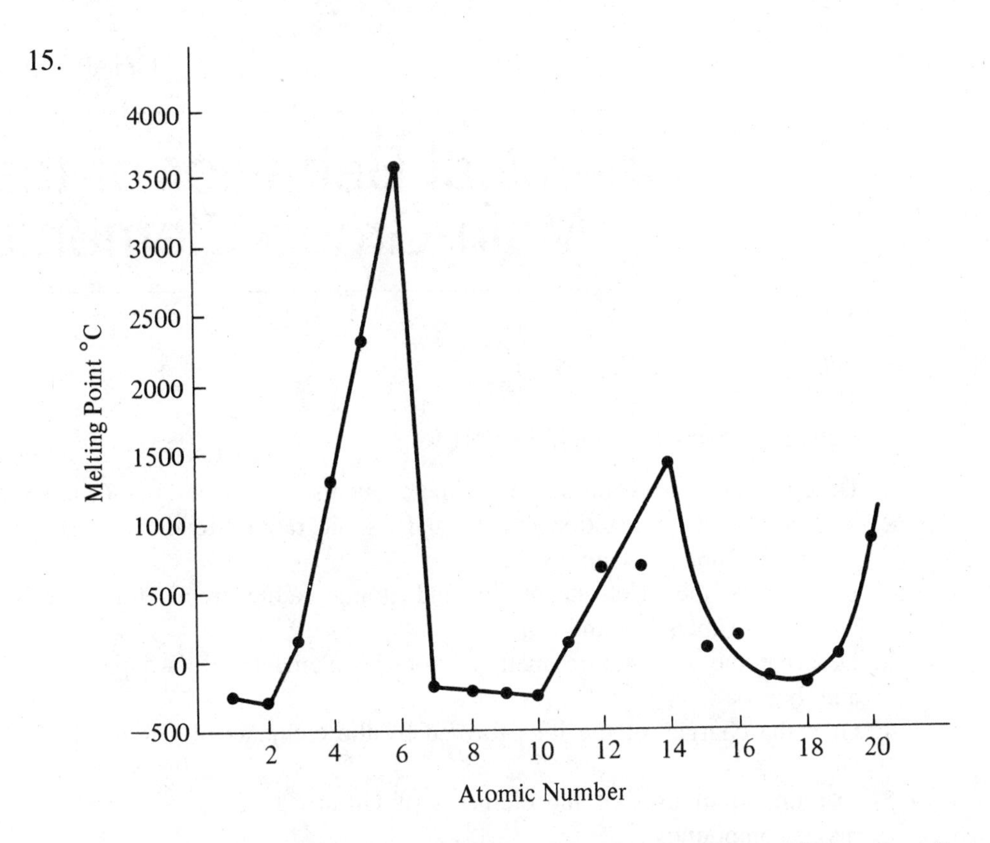

The two peaks are Group 4 elements, C and Si, while the minimums are Group 8 elements, He, Ne, and Ar. Al and P appear to have lower than expected melting points while those of Mg and S appear to be higher.

16. The elements of similar properties are placed in vertical groups of the periodic table. The missing letters are assigned to horizontal periods in such a way that the alphabet is completed, in order, between the groups.

A							B
C	D	E					F
G	H	I					J
K	L	M	N	O			P
Q	R	S	T	U	V	W	X
Y	Z						

17. a. Np (neptunium): from planet, Neptune.
    b. Po (polonium): from nation, Poland. Discovered by Marie Curie, a native of Poland.
    c. Fr (francium): from nation, France. Discovered by Marguerite Perey in Paris.

Chemical Behavior of the Main-Group Elements

I. Basic Skills

A. Qualitative Students should be able to:

1. Describe the general properties of the elements in Groups 1, 2, 6, and 7.
2. a. Describe the ionization energy and atomic radius trends for Groups 1 and 2 (Tables 8.1 and 8.2).
   b. Describe the electronegativity and atomic radius trends for Groups 6 and 7 (Tables 8.3 and 8.4).
3. Describe the preparation methods for the elements of Groups 1, 2, 6, and 7.
4. Give the charges of the ions formed by the elements of Groups 1, 2, 6, and 7.
5. List important uses of the elements of Groups 1, 2, 6, and 7 and their major compounds.
6. Explain what is meant by allotropy and give an example.

B. Quantitative Students should be able to:

1. Write the formulas of the compounds formed between the metals of Groups 1 and 2 and the non-metals of Groups 6 and 7.
2. Write balanced chemical equations for the:
   a. Preparation of Group 1 or 2 metal by electrolysis (Equations 8.1 and 8.9).
   b. Reaction of a Group 1 or 2 metal with oxygen (Equations 8.2 and 8.10); with water (Equations 8.3 and 8.11).
   c. Reaction of $NaHCO_3$ with an acid (Equation 8.4).
   d. Decomposition of a Group 2 carbonate by heat (Equation 8.13).
   e. Reaction of a Group 2 oxide with water (Equation 8.14).
   f. Reaction of a Group 2 hydroxide with CO_2 (Equation 8.15).
   g. Reaction of a Group 6 element with a metal (Equation 8.17).
   h. Reaction of sulfur with oxygen (Equation 8.19).
   i. Preparation of the halogens (Equations 8.1, 8.22, 8.23, and 8.25).
   j. Reaction of halogens with metals (Equations 8.26 and 8.27).
   k. Reaction of halogens with hydrogen (Equation 8.28).
   l. Preparation of hydrogen halides from halide solutions (Equation 8.29).

II. Chapter Development

1. Chapter 8 builds on the material presented previously in Chapter 7, The Peri-

odic Table. Its purpose is to emphasize the family relationships which exist between elements of the same group. The descriptive chemistry of four important groups, 1, 2, 6, and 7, is covered. Group 8, the noble gases, was discussed previously in Chapter 6.

2. Students should recognize how the atomic radius, ionization energy, and electronegativity trends correlate with the change in properties of the elements within a group. The reason for the larger changes between groups is brought out in Chapter 9, Electronic Structure of Atoms.
3. The descriptive chemistry is purposely at a basic level. As a student learns reaction principles the depth and breadth of the chemistry will be increased. Do not rush into more complicated reactions, they will come later.

III. Problem Areas

1. There is a minor problem here, mostly for teachers, in avoiding the subject of electronic structure when dealing with the chemistry of the elements. Keep in mind that Mendeleev didn't have the first idea about structure. He was led to the periodic table by his thorough knowledge of the properties (physical and chemical) of the elements and their compounds.
2. Equation writing and balancing should be becoming less of a problem. Stress family relationships for reactions such as those shown in Examples 8.4, 8.5, and 8.7.

IV. Suggested Activities

1. Group Reactions: **Teacher Demonstrations Only**
   Demonstrate characteristic group reactions. Limit yourself to those which are not performed during Experiment 14. Possible reactions are:
   a. burn a magnesium ribbon.
   b. heat a mixture of copper wool and sulfur in a large test tube.
   c. add some $NaHCO_3$ to a dilute acid.
   d. add a seltzer tablet to water.
   Have students make observations and write equations for each reaction.
2. Predicting Properties:
   Have students predict the physical properties of radium using the data given in Table 8.2. (Actual data: Density, 5.0 g/cm^3; Atomic Radius, 0.220 nm; I.E., 121 kcal; MP, 700°C; BP, 1140°C.) Point out the obvious risk of making predictions based on extrapolations.
3. Flame Tests of Group 1 and 2 Ions: **Teacher Demonstration Only**
   Demonstrate the flame colors given off by the heating of solutions of Group 1 and 2 ions. Point out that the procedure could be used as a qualitative test.

V. Answers to Questions

1. a. Group 1, +1 b. Group 2, +2 c. Group 6, −2 d. Group 7, −1
2. Of the listed groups, only oxygen and sulfur in Group 6 occur in the uncombined state.
3. a. Group 1, Na b. Group 2, Ca c. Group 6, O d. Group 7, Cl
4. a. Group 1, Fr b. Group 2, Ra c. Group 6, Po d. Group 7, At
5. a. Group 1, Fr b. Group 2, Ra c. Group 6, Po d. Group 7, At
6. a. Group 1, Li b. Group 2, Be c. Group 6, O d. Group 7, F
7. Due to their high chemical reactivity, Na, Mg, Ca, Cl and F are obtained by electrolysis. Sulfur is primarily obtained from natural deposits of the element.
8. Stable diatomic molecules are formed by (c) O, (d) F, (e) Cl, and (f) I.
9. The Group 2 elements have higher densities than the Group 1 elements because they have a larger mass and a smaller atomic radius.
10. a. Na: used for making sodium compounds and as a heat-transfer medium in nuclear reactors.
    b. NaCl: used for making chlorine and as a necessary ingredient of diets.
    c. NaOH: used in making soap and to dissolve fats and grease.
    d. Na_2CO_3: used in making glass and as a water softener.
    e. $NaHCO_3$: used as an ingredient of baking powder and in some antacids.
11. Calcium carbonate, $CaCO_3$, is the principal component of limestone, marble, pearls and seashells.
12. Hard water is caused primarily by Ca^{2+} and Mg^{2+} ions. These ions react with the negative ions in water to form insoluble compounds. The precipitates form a "scale" which tends to clog hot water systems. The Mg^{2+} and Ca^{2+} ions also react with soap to form insoluble curds which stain clothes and produce the "ring-around-the-tub."
13. Magnesium is used in aircraft parts because of its low density. Because of its high reactivity, magnesium is a hazard in aircraft fires.
14. Hot water, under pressure, is pumped underground to melt the sulfur deposits. Compressed air is then blown through the liquid sulfur. The mixture rises to the surface through another pipe.
15. Allotropy refers to the ability of some elements to exist in different forms in the same physical state. Oxygen allotropes are O_2 and O_3, ozone. Sulfur allotropes are rhombic and monoclinic crystal forms.
16. Fluorine and chlorine.
17. a. Highest electronegativity, F b. Highest density, I
    c. Highest radius, I d. Highest melting point, I
18. Yes. Br_2 has a stronger attraction for electrons than I_2.
19. a. NaOH b. Na_2CO_3 c. $NaHCO_3$
20. a. $Ca(OH)_2$ b. $CaCO_3$ c. CaO
21. a. MgS b. SO_2
22. a. KCl b. $MgCl_2$ c. HCl

VI. Solutions to Problems

1. a. $4\ Na(s) + O_2(g) \rightarrow 2\ Na_2O(s)$

   b. $2\ Na(s) + 2\ H_2O(l) \rightarrow H_2(g) + 2\ Na^+(aq) + 2\ OH^-(aq)$

2. $Mg(s) + H_2O(g) \rightarrow MgO(s) + H_2(g)$

3. a. $CaCO_3(s) \rightarrow CaO(s) + CO_2(g)$

   b. $CaO(s) + H_2O(l) \rightarrow Ca^{2+}(aq) + 2\ OH^-(aq)$

   c. $Ca(OH)_2(s) + CO_2(g) \rightarrow CaCO_3(s) + H_2O(l)$

4. a. $Mg(s) + S(s) \rightarrow MgS(s)$

   b. $Fe(s) + S(s) \rightarrow FeS(s)$

   c. $S(s) + O_2(g) \rightarrow SO_2(g)$

5. a. $Cl_2(g) + 2\ Br^-(aq) \rightarrow Br_2(l) + 2\ Cl^-(aq)$

   b. $Cl_2(g) + 2\ I^-(aq) \rightarrow I_2(s) + 2\ Cl^-(aq)$

6. a. $MnO_2(s) + 4\ H^+(aq) + 2\ Cl^-(aq) \rightarrow Mn^{2+}(aq) + Cl_2(g) + 2\ H_2O$

   b. $2\ NaCl(l) \rightarrow 2\ Na(l) + Cl_2(g)$

7. a. $4\ Li(s) + O_2(g) \rightarrow 2\ Li_2O(s)$

   b. $2\ Li(s) + 2\ H_2O(l) \rightarrow H_2(g) + 2\ Li^+(aq) + 2\ OH^-(aq)$

8. a. $2\ KCl(l) \rightarrow 2\ K(l) + Cl_2(g)$

   b. $SrCl_2(l) \rightarrow Sr(s) + Cl_2(g)$

9. a. $2\ Ba(s) + O_2(g) \rightarrow 2\ BaO(s)$

   b. $Ba(s) + 2\ H_2O(l) \rightarrow H_2(g) + Ba^{2+}(aq) + 2\ OH^-(aq)$

10. a. $Mg(s) + Te(s) \rightarrow Mg\ Te(s)$

    b. $Te(s) + O_2(g) \rightarrow TeO_2(s)$

11. a. $Mg(s) + I_2(s) \rightarrow MgI_2(s)$

    b. $2\ Na(s) + I_2(s) \rightarrow 2\ NaI(s)$

12. a. $2\ Rb(s) + 2\ H_2O(l) \rightarrow H_2(g) + 2\ Rb^+(aq) + 2\ OH^-(aq)$

    b. $Ca(s) + 2\ H_2O(l) \rightarrow H_2(g) + Ca^{2+}(aq) + 2\ OH^-(aq)$

    c. $2\ Sr(s) + O_2(g) \rightarrow 2\ SrO(s)$

13. a. $H_2(g) + Br_2(l) \rightarrow 2\ HBr(g)$

    b. $Br_2(l) + 2\ I^-(aq) \rightarrow 2\ Br^-(aq) + I_2(s)$

    c. $MnO_2(s) + 2\ Br^-(aq) + 4\ H^+(aq) \rightarrow Mn^{2+}(aq) + Br_2(l) + 2\ H_2O$

14. From Eqn. 8.3: 2 mol Na $\cong$ 1 mol H_2

a. $\text{moles } H_2 = 0.120 \text{ g Na} \times \frac{1 \text{ mol Na}}{23.0 \text{ g Na}} \times \frac{1 \text{ mol } H_2}{2 \text{ mol Na}} = 2.61 \times 10^{-3} \text{ mol}$

b. $\text{volume } H_2 = 0.00261 \text{ mol } H_2 \times \frac{22.4\ \ell}{1 \text{ mol } H_2} = 0.0585\ \ell$

15. From Eqn. 8.13: 1 mol $CaCO_3$ $\cong$ 1 mol CO_2

$$\text{moles } CaCO_3 = 2.00\ \ell\ CO_2 \times \frac{1 \text{ mol } CO_2}{22.4\ \ell\ CO_2} \times \frac{1 \text{ mol } CaCO_3}{1 \text{ mol } CO_2}$$

$$= 8.93 \times 10^{-2} \text{ mol}$$

$$\text{mass } CaCO_3 = 0.0893 \text{ mol } CaCO_3 \times \frac{100.1 \text{ g } CaCO_3}{1 \text{ mol } CaCO_3} = 8.94 \text{ g}$$

16. From Eqn. 8.26: 1 mol MnO_2 $\cong$ 1 mol Cl_2

$$\text{volume/mole at } 25°\text{C} = 22.4\ \ell \times \frac{298 \text{ K}}{273 \text{ K}} = 24.5\ \ell$$

$$\text{mass } MnO_2 = 1.00\ \ell\ Cl_2 \times \frac{1 \text{ mol } Cl_2}{24.5\ \ell\ Cl_2} \times \frac{1 \text{ mol } MnO_2}{1 \text{ mol } Cl_2} \times \frac{86.9 \text{ g } MnO_2}{1 \text{ mol } MnO_2}$$

$$= 3.55 \text{ g}$$

17. From Eqn. 8.29: 1 mol I_2 $\cong$ 1 mol MgI_2

$$\text{mass } I_2 = 1.00 \text{ g } MgI_2 \times \frac{1 \text{ mol } MgI_2}{278.1 \text{ g } MgI_2} \times \frac{1 \text{ mol } I_2}{1 \text{ mol } MgI_2} \times \frac{253.8 \text{ g } I_2}{1 \text{ mol } I_2}$$

$$= 0.913 \text{ g}$$

18. Beryl: $Be_3Al_2Si_6O_{18}$

FM = 3(9.01) + 2(26.98) + 6(28.09) + 18(16.00) = 537.5

$$\% \text{ Be} = \frac{27.03}{537.5} \times 100 = 5.029 \qquad \% \text{ Al} = \frac{53.96}{537.5} \times 100 = 10.04$$

$$\% \text{ Si} = \frac{168.5}{537.5} \times 100 = 31.35 \qquad \% \text{ O} = \frac{288.0}{537.5} \times 100 = 53.58$$

19. 1 mol NaCl $\cong$ 1 mol NaOH; 58.5 g NaCl $\cong$ 40.0 g NaOH

$$\text{mass NaCl} = 1.0 \times 10^{10} \text{ kg NaOH} \times \frac{58.5 \text{ kg NaCl}}{40.0 \text{ kg NaOH}} = 1.5 \times 10^{10} \text{ kg NaCl}$$

20. From Eqn. 8.10: 2 mol Sr $\cong$ −282 kcal

$$\Delta H = 1.00 \text{ g Sr} \times \frac{1 \text{ mol Sr}}{87.6 \text{ g Sr}} \times \frac{-282 \text{ kcal}}{2 \text{ mol Sr}} = -1.61 \text{ kcal}$$

21. density $O_3 = \frac{48.0 \text{ g } O_3}{22.4\ \ell} = 2.14 \text{ g}/\ell$

From Table 8.5: Density $O_3 = 2.145 \text{ g}/\ell$

22. density $Cl_2 = \frac{70.9 \text{ g } Cl_2}{22.4\ \ell} = 3.17 \text{ g}/\ell$

From Table 8.6: Density $Cl_2 = 3.21 \text{ g}/\ell$

23. $C_2H_4Br_2(l) + (C_2H_5)_4\,Pb(l) + 16\ O_2(g) \rightarrow PbBr_2(s) + 10\ CO_2(g) + 12\ H_2O$

24. From Eqn. 8.15: 1 mol $Ca(OH)_2 \simeq$ 1 mol $CaCO_3$

$$\text{mass } CaCO_3 = 20 \text{ kg mortor} \times \frac{40 \text{ kg } Ca(OH)_2}{100 \text{ kg mortor}} \times \frac{100 \text{ kg } CaCO_3}{74 \text{ kg } Ca(OH)_2} = 11 \text{ kg}$$

25. a. $\text{mass} = 1 \text{ Na atom} \times \frac{1 \text{ mol Na}}{6.022 \times 10^{23} \text{ atoms}} \times \frac{22.99 \text{ g Na}}{1 \text{ mol Na}}$

$= 3.818 \times 10^{-23}$ g

b. $\text{radius} = 0.186 \text{ nm} \times \frac{1 \times 10^{-7} \text{ cm}}{1 \text{ nm}} = 1.86 \times 10^{-8}$ cm

c. $V = 4/3\ \pi r^3$

$V = 4/3\ \pi(1.86 \times 10^{-8} \text{ cm})^3 = 2.70 \times 10^{-23} \text{ cm}^3$

d. $\text{Density} = \frac{3.82 \times 10^{-23} \text{ g}}{2.70 \times 10^{-23} \text{ cm}^3} = 1.41 \text{ g/cm}^3$

NOTE: The density of an individual Na atom is greater than that of Na metal (0.971 g/cm^3) because the metal contains voids when the atoms are packed.

CHAPTER 9

Electronic Structure of Atoms

I. Basic Skills

A. Qualitative   Students should be able to:

1. Explain the quantum theory as it applies to the hydrogen electron.
2. Explain how Bohr fit the quantum theory to the hydrogen atom.
3. Give the probability description of the hydrogen electron.
4. Explain what is meant by:
   a. energy levels   b. sublevels   c. electron configurations
5. Explain what is meant by an orbital and describe s and p orbitals.
6. Correlate electron configurations with the structure of the periodic table.
7. Explain what is meant by valence electrons.

B. Quantitative   Students should be able to:

1. a. Calculate the energy of the hydrogen electron in any energy level (Equation 9.2).
   b. Calculate the energy change when the hydrogen electron changes its energy level (Example 9.1).
2. a. Give the relative energies of levels and sublevels (Example 9.2).
   b. Give the electron capacities of levels and sublevels (Example 9.2).
3. Write the electron configuration of an atom (Examples 9.3 and 9.4).
4. Give the electron capacity of an orbital and the number of orbitals in a sublevel (Example 9.5).
5. Write the outer electron configuration of an atom using the periodic table (Example 9.6).
6. Give the number of valence electrons for any of the main group elements.

II. Chapter Development

1. The quantum theory and the Bohr equation are applied to the hydrogen atom. The idea of energy levels is first developed for this simplest case. Energy changes are related to atomic spectra in an optional section. Even here, physics is held to a minimum. The ΔE is related to wavelength but not to frequency.
2. Probability descriptions are used to develop the idea that the location of an electron cannot be exactly known. Probability pictures of the hydrogen electron are used to further the image of an electron cloud.
3. The chapter's primary objective is to develop the student's ability to under-

stand and to write electron configurations of atoms. The rules are developed for this purpose only and no attempt is made to present their deeper meaning. Electron configurations will be used in Chapters 10 and 11 in discussions of chemical bonding.

4. The order of filling orbitals within a sublevel (Hund's Rule) is presented in an optional section. The concepts of electron spin and unpaired electrons (paramagnetism) are not introduced.

III. Problem Areas

1. Some students will experience difficulty in calculating the electron's energy change for an energy level change. (They will solve for E, rather than ΔE.) Compare ΔE to ΔH for chemical reactions and show that both E and H have initial and final states. An energy diagram can be used to emphasize that $\Delta E = E_{final} - E_{initial}$.
2. The concept of probability can be further developed by using a deck of 52 cards. Ask students for the probability of locating an ace by selecting one card ($1/13$), of locating a face card ($3/13$), or of locating a heart ($1/4$). Of course, the probability of locating an electron is much, much smaller.
3. The writing of electron configurations tends to be mechanical until the first overlap in the fourth period is reached. Students are helped to recall this fact by relating electronic structure to chemical properties. Point out that K displays all the properties of an alkali metal, Ca displays all the properties of an alkaline earth metal, and neither behaves as a transition metal. Ideally, students should be able to write most configurations directly from the periodic table. They should not be responsible for the exceptions to the general rules for the filling of sublevels (Cr, Cu, etc.).

IV. Suggested Activities

1. Atomic Spectra: **Teacher Demonstration Only**
   Use a spectroscope to view the atomic spectra of elements. Use emitted light from gas discharge tubes or flame tests. Point out the discontinuous nature of the spectra in contrast with the continuous spectrum of white light. The hydrogen visible spectrum can be related to specific electronic transitions to the $n = 2$ level.
2. The Bohr Atom:
   Have students draw a one-dimensional graph of energy levels in the Bohr atom. Begin at the origin for the $n = 1$ level and use a scale of 1 cm = 20 kcal. Calculate E values for $n = 1$ to $n = 5$ and complete with the value for $n = \infty$. Point out how the energy levels become closer as the value of n increases. Have students explain the significance of the $n = \infty$ value.
3. Electron Clouds and Orbitals:

Use styrofoam balls or commercial models to discuss s and p orbitals. Discuss the geometry of the different orbitals and how they are related to the probability description of an electron. Relate the number of the principle energy level to the size of the orbital. Explain how the three p orbitals are at 90° angles to each other.

V. Answers to Questions

1. When an electron moves from a lower to a higher energy level, energy is absorbed and ΔE is positive. For the reverse direction, energy is evolved and ΔE is negative.
2. The hydrogen electron in its ground state (lowest state) is in the $n = 1$ energy level. When the electron is removed from the atom, it is in the $n = \infty$ level, with $E = 0$.
3. a. The spacings between the rungs of a ladder are equal. The energy levels in the hydrogen atom are not equally spaced, but become increasingly closer together as their n value increases.
   b. The earth revolves in a fixed and known orbit about the sun. The electron in the hydrogen atom does not move in a fixed orbit. Its exact location, relative to the nucleus, cannot be known. We can only speak in terms of the "probability" of locating the electron in a particular region of space.
4. Some examples of quantization which students might think of are ladders, steps, elevators, three-way light bulbs, and digital watches. All except the light bulbs have equal spacing. Teachers might think of quantized salary schedules.
5. The electron does not occupy a fixed position nor does it move in a fixed orbit. Instead, the electron has probable locations which are independent of direction from the nucleus, but whose density decreases with distance from the nucleus.
6.

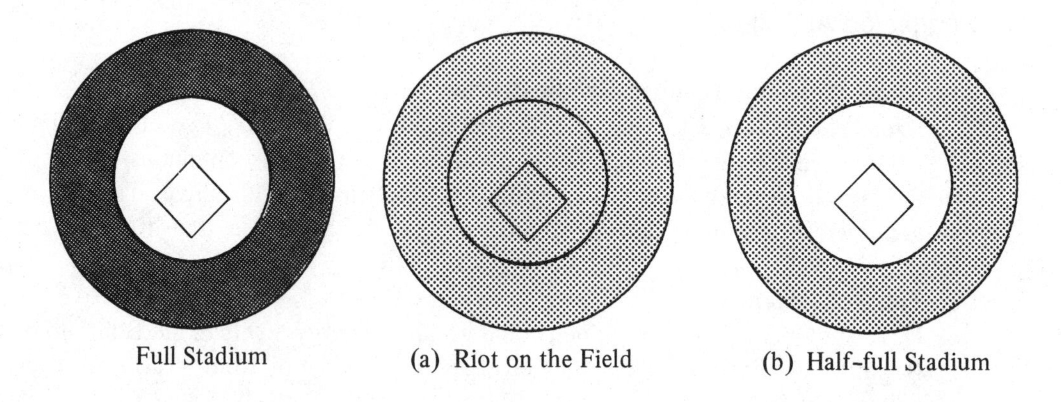

Full Stadium (a) Riot on the Field (b) Half-full Stadium

The shading in (a) extends onto the field showing a probability of locating fans throughout the stadium during a riot. The shading in (b) is decreased showing a decreased probability of locating fans in the stands of a half-full stadium as compared to a full stadium.

7. a. $n = 1, 2$ b. $n = 2, 8$ c. $n = 3, 18$ d. $n = 4, 32$
   e. $n = 5, 50$
8. a. $n = 2$ b. 2s   c. 2p
9. a. 1s   b. 2p   c. 3d   d. 4f
10. a. $s = 2$ b. $p = 6$ c. $d = 10$ d. $f = 14$
11. a. 2   b. 2   c. 6   d. 10   e. 2   f. 18
12. Sublevels 1p, 2d, and 3f do not exist.
13. $1s < 2s < 2p < 3s < 3p$
14. In the third principal energy level, the total electron capacity is 18. There are three different sublevels in this level. In order of increasing energy, they are 3s, 3p, and 3d. The 3d sublevel is higher in energy than the 4s sublevel of the fourth level.
15.

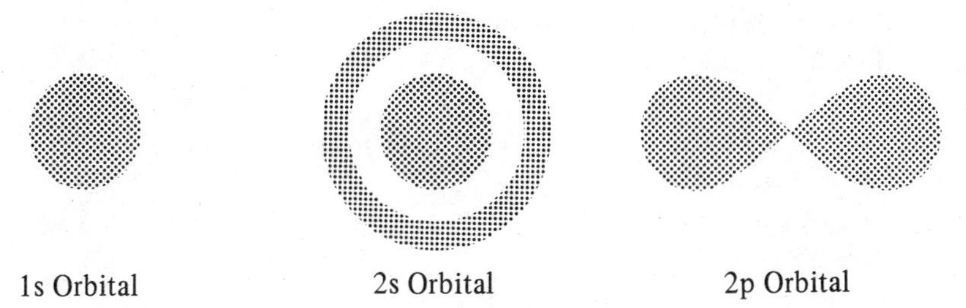

1s Orbital   2s Orbital   2p Orbital

The 1s orbital is spherical with equal probability of locating the electron in any direction, and decreasing probability with increasing distance from the nucleus. In the 2s orbital, there are two regions of high probability, one near the nucleus and one farther away. Between these two regions, there is a low probability region. The 2s orbital is spherical and larger than the 1s orbital. The 2p orbital is dumbbell-shaped and is symmetrical about an axis. The probability of locating a p electron is highest along the axis and is lowest in the region perpendicular to the axis.

16. a. $s = 1$ b. $p = 3$ c. $d = 5$ d. $f = 7$
17. a. $n = 1, 1$ b. $n = 2, 4$ c. $n = 3, 9$ d. $n = 4, 16$
18. a. Ar, at. no. = 18, has filled 9 orbitals.
    b. Zn, at. no. = 30, has filled 15 orbitals.
19. a. Group 3, ns^2np^1 b. Group 5, ns^2np^3
    c. Group 7, ns^2np^5 d. Group 8, ns^2np^6
20. a. Group 2   b. Group 4   c. Group 7
21. a. The first period of the periodic table contains the elements of the first energy level. Using the rule of $2n^2$, the $n = 1$ level is complete with two elements.
    b. The second period contains the elements of the second energy level. This level is complete with eight elements.
    c. The elements in the fourth period of the table fill the 4s and 4p orbitals of the fourth level and the 3d orbitals of the third level. The total of 9 orbitals provides for a maximum of 18 elements.
    d. The elements in the sixth period fill the 6s and 6p orbitals, the 4f orbitals and the 5d orbitals. The total number of orbitals is 16 which provides for a maximum of 32 elements.

22. The number of valence electrons of a main-group element is equal to its group number. Hence: a. 1 b. 3 c. 4 d. 6 e. 7
23. a. Group 2 b. Group 5 c. Group 6

VI. Solutions to Problems

1. $En = \frac{-313.6}{n^2}$ kcal/mol

   a. $n = 1, E = -313.6$ kcal/mol b. $n = 2, E = -78.4$ kcal/mol
   c. $n = 3, E = -34.8$ kcal/mol d. $n = 4, E = -19.6$ kcal/mol
   e. $n = 5, E = -12.5$ kcal/mol
2. Use the values obtained in Problem 1 for En.
   $\Delta E = E(\text{final}) - E(\text{initial})$
   a. For $n = 1$ to $n = 3$:
      $\Delta E = -34.8 - (-313.6) = 278.8$ kcal/mol
   b. For $n = 3$ to $n = 2$:
      $\Delta E = -78.4 - (-34.8) = -43.6$ kcal/mol
   c. For $n = 2$ to $n = 5$:
      $\Delta E = -12.5 - (-78.4) = 65.9$ kcal/mol
3. a. $_5B$: $1s^22s^22p^1$

   b. $_{10}Ne$: $1s^22s^22p^6$

   c. $_{13}Al$: $1s^22s^22p^63s^23p^1$

   d. $_{17}Cl$: $1s^22s^22p^63s^23p^5$
4. a. $_{20}Ca$: [Ar] $4s^2$

   b. $_{21}Sc$: [Ar] $4s^23d^1$

   c. $_{26}Fe$: [Ar] $4s^23d^6$

   d. $_{30}Zn$: [Ar] $4s^23d^{10}$
5. a. $_{33}As$: $4s^24p^3$ b. $_{37}Rb$: $5s^1$

   c. $_{50}Sn$: $5s^25p^2$ d. $_{53}I$: $5s^25p^5$
6. a. P b. K c. At d. Te
7. For $n = 1$ to $n = \infty$:
   $\Delta E = 0 - (-313.6) = 313.6$ kcal/mol
8. a. Li b. Si c. Ti
9. a. Al: $1s^22s^22p^63s^23p^1$

   b. Ca: $1s^22s^22p^63s^23p^64s^2$

   c. Cr: $1s^22s^22p^63s^23p^64s^13d^5$
10. Symbol Electron Configuration

    Na $1s^22s^22p^63s^1$

Mg	$1s^22s^22p^63s^2$
Al	$1s^22s^22p^63s^23p^1$
Si	$1s^22s^22p^63s^23p^2$
P	$1s^22s^22p^63s^23p^3$
S	$1s^22s^22p^63s^23p^4$
Cl	$1s^22s^22p^63s^23p^5$
Ar	$1s^22s^22p^63s^23p^6$

11. Electron configuration: $1s^22s^22p^63s^23p^3$
    Number of electrons = 2 + 2 + 6 + 2 + 3 = 15
12. P: $1s^22s^22p^63s^23p^3$
13. From Problem 5: As = 5, Rb = 1, Sn = 4, I = 7
    From Problem 6: P = 5, K = 1, At = 7, Te = 6
14. Ground state configurations: (a) and (e)
    Excited state configurations: (c) and (d). These configurations are excited because lower sublevels are available.
    Impossible configurations: (b) and (f). These configurations are impossible because they violate the rules for allowable sublevels.
15. Use the values of E_n obtained in Problem 1.
    For n = 3 to n = 2:
    $\Delta E = -78.4 - (-34.8) = -43.6$ kcal/mol

$$\lambda = \frac{2.858 \times 10^4 \text{ kcal} \cdot \text{nm}}{\Delta E}$$

$$\lambda = \frac{2.858 \times 10^4 \text{ kcal} \cdot \text{nm}}{43.6 \text{ kcal}} = 656 \text{ nm}$$

16. $$\Delta E = \frac{2.858 \times 10^4 \text{ kcal} \cdot \text{nm}}{\lambda}$$

$$\Delta E = \frac{2.858 \times 10^4 \text{ kcal} \cdot \text{nm}}{102.5 \text{ nm}} = 278.8 \text{ kcal}$$

From Figure 9.3: The electron transition is from n = 3 to n = 1.

17. a. Tc: [Kr] $5s^24d^5$
    b. I: [Kr] $5s^24d^{10}5p^5$
    c. La: [Xe] $6s^25d^1$
18. Al 1, 0, 0 S 2, 1, 1
    Si 1, 1, 0 Cl 2, 2, 1
    P 1, 1, 1 Ar 2, 2, 2
19. Using Hund's Rule, electrons will occupy an empty orbital in the same sublevel before pairing up. a. V = 3 b. Cr = 6 c. Co = 3

CHAPTER 10

Ionic Bonding

I. Basic Skills

A. Qualitative   Students should be able to:

1. a. Describe the necessary conditions for the forming of an ionic bond.
   b. Give examples of typical metals and nonmetals which form ionic bonds.
2. Describe the general properties of ionic compounds.
3. Explain the significance of the noble-gas structure in the forming of ions.
4. Explain what is meant by a hydrate and be able to interpret a hydrate formula.

B. Quantitative   Students should be able to:

1. Apply Hess' Law to obtain the ΔH for the formation of an ionic compound from its elements.
2. Describe how the ionic radius compares in size to the atomic radius of an element (Example 10.1).
3. Write the electron configurations of ions.
4. a. Determine the charges of monatomic ions with noble-gas structures (Table 10.1 and Example 10.2).
   b. Give the charges of the more important transition and post-transition metal ions (Table 10.2).
   c. Give the formulas of the more important polyatomic ions (Table 10.3).
5. Write the formulas of ionic compounds using the principle of electrical neutrality (Example 10.4).
6. Write balanced chemical equations for the formation of ionic compounds (Example 10.5).
7. Calculate the mass percentage of water in a hydrate (Example 10.6).
8. Name ionic compounds from their formulas (Example 10.7).

II. Chapter Development

1. The concepts of ionization energy and electronegativity are used to explain the formation of ionic bonds. An energy analysis of an electron-transfer reaction is given to show why ionic bond formation is an exothermic process. While the basic properties of ionic compounds are given, additional material on this subject will come in Chapter 12 under the study of liquids and solids.

2. The majority of the material is directed toward the study of ionic nomenclature, names and formulas, and the writing of equations. While the formation of an ionic compound from elements is an oxidation-reduction reaction, it is not advisable to begin writing half-equations at this stage. The given reactions are simple enough that the equations can be handled without going through the mechanics of the electron-transfer process. The concept of oxidation number will be used in Chapter 23.
3. Students are not expected to be able to deduce the electron configurations of transition metal ions and post-transition metal ions. Students should be able to deduce the charge of these ions from an ionic formula.

III. Problem Areas

1. In working on the student's ability to write correct formulas for ions, relate the ionic charge to the atom's position on the periodic table. A certain amount of memorization will be necessary for non-noble gas structure ions and for polyatomic ions.
2. In writing formulas of ionic compounds some students persist in reversing subscripts: Mg_2Cl instead of $MgCl_2$ or $Al_3(SO_4)_2$ instead of $Al_2(SO_4)_3$. If they can't see electrical neutrality any other way, have them arrive at the correct formula by transposing the ionic charges to obtain the subscripts.

IV. Suggested Activities

1. Nomenclature Exercise:
   Make a grid with metal ions (main-groups, transition and post-transition) on the left side and nonmetal ions (monatomic and polyatomic) across the top. Have students complete the grid by writing the name and formula of the compound formed by each ion-pair.
2. Conductivity and Color of Ionic Solutions: **Teacher Demonstration Only**
   a. Demonstrate the low conductivity of distilled water with a conductivity tester. Add a small amount of an ionic salt, stir thoroughly, and show the solution's conductivity. Emphasize that the change is due to the presence of positive and negative ions. A solution of a covalent solid (sucrose) can also be tested for comparison.
   b. Exhibit solutions of KNO_3, $Cr(NO_3)_3$, and $KMnO_4$. Have students deduce the color of each positive and negative ion in solution. Show white (anhydrous) $CuSO_4$, brown $CuBr_2$, and green $CuCl_2$ solids. Make separate solutions of each salt and have students explain the results.

V. Answers to Questions

1. Lithium forms the Li^{+} ion and nickel forms the Ni^{2+} ion. Sulfur forms the S^{2-} ion. Carbon does not ordinarily form ions.
2. Aluminum is more metallic than boron because of its smaller ionization energy.
3. The small nitrogen atom has a high ionization energy and a high electronegativity. It can attain a noble gas configuration by forming the N^{3-} ion. The much larger bismuth atom has the characteristics of a metal. Due to its low ionization energy bismuth can lose electrons and form the Bi^{3+} ion.
4. Ions are attracted to ions of opposite charge and repelled by ions of like charge. For this reason, an arrangement in which ions of opposite charge are near each other produces the most stable structure.
5. The electrical forces between oppositely charged ions in ionic solids are very strong. The melting points are high because those forces must be overcome before melting will occur.
6. The ions in an ionic solid are not free to move and cannot transfer a charge. The ions are free to move in the liquid state.
7. a. An atom is larger than its positive ion. The greater nuclear charge causes the electron cloud to decrease in size.
   b. A negative ion is larger than its corresponding atom. The greater electron charge causes the electron cloud to increase in size.
   c. The metal atom is larger than a nonmetal of the same period. Atomic radius decreases going from left-to-right within a period due to the increasing nuclear charge.
   d. Nonmetal ions are negative and larger than the positive metal ions of the same period.
8. a. Na^{+}: $1s^2 2s^2 2p^6$ b. Ca^{2+}: $1s^2 2s^2 2p^6 3s^2 3p^6$

   c. Al^{3+}: $1s^2 2s^2 2p^6$ d. Cl^{-}: $1s^2 2s^2 2p^6 3s^2 3p^6$

   e. O^{2-}: $1s^2 2s^2 2p^6$
9. a. K^{+} b. Sc^{3+} c. Sr^{2+} d. Li^{+} e. Al^{3+}
10. a. S^{2-} b. F^{-} c. N^{3-} d. I^{-} e. O^{2-}
11. Cs^{+} and Be^{2+} have noble gas configurations.
12. S^{2-}, Cl^{-}, K^{+}, Ca^{2+}, and Sc^{3+} have the electron configuration of argon.
13. a. Ag^{+} b. Zn^{2+} c. Cu^{2+} or Cu^{+} d. Fe^{2+} or Fe^{3+}
14. a. NH_4^{+}, ammonium b. CO_3^{2-}, carbonate
    c. NO_3^{-}, nitrate d. ClO_3^{-}, chlorate
15. a. sulfate, SO_4^{2-} b. phosphate, PO_4^{3-}
    c. hydrogen carbonate, HCO_3^{-} d. chromate, CrO_4^{2-}
16. Li^{+}, Na^{+}, K^{+}, Rb^{+}, Cs^{+}, Ag^{+} and NH_4^{+}
17. H^{-}, F^{-}, Cl^{-}, Br^{-}, I^{-}, OH^{-}, NO_3^{-}, ClO_3^{-}, ClO_4^{-}, MnO_4^{-}, and HCO_3^{-}
18. a. AX b. BX_2 c. CX_3 d. A_2Y e. BY f. C_2Y_3

VI. Solutions to Problems

1. The correct choice is (d), 0.072 nm and 0.184 nm. The radius of a positive ion is less than, while that of the negative ion is larger than, that of the neutral atom.
2. FM $CrSO_4$ = 148.1

$$\% \text{ Cr} = \frac{52.00}{148.1} \times 100 = 35.11$$

FM $Cr_2(SO_4)_3$ = 392.2

$$\% \text{ Cr} = \frac{104.0}{392.2} \times 100 = 26.52$$

3. FM $Na_2CO_3 \cdot 10\ H_2O$ = 286.2

$$\% \text{ H}_2\text{O} = \frac{180.2}{286.2} \times 100 = 62.96$$

4. a. CaI_2 b. K_2S c. Al_2O_3 d. SrO e. RbCl f. AlF_3
5. a. $NiCl_2$ b. NiO c. NiS d. $NiBr_2$
6. a. AgCl b. $CuCl_2$ or CuCl c. $FeCl_2$ or $FeCl_3$ d. $ZnCl_2$
7. a. $Ca(s) + I_2(s) \rightarrow CaI_2(s)$

   b. $2\ K(s) + S(s) \rightarrow K_2S(s)$

   c. $4\ Al(s) + 3\ O_2(g) \rightarrow 2\ Al_2O_3(s)$

   d. $2\ Sr(s) + O_2(g) \rightarrow 2\ SrO(s)$

   e. $2\ Rb(s) + Cl_2(g) \rightarrow 2\ RbCl(s)$

   f. $2\ Al(s) + 3\ F_2(g) \rightarrow 2\ AlF_3(s)$
8. a. $Ni(s) + Cl_2(g) \rightarrow NiCl_2(s)$

   b. $2\ Ni(s) + O_2(g) \rightarrow 2\ NiO(s)$

   c. $Ni(s) + S(s) \rightarrow NiS(s)$

   d. $Ni(s) + Br_2(l) \rightarrow NiBr_2(s)$
9. a. calcium iodide d. strontium oxide
   b. potassium sulfide e. rubidium chloride
   c. aluminum oxide f. aluminum fluoride
10. a. iron (II) chloride d. cobalt (III) fluoride
    b. iron (III) chloride e. copper (I) nitrate
    c. cobalt (II) bromide
11. a. $KHCO_3$ b. $MgSO_4$ c. $Co(NO_3)_3$
12. $Fe^{3+} < Fe^{2+} < Fe$
13. a. Na_2S b. $Fe(NO_3)_2$ c. $ZnSO_4$ d. $BaCO_3$

14. a. $2\ K(s) + F_2(g) \rightarrow 2\ KF(s)$

 b. $2\ K(s) + Cl_2(g) \rightarrow 2\ KCl(s)$

 c. $2\ K(s) + I_2(s) \rightarrow 2\ KI(s)$

 d. $4\ K(s) + O_2(g) \rightarrow 2\ K_2O(s)$

 e. $2\ K(s) + S(s) \rightarrow K_2S(s)$

15. a. $4\ Li(s) + O_2(g) \rightarrow 2\ Li_2O(s)$

 b. $2\ Zn(s) + O_2(g) \rightarrow 2\ ZnO(s)$

 c. $4\ Sc(s) + 3\ O_2(g) \rightarrow 2\ Sc_2O_3(s)$

 d. $4\ Bi(s) + 3\ O_2(g) \rightarrow 2\ Bi_2O_3(s)$

16. a. $2\ Rb(s) + S(s) \rightarrow Rb_2S(s)$

 b. $2\ Sr(s) + O_2(g) \rightarrow 2\ SrO(s)$

 c. $Cu(s) + Cl_2(g) \rightarrow CuCl_2(s)$

 d. $4\ Ag(s) + O_2(g) \rightarrow 2\ Ag_2O(s)$

 e. $Zn(s) + Br_2(l) \rightarrow ZnBr_2(s)$

17. a. potassium phosphate d. chromium (III) chloride
 b. iron (III) nitrate e. aluminum sulfide
 c. ammonium carbonate

18. a. FM $NaNO_3 = 85.00$

$$\% \text{ Na} = \frac{22.99}{85.00} \times 100 = 27.05 \qquad \% \text{ N} = \frac{14.01}{85.00} \times 100 = 16.48$$

$$\% \text{ O} = 100.00 - 27.05 - 16.48 = 56.47$$

 b. FM $Ag_2O = 231.7$

$$\% \text{ Ag} = \frac{215.7}{231.7} \times 100 = 93.09 \qquad \% \text{ O} = 100.0 - 93.09 = 6.91$$

 c. FM $(NH_4)_3PO_4 = 149.1$

$$\% \text{ N} = \frac{42.02}{149.1} \times 100 = 28.18 \qquad \% \text{ H} = \frac{12.10}{149.1} \times 100 = 8.12$$

$$\% \text{ P} = \frac{30.97}{149.1} \times 100 = 20.77 \qquad \% \text{ O} = \frac{64.00}{149.1} \times 100 = 42.92$$

19. a. FM $Al_2(SO_4)_3 = 342.1$

$$\% \text{ Al} = \frac{53.96}{342.1} \times 100 = 15.77$$

 b. FM $BaCl_2 \cdot 2\ H_2O = 244.3$

$$\% \text{ Ba} = \frac{137.3}{244.3} \times 100 = 56.20$$

20. a. FM $CaCl_2$ = 111.0

$$\% \text{ Cl} = \frac{70.91}{111.0} \times 100 = 63.88$$

b. FM $CaCl_2 \cdot 6\ H_2O$ = 219.1

$$\% \text{ Cl} = \frac{70.91}{219.1} \times 100 = 32.36$$

c. FM $Ca(ClO_3)_2$ = 207.0

$$\% \text{ Cl} = \frac{70.91}{207.0} \times 100 = 34.26$$

21. FM $CuSO_4$ = 159.6

FM $CuSO_4 \cdot 5\ H_2O$ = 249.7

$$\text{mass CuSO}_4 = 1.00 \text{ g hydrate} \times \frac{159.6 \text{ g CuSO}_4}{249.7 \text{ g hydrate}} = 0.639 \text{ g}$$

22. a. $4\ Sc(s) + 3\ O_2(g) \rightarrow 2\ Sc_2O_3(s)$

b. $$\text{mass O}_2 = 1.00 \text{ g Sc} \times \frac{3(32.0) \text{ g O}_2}{4(45.0) \text{ g Sc}} = 0.533 \text{ g}$$

c. $$\text{volume O}_2 = 0.533 \text{ g O}_2 \times \frac{1 \text{ mol O}_2}{32.0 \text{ g O}_2} \times \frac{22.4\ \ell}{1 \text{ mol O}_2} = 0.373\ \ell$$

23. Adding the three given equations gives the desired equation:

$Li(s) + 1/2\ F_2(g) \rightarrow LiF(s)$.

The ΔH of the reaction then equals the sum of the individual heat effects:

$\Delta H = +55 \text{ kcal} + 43 \text{ kcal} - 244 \text{ kcal} = -146 \text{ kcal}$

24. In order to arrive at the equation for lattice energy, the second and third reactions must be reversed:

$Ag^+(g) + Cl^-(g) \rightarrow Ag(g) + Cl(g)$; $\Delta H = -87$ kcal

$Ag(g) + Cl(g) \rightarrow Ag(s) + 1/2\ Cl_2(g)$; $\Delta H = -90$ kcal

$Ag(s) + 1/2\ Cl_2(g) \rightarrow AgCl(s)$; $\Delta H = -30$ kcal

$Ag^+\ (g) + Cl^-(g) \rightarrow AgCl(s)$

$\Delta H = -87 \text{ kcal} - 90 \text{ kcal} - 30 \text{ kcal} = -207 \text{ kcal}$

25. $$\text{moles O} = \frac{60.7 \text{ g O}}{16.0 \text{ g O/mol}} = 3.79 \qquad \text{moles N} = \frac{17.7 \text{ g N}}{14.0 \text{ g N/mol}} = 1.26$$

$$\text{moles C} = \frac{15.2 \text{ g C}}{12.0 \text{ g C/mol}} = 1.27 \qquad \text{moles H} = \frac{6.37 \text{ g H}}{1.01 \text{ g H/mol}} = 6.31$$

$$\frac{\text{moles O}}{\text{mole N}} = \frac{3.79}{1.26} = 3.01 \qquad \frac{\text{moles H}}{\text{mole N}} = \frac{6.31}{1.26} = 5.01$$

$$\frac{\text{mole C}}{\text{mole N}} = \frac{1.27}{1.26} = 1.01$$

Simplest formula is NH_5CO_3; Ionic formula is NH_4HCO_3

Ions present are $NH_4{}^+$ and $HCO_3{}^-$

26. Assuming 1.000 g of $CoCl_2 \cdot X\ H_2O$:

mass $CoCl_2$ = 0.643 g

mass H_2O = 1.000 g − 0.643 g = 0.357 g

$$\text{moles } CoCl_2 = \frac{0.643 \text{ g } CoCl_2}{130 \text{ g } CoCl_2/\text{mol}} = 4.95 \times 10^{-3}$$

$$\text{moles } H_2O = \frac{0.357 \text{ g } H_2O}{18.0 \text{ g } H_2O/\text{mol}} = 1.98 \times 10^{-2}$$

$$\frac{\text{moles } H_2O}{\text{mol } CoCl_2} = \frac{1.98 \times 10^{-2}}{4.95 \times 10^{-3}} = 4.00$$

Hydrate formula is $CoCl_2 \cdot 4\ H_2O$

CHAPTER 11

Covalent Bonding

I. Basic Skills

A. Qualitative Students should be able to:

1. Describe the necessary conditions for the forming of a covalent bond.
2. Describe covalent bonding using the orbital overlap model.
3. Distinguish among single, double, and triple bonds.
4. Describe the general properties of molecular compounds.
5. Explain what is meant by:
   a. The Lewis structure of an atom or molecule.
   b. The skeleton structure of a molecule.
6. Describe the octet rule.
7. Explain the electron pair repulsion principle and how it affects molecular geometry.
8. Describe linear, bent, pyramidal, and tetrahedral geometries.
9. Describe what is meant by bond polarity and molecular polarity.

B. Quantitative Students should be able to:

1. Write the Lewis structure for:
   a. any of the main-group elements (Example 11.1).
   b. simple molecules (Example 11.2).
   c. complex molecules, if given the skeleton structure.
   d. molecules containing multiple bonds (Example 11.3).
2. Predict the geometry of simple molecules (Example 11.4).
3. Predict the polarity of covalent bonds (Example 11.5) and the polarity of molecules (Example 11.6).
4. Name binary molecular compounds (Example 11.7) and oxygen-containing acids (Example 11.8).

II. Chapter Development

1. The nature of the covalent bond is briefly introduced by discussing the orbital overlap of half-filled orbitals. This is then extended to the idea that nonmetal atoms reach a noble-gas structure by sharing electrons.
2. The octet rule is used to guide the writing of correct Lewis structures. Exceptions to the octet rule are covered in an optional section.
3. Students are not exposed to the idea of ''bonding capacity'' which implies that an element of a particular group forms just so many bonds. (For example, Gp 7 atoms are said to have a bonding capacity of 1.) We believe that there are too many exceptions to these general rules and that students must eventually

unlearn them.

4. The electron pair repulsion principle is used as a simple and logical method for explaining the geometry of a molecule. The concepts of electron promotion and hybridization of orbitals are not believed to be necessary.
5. The geometry of multiple-bonded molecules is not included in this chapter. Questions on this subject may naturally arise, however (CO_2, C_2H_4, etc.). Their geometry is easily predicted by considering the multiple bond as a single bond and applying the principle of electron repulsion as before. (CO_2 with 2 double bonds is treated as having 2 single bonds and is therefore linear; C_2H_4 with 2 single bonds and a double bond is treated as having 3 single bonds and is therefore triangular.) Some applications of this type are included in Chapter 14 when discussing hydrocarbons.
6. The Lewis structures and the geometry of polyatomic ions are not specifically covered. Most students will be able to deduce them using the same principles used for neutral molecules.
7. The concept of resonance with regard to molecules like SO_2 and SO_3 is not introduced at this time. Resonance is covered later in Chapter 14 when discussing the structure of benzene.

III. Problem Areas

1. Drawing Lewis structures can be difficult for some students (especially those who can't count to eight). Some will have difficulty in arranging the atoms correctly. (Why not Cl—Cl—C—Cl—Cl instead of the following?)

```
      Cl
      |
Cl — C — Cl
      |
      Cl
```

   Many cases, like the above, can be handled by the generalization that multiple atoms of the same type are usually bonded individually to a central nonmetal atom. In other cases, the student cannot be expected to select the correct structure and should be given the skeleton structure.
2. In discussing the polarity of molecules, use the idea that a polar bond is like a force which can cause a molecule to turn in an electrical field. If these forces cancel, the molecule will not turn and is said to be nonpolar. If the forces do not cancel, the molecule will turn and is said to be polar.
3. There will be some confusion in naming molecular compounds as contrasted to ionic compounds of Chapter 10. While SCl_2 is sulfur dichloride, $CaCl_2$ is not calcium dichloride but simply calcium chloride. As the rules differ between molecular and ionic compounds, the student must first decide which type of compound is under consideration.

IV. Suggested Activities

Molecular Geometry:
Build models of simple molecules with a ball and stick model kit. Contrast the shape and bond angles of linear, bent, pyramidal, and tetrahedral molecules which obey the octet rule. Use sticks to represent nonbonding electron pairs. This helps in relating the structure (as defined by the nuclei) to the electron pair repulsion principle. Show the Lewis structure for each molecule simultaneously with its model. (Most ball and stick model kits do not allow the building of linear AX_2 or triangular AX_3 molecules. These, and all other geometries, can be made using styrofoam balls and toothpicks.)

V. Answers to Questions

1. Charlie Brown meant that atoms achieve "happiness" by sharing a pair of electrons so as to form a covalent bond.
2. a. A covalent bond is the sharing of an electron pair between two atoms.
   b. A single bond is the sharing of one electron pair between two atoms.
   c. Marital bonds exist between husband and wife and are not normally covered in a high school chemistry course.
   d. A double bond is the sharing of two electron pairs between two atoms.
   e. A triple bond is the sharing of three electron pairs between two atoms.
   f. An ionic bond is the attractive force between positive and negative ions. Electrons are not shared, but transferred from the metal atom to the nonmetal atom.
3. Hydrogen atoms have a half-filled 1s orbital. Their overlap produces a stable molecule in which the attractive forces are greater than the repulsive forces. In contrast, helium atoms have a filled 1s orbital. The overlap of these orbitals would produce an unstable combination.
4. The Li_2 molecule would be more stable than the Be_2 molecule. The Li_2 molecule is formed by the overlap of half-filled 2s orbitals while the 2s orbitals of Be are filled.
5. CCl_4 is a molecular compound while NaCl is an ionic compound. Molecular compounds boil readily because there are only weak forces (Chapter 12) between the molecules. In order to boil an ionic compound, the strong ionic

bonds must be broken.

6. Molecular liquids, such as CCl_4, do not conduct electricity as no ions are present. Molten ionic compounds, such as NaCl, conduct because their ions are free to move in the liquid state.
7. The skeleton structure shows the correct bonding arrangement of the atoms in a molecule. The Lewis structure shows all valence electrons around the atoms and distinguishes between single, double, and triple bonds. The skeleton structure for H_2O is H—O—H, while its Lewis structure is $H-\underset{\cdot\cdot}{\ddot{O}}-H$. The structures for CH_4 are the same,

$$\begin{array}{ccc} & H & \\ & | & \\ H- & C & -H \\ & | & \\ & H & \end{array}$$

   because all valence electrons form single bonds.
8. a. CH_4, 4 single bonds
   b. C_2H_6, 7 single bonds
   c. C_2H_4, 4 single bonds and 1 double bond
   d. C_2H_2, 2 single bonds and 1 triple bond
9. The Lewis structure of an atom consists of its chemical symbol surrounded by the correct number of valence electrons shown as dots. For the main-group elements, the number of valence electrons is equal to the Group number in the Periodic Table.
10. In most stable molecules, the atoms acquire the structure of the nearest noble gas by surrounding themselves with eight valence electrons. The nearest noble gas to hydrogen is helium with two valence electrons.
11. The octet rule is followed in structures (a) and (d).
12. A deficiency of 4 electrons can be accounted for by the presence of 1 triple bond or 2 double bonds.
13. An AX_2 molecule can have a linear structure, X—A—X, or a bent structure,

   The linear structure would be nonpolar, while the bent structure would be polar.
14. The electron pairs around an atom in a molecule are directed so as to be as far apart as possible. This arrangement minimizes the repulsive forces between electron pairs. In the case of CH_4, the four electron pairs are directed at tetrahedral angles, 109.5°C.
15. a. tetrahedral b. pyramidal c. bent
16. A regular tetrahedron has 4 sides, each of which is an equilateral triangle. It has 4 corners. The tetrahedral angle is 109.5°C.
17. If there is a separation of positive and negative charges within a molecule, the molecule is said to be polar; if not, the molecule is nonpolar. A separation of charges exists when the center of all positive charges does not coincide with the center of all negative charges. Polar molecules line up in an electrical field so that the positive end of the molecule is directed toward the negative

plate and the negative end is directed toward the positive plate.

18. If a molecule is symmetrical, the polar bonds may cancel each other making the molecule nonpolar.
19. Structures (a) and (c) would give polar molecules. A molecule with structure (b) would be nonpolar because the molecule is symmetrical and the polar bonds would cancel each other.
20. The H_2O molecule is polar because its structure is bent. The BeF_2 molecule is nonpolar because it has a symmetrical linear structure.
21. a. hydrochloric acid b. hydrobromic acid
    c. hydrofluoric acid d. hydriodic acid
22. Carbon dioxide and dinitrogen oxide are systematic names. Water and nitrous oxide, for N_2O, are common names.
23. a. nitric acid b. sulfuric acid c. boric acid
24. Hypophosphorus acid, H_3PO_2, contains the least number of oxygen atoms. Phosphoric acid, H_3PO_4, contains the largest number of oxygen atoms.

VI. Solutions to Problems

1. a. ·Ẋ· b. :Ẍ· (with pair below) c. ·Ẍ· (with dot below) d. ·Ẍ· (with pair below)

2. a. H—C̈l: (with pair below) b. H—P̈—H, with H below P c. H—S̈—H (with pair below) d. H—Si—H, with H above and below Si

3. 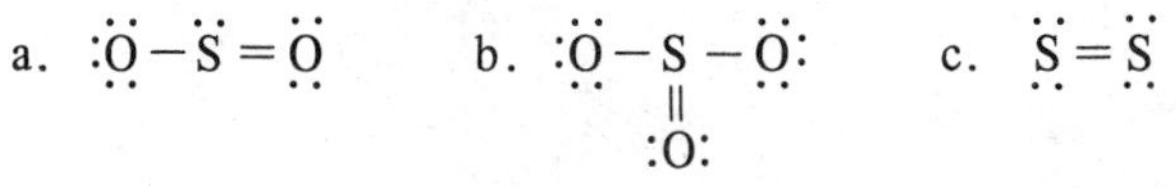 a. :Ö—S̈=Ö (with pairs below O) b. :Ö—S—Ö: (with pairs below O), with =O: below S c. S̈=S̈ (with pairs below)

4. a. HCl: linear b. PH_3: pyramidal c. H_2S: bent
   d. SiH_4: tetrahedral
5. a. N_2: nonpolar b. HI: polar c. ICl: polar
   d. H_2: nonpolar
6. HCl, PH_3 and H_2S are polar molecules; SiH_4 is nonpolar.
7. a. dihydrogen oxide b. phosphorus pentachloride
   c. dinitrogen trioxide d. disulfur dichloride
8. H_3PO_4: phosphoric acid PO_4^{3-}: phosphate ion Na_3PO_4: sodium phosphate
   H_3PO_3: phosphorous acid PO_3^{3-}: phosphite ion Na_3PO_3: sodium phosphite
   H_3PO_2: hypophosphorous acid PO_2^{3-}: hypophosphite ion

9. a. ·Ṡi· (with dot below) b. ·Äs· (with dot below) c. :K̈r: (with pair below) d. :B̈r· (with pair below)

10. a. :C̈l—P̈—C̈l: (with pairs below Cl), with :C̈l: below P b. :F̈—Ö—F̈: (with pairs below) c. :Ï—C̈l: (with pairs below)

11. a. PCl_3: pyramidal b. OF_2: bent c. ICl: linear
12. PCl_3, OF_2 and ICl are polar molecules.

```
                                                              H
                                                              |
13. a.  H—N̈—N̈—H        b.  H—Ö—Ö—H            c.  H—C—Ö—H
            |  |                                              |
            H  H                                              H

14. a.  Ö=C=Ö          b.  H—C≡N:             c.  :F̤—N̈=N̈—F̤:

            H  H                   H  H                       H        H
            |  |                   |  |                       |        |
15. a.  H—C—C—H        b.  H—C—C—Ö—H     or     H—C—Ö—C—H
            |  |                   |  |                       |        |
            H  H                   H  H                       H        H

            H  H
            |  |
    c.  H—C—C=Ö
            |
            H
```

```
        H       H
         \     /
16.       C—O
         / \
        H    H
```

The carbon atom is the center of a tetrahedron and all bond angles are about 109°.

17. The CH_3Cl, CH_2Cl_2 and $CHCl_3$ molecules are polar. The CH_4 and CCl_4 molecules are nonpolar because they are symmetrical.
18. a. NH_4^+; tetrahedral b. SO_4^{2-}: tetrahedral c. ClO_2^-: bent

```
                                                              :C̤l:
                                                                |
19. a.  :C̤l—Ö—C̤l:       b.  H—N̈—H             c.  :C̤l—Si—C̤l:
                                  |                             |
                                :C̤l:                          :B̤r:
        bent, polar
                          pyramidal, polar        tetrahedral, polar
```

```
20. H—Ö—N—Ö:
          ||
          :O:
```

21. a. SCl_2 b. P_4O_6 c. N_2O_5
22. a. selenic acid b. selenous acid
    c. diarsenic pentoxide d. dihydrogen dioxide (hydrogen peroxide)

```
23. :Ö—N̈=Ö        :Ö—C̤l—Ö:
```

NOTE: In each of these structures, the least electronegative atom is short one electron. Both molecules have a bent structure.

24. a. $:\ddot{\underset{..}{O}}-H^-$ b. $:\ddot{\underset{..}{Cl}}-\ddot{\underset{..}{O}}:^-$ c. $\left[\begin{array}{c} :O: \\ \| \\ :\ddot{\underset{..}{O}}-C-\ddot{\underset{..}{O}}: \end{array}\right]^{2-}$

(8 Val. e$^-$) (14 Val. e$^-$) (24 Val. e$^-$)

25. $$\text{moles H} = \frac{10.1 \text{ g H}}{1.01 \text{ g H/mol}} = 10.0$$

$$\text{moles C} = \frac{89.9 \text{ g C}}{12.0 \text{ g C/mol}} = 7.49$$

$$\frac{\text{moles H}}{\text{mole C}} = \frac{10.0}{7.49} = 1.33 \text{ or } \frac{4}{3}$$

Simplest formula is C_3H_4. As MM is 40, the molecular formula is also C_3H_4. A possible structure is:

$$\begin{array}{c} \quad\quad\quad\quad\quad\;\; H \\ \quad\quad\quad\quad\quad\;\; | \\ H-C\equiv C-C-H \\ \quad\quad\quad\quad\quad\;\; | \\ \quad\quad\quad\quad\quad\;\; H \end{array}$$

26.

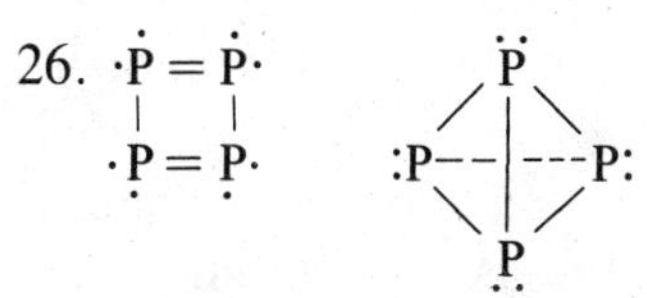

planar tetrahedral (true structure)

CHAPTER 12

Liquids and Solids

I. Basic Skills

A. Qualitative Students should be able to:

1. Describe how the condensed states differ from the gas state with regard to distance between particles, molar volume, and attractive forces.
2. Describe the relationship between attractive forces and the heats of fusion and vaporization of a substance.
3. a. Explain what is meant by the dynamic equilibrium between a liquid and its vapor.
   b. Explain what is meant by the equilibrium vapor pressure of a liquid.
4. Explain how a liquid's vapor pressure is related to the nature of the liquid and to its temperature.
5. a. Explain what is meant by the boiling of a liquid and its boiling point.
   b. Explain the effect of pressure on the boiling point.
6. Describe Van der Waals forces and how they are related to molecular mass.
7. Describe hydrogen bonding and the necessary conditions for its presence.
8. Explain what is meant by the melting of a solid and its melting point.
9. Explain the effect of attractive forces on the melting of a solid and the boiling of a liquid.
10. List the types of ionic compounds which tend to decompose when heated.
11. Explain what is meant by sublimation and give an example of a substance which sublimes.
12. Describe the "electron sea" model for metals.
13. Describe the bonding in covalent network solids.
14. Describe the general properties of ionic, molecular, metallic, and covalent network solids (Table 12.4). Give an example of each type.

B. Quantitative Students should be able to:

1. Determine the vapor pressure of water at a particular temperature (Table 12.2).
2. Calculate the relative humidity of air given the vapor pressure of water at a given temperature (Equation 12.2 and Example 12.1).
3. Determine the boiling point of water at a particular pressure (Example 12.2).
4. Predict the order of boiling point for molecular substances of similar

structure (Example 12.3).
5. Distinguish by their structural formulas which molecules can exhibit hydrogen bonding (Example 12.4).
6. Predict the type of solid based on its physical properties (Example 12.6).

II. Chapter Development

1. No simple physical laws, like the gas laws, are known which describe the solid and liquid states. The condensed states are distinguished from the gas state by the close distance between particles and the resulting strength of the attractive forces. In contrast to the gas state, their behavior is strongly dependent upon their composition.
2. A kinetic-molecular approach is used to describe vapor pressure, boiling, and melting. These phenomena are related to the substance's attractive forces. No particular emphasis is placed on the effect of dipole forces between polar molecules. The effect of these forces on physical properties is small.
3. Solids are classified into four types which are distinguished by their properties. The solubility characteristics of solids will be covered in Chapter 13, Solutions. Molecular substances will be illustrated in the organic chemistry of Chapters 14 and 15. Covalent network solids are basic to the discussion of polymers in Chapter 16. Additional material on metallic solids is included in Chapter 25.
4. No effort is made to cover the different types of crystal systems and their unit cells.

III. Problem Areas

1. Students are often confused as to why the volume of liquid or its container has no effect on its equilibrium vapor pressure. Explain that changing either of these volumes creates a temporary disturbance which causes a temporary change in the vapor pressure. Then have them explain how the system would adjust to the disturbance and regain its original equilibrium vapor pressure.
2. When discussing the effect of pressure (or altitude) on boiling, sometimes we confuse the boiling point with the time required to cook something. A student will say that "it takes longer to boil water at a higher altitude" when what is meant is that "it takes longer to cook an egg in the water." Separate the physical change of boiling water, which is faster at high altitudes, from the chemical cooking of the egg, which is slower.

IV. Suggested Activities

1. The Effect of Pressure on the Boiling Point: **Teacher Demonstration Only** Fill a 250 cm^3 round-bottomed Florence flask about one-third with water.

CAUTION: Make certain that the flask contains no cracks so that it will not explode. Fit the flask with a stopper-thermometer assembly so that the bulb extends into the water. Loosen the stopper and gradually heat the water. Relate the temperature increase to the increasing energy of the water and to its increasing vapor pressure. As the water reaches its boiling point, relate its vapor pressure to the atmospheric pressure. (Read the barometer.) After the water has boiled for a minute, remove the burner and (unnoticed by the class) seal the flask with the stopper. After the boiling has momentarily stopped, hold the flask under running cold water. The water will begin to boil vigorously at the reduced pressure caused by the condensation of steam. Ask students to explain their observations.

2. Diffusion in Liquids:
   Compare the rate of diffusion in water samples of different temperature. Add one drop of food coloring to separate beakers of cold water and hot water. Relate the diffusion rate to the energy and velocity of the water molecules.
3. Types of Solids:
   Display large models of ionic, molecular, metallic, and covalent network solids. If possible, display chemical samples of the same solids (NaCl, CO_2 or H_2O, Cu or Mg, graphite or silica). Discuss each solid from the viewpoint of its building unit, attractive forces, and properties. (See Table 12.4.)

V. Answers to Questions

1. Students during a lecture period are comparable to the solid state; they are restricted to definite locations (seats) in the classroom. Students during a laboratory period are comparable to the liquid state; they have more energy and have limited freedom to move about the room. Students at dismissal are comparable to the gaseous state; they leave the room and become distributed throughout the school.
2. a. Particles of matter are farthest apart in the gaseous state. The distances are much less in the solid and liquid states and are comparable to each other.
   b. The attractive forces are strongest in the solid state and weakest in the gaseous state.
   c. Particles can move past each other in the liquid and gaseous states.
3. In melting or vaporizing, heat must be absorbed to break down the attractive forces between particles.
4. a. Evaporation is the movement of molecules from the liquid state to the gaseous state at temperatures below the boiling point.

b. Dynamic equilibrium is attained when opposing processes take place at equal rates.
c. The equilibrium vapor pressure is the pressure exerted by a vapor in equilibrium with its liquid. The system must be closed and at constant temperature.

5. a. If more water is added, the pressure remains the same.
   b. If the temperature drops, the pressure drops.
   c. If the volume is increased, the pressure remains the same. More water will vaporize to maintain a constant pressure.
6. A relative humidity of 50% means that the air contains 50% of the water that it is capable of holding. That is, the pressure of the water in the air is 50% of the vapor pressure of water at the same temperature.
7. The vapor pressures of water at 0°C and 20°C are 4.6 and 17.5 mm Hg, respectively. At 0°C and 80% RH, the water in the air has a pressure of 0.80×4.6 mm Hg or 3.7 mm Hg. This same air when heated to 20°C has an RH of $\frac{3.7 \text{ mm Hg}}{17.5 \text{ mm Hg}} \times 100$ or 21%. This accounts for the dryness of the air in homes heated by forced-air systems.
8. Boiling is the rapid vaporization with formation of vapor bubbles, which takes place at a liquid's boiling point. Evaporation is the gradual vaporization which takes place at temperatures below the boiling point.
9. The boiling point of a liquid decreases as the opposing pressure decreases.
   a. At 720 mm Hg, the BP is less than 80°C.
   b. At 760 mm Hg, or one atmosphere, the BP is 80°C.
   c. At 800 mm Hg, the BP is greater than 80°C.
10. At high altitudes, atmospheric pressure is quite low. A pressure cooker is a closed system which operates at a high pressure. This will force the water to boil at a higher temperature so that the food will cook at a faster rate.
11. When a series of similar molecules is studied, such as F_2, Cl_2, Br_2, and I_2, the melting points and boiling points are observed to increase with molecular mass. The increase in these properties is attributed to an increase in Van der Waals forces.
12. A hydrogen bond is the attractive force between the hydrogen atom of one molecule and a small, highly electronegative atom of another molecule. Hydrogen bonds between like molecules are only formed when the hydrogen atom is bonded to an atom of N, O, or F. Examples are NH_3, H_2O and HF.
13. $KClO_3$, (b), releases oxygen gas when heated. $MgCO_3$, (d), releases carbon dioxide gas when heated.
14. Ionic solids consist of a network of ions held together by strong electrical forces. Molecular solids consist of individual molecules held together by relatively weak Van der Waals forces.
15. Snow and ice sublime outdoors when the temperature is below freezing. Freeze-dried foods are prepared by reducing the temperature below 0°C and lowering the pressure. Under these conditions, the water in the food is removed by sublimation.
16. As ice is less dense than liquid water (It floats!), water expands to a larger volume when it freezes. This expansion in a pipe can easily cause the pipe

to break. When the ice melts the water leaks out through the break.

17. In a metallic solid the lattice sites are occupied by metal ions and the ions are surrounded by a "sea" of valence electrons. Using the electron sea model, potassium can be diagrammed as:

```
          e−     e−     e−
      K+     K+     K+     K+
   e−     e−     e−     e−     e−
      K+     K+     K+     K+
   e−     e−     e−     e−     e−
      K+     K+     K+     K+
```

18. Metals are good conductors because their valence electrons are not tied down to a particular ion but are free to move in an electrical field. The fact that the electrons are relatively free also accounts for their low ionization energy.
19. A covalent network solid consists of a network of atoms held together by covalent bonds. An ionic solid is a network of ions held together by the electrical forces of ionic bonds. Molecular solids consist of individual molecules held together by weak Van der Waals forces.
20. Covalent network solids have high melting points because the bonds in the network must be broken for melting to occur.
21. The carbon atoms in diamond are covalently bonded to each other to form a hard, 3-dimensional, tetrahedral structure. The carbon atoms in graphite form a planar hexagonal structure. As the forces between the layers are weak Van der Waals forces, layers of graphite can be easily separated.
22. a. Covalent network solids have the highest melting points.
    b. With the exception of the covalent network solid, graphite, only metallic solids conduct electricity.
23. Simple tests which could be used to identify an unknown white solid are solubility, conductivity and melting point.
    a. Many ionic solids are soluble in water. A few molecular solids are soluble if hydrogen bonding is present.
    b. Solutions of ionic compounds conduct electricity. Only metallic solids conduct in the solid state.
    c. Molecular solids have low melting points.

VI. Solutions to Problems

1. $P_{H_2O} = 10.5 \text{ mm Hg at } 12°\text{C}$

$$RH = \frac{6.5 \text{ mm Hg}}{10.5 \text{ mm Hg}} \times 100 = 62\%$$

2. From Table 12.2:

   $P_{H_2O} = 300$ mm Hg at about 75°C

   Therefore, water boils at about 75°C at 300 mm Hg pressure.
3. From Figure 12.4:

   $P_{H_2O} = 300$ mm Hg at about 77°C

   The boiling point obtained from the graph is more accurate than that obtained from the table. Interpolation of the table assumes a linear relationship between P and T, while the actual relationship is exponential.
4. Increasing boiling point: $CH_4 < C_2H_6 < C_3H_8 < C_4H_{10}$
5. Due to their equal molecular masses, N_2, (b), and CO, (c), are expected to have similar boiling points. Actual: N_2 (−196°C), CO (−192°C).
6. Hydrogen peroxide, (a), urea, (b), and acetic acid, (d), form hydrogen bonds. The H atom must be bonded to an N, O or F atom.
7. a. $Ni(OH)_2(s) \rightarrow NiO(s) + H_2O(g)$

   b. $2\ AgOH(s) \rightarrow Ag_2O(s) + H_2O(g)$

   c. $NiCO_3(s) \rightarrow NiO(s) + CO_2(g)$

   d. $Ag_2CO_3(s) \rightarrow Ag_2O(s) + CO_2(g)$
8. The element is a metallic solid. Only metals conduct in either the solid or liquid state.
9. The element is a covalent network solid. Ionic and metallic solids conduct in the liquid state.
10. $\Delta H_{fus} = 23.0\ \frac{cal}{g} \times \frac{1\ kcal}{10^3\ cal} \times \frac{70.9\ g}{1\ mol} = 1.63\ kcal/mol$

    $\Delta H_{vap} = 67.4\ \frac{cal}{g} \times \frac{1\ kcal}{10^3\ cal} \times \frac{70.98}{1\ mol} = 4.78\ kcal/mol$
11. a. $\Delta H = 1.50\ mol\ Hg \times \frac{14.2\ kcal}{1\ mol} = 21.3\ kcal$

    b. $\Delta H = 1.29\ g\ C_6H_6 \times \frac{94\ cal}{g} = 120\ cal$
12. a. BP of H_2O at 100 mm Hg is about 51°C

    b. P_{H_2O} at 105°C is about 910 mm Hg
13. P_{H_2O} at 22°C = 19.8 mm Hg

    P_{H_2O} (in air) = 0.42 (19.8 mm Hg) = 8.3 mm Hg
14. The temperature at which water's vapor pressure is about 8.3 mm Hg (Problem 13) is about 8°C. This is called the dew point because the air would be saturated (RH = 100%) at these conditions.

15. a. $N_2O > N_2$: greater molecular mass

    b. $N_2O_4 > N_2O$: greater molecular mass

    c. $H_2O > CH_4$: hydrogen bonding in H_2O

16. Only HOCl can form hydrogen bonds.

17. a. $2\ NaClO_3(s) \rightarrow 2\ NaCl(s) + 3\ O_2(g)$

    b. $Ca(ClO_3)_2(s) \rightarrow CaCl_2(s) + 3\ O_2(g)$

18. a. Cr: metallic b. I_2: molecular
    c. C: covalent network d. $BaCl_2$: ionic
    e. SiO_2: covalent network f. Al: metallic
    g. H_2O: molecular h. $Al(NO_3)_3$: ionic

19. The unknown solid is most likely (b), LiCl, as only ionic compounds behave in this way.

20. The solid is either $BaCl_2 \cdot 2\ H_2O$, (a), $Mg(OH)_2$, (c), which decomposes to MgO when heated.

21. a. $MgO > I_2$: ionic vs molecular

    b. $SiO_2 > CO_2$: covalent network vs molecular

    c. $Fe > H_2O$: metallic vs molecular

    d. $MgO > NaF$: stronger ionic forces in MgO due to $+2$, -2 charges.

22. $Hg(s) \rightarrow Hg(l)$ $\Delta H_{fus} = 0.56$ kcal

    $Hg(l) \rightarrow Hg(g)$ $\Delta H_{vap} = 14.2$ kcal

    $Hg(s) \rightarrow Hg(g)$ $\Delta H_{sub1} = \Delta H_{fus} + \Delta H_{vap}$

    $\Delta H_{sub1} = 0.56 + 14.2 = 14.8$ kcal

23. $$\text{molecules (l)} = 1.00\ \text{cm}^3 \times \frac{1.00\ \text{g}}{1.00\ \text{cm}^3} \times \frac{6.02 \times 10^{23}\ \text{molecules}}{18.0\ \text{g}} = 3.34 \times 10^{22}$$

    $$\text{molecules (g)} = 1.00\ \text{cm}^3 \times \frac{1\ \ell}{10^3\ \text{cm}^3} \times \frac{6.02 \times 10^{23}\ \text{molecules}}{22.4\ \ell} = 2.69 \times 10^{19}$$

CHAPTER 13

Water Solutions

I. Basic Skills

A. Qualitative Students should be able to:

1. a. Explain what is meant by solution, solvent, and solute.
   b. Describe the types of solutions and give an example of each (Table 13.1).
2. a. Describe what is meant by the concentration of a solution.
   b. Define the mass percent and the molarity of a solution.
3. a. Explain how the nature of the solute affects its solubility in water.
   b. Explain the conditions which favor the solubility of molecular solutes.
4. Describe the process of solution equilibrium.
5. Distinguish among saturated, unsaturated, and supersaturated solutions.
6. Explain the effect of temperature changes on the solubility of solutions containing solid or gaseous solutes.
7. Define electrolyte and nonelectrolyte. Explain how these types of solutes affect the electrical conductivity of solutions.
8. a. Explain what is meant by a colligative property.
   b. Explain the effect of concentration on the vapor pressure, boiling point, and freezing point of a solution.

B. Quantitative Students should be able to:

1. a. Calculate the mass percent of a solution or the amount of components present in a solution of a given mass percent (Example 13.1).
   b. Describe how to make a solution of a given mass percent (Example 13.2).
2. a. Calculate the molarity of a solution (Example 13.3).
   b. Determine the moles of a solute in a particular volume or the volume which contains a particular number of moles for a solution of known molarity (Example 13.4).
   c. Describe how to make a solution of given molarity (Example 13.5).
3. Write equations showing how ionic solutes dissolve in water (Example 13.6).
4. Determine the solubility of solutions using solubility curves (Figure 13.6 and Example 13.7).
5. Determine the relative order of freezing points or boiling points of solutions containing either molecular or ionic solutes (Examples 13.8 and 13.9).

II. Chapter Development

1. While the principles can be applied to other solvents, the chapter is essentially limited to water solutions.
2. Methods of expressing concentration are limited to mass percent, still commonly used, and molarity. Molality is introduced in an optional section at the end of the chapter. Students may ask about volume percent, having read the term on the labels of beverages. Volume percent is used in this context in Chapter 15.
3. Strong acids are used as examples of molecular electrolytes. Weak electrolytes (weak acids) are not discussed until Chapter 19.
4. Solution equilibrium is presented in a manner similar to that of vapor pressure equilibrium in Chapter 12. These examples of physical equilibrium should prepare the student for chemical equilibrium applications in Chapters 18–20.
5. The colligative properties of water solutions are discussed from a qualitative viewpoint. Students are expected to distinguish between the effects of molecular and ionic solutes. A quantitative approach to freezing points, and its use in determining molar mass, is presented in an optional section.

III. Problem Areas

1. Students will have previously used molar solutions in their laboratory work. This will help in selling the molarity system. Demonstrate its utility by showing how a volume measurement, instead of a mass measurement, is used to obtain a particular number of moles.
2. Students frequently mess up molarity calculations by using milliliters or cubic centimeters instead of liters for the volume. Stress the basic definition (mols/ℓ) and show how units must cancel in molarity calculations.

IV. Suggested Activities **Teacher Demonstrations Only**

1. Supersaturation:
   Add 25 ml of water to about 200 ml of sodium acetate hydrate in a Florence flask. Heat the mixture slowly on a hot plate until all solute is dissolved. Stopper the flask and allow to cool to room temperature. Handle the flask carefully at demonstration time but clearly show that the contents are liquid. Remove the stopper and add one crystal of sodium acetate. The supersaturated solution will immediately crystallize into an apparent solid. The contents can be dissolved again and reused after cooling.
2. Electrical Conductivity:
   Use a conductivity tester and show the low conductivity of distilled water.

Test the conductivity of molecular and ionic solutions. Include solutions of molecular solutes which ionize (HCl and H_2SO_4). Explain the conductivity in terms of mobile ions which carry a charge between the electrodes. (NOTE: Do this demonstration only if not done by students in Experiment 19.)

3. Colligative Properties:
   Prepare two 600-ml beakers of distilled water in advance, one boiling and one with crushed ice. Measure the boiling point and the freezing point of the water. Add about 100 g of NaCl (rock salt) to each beaker and stir until dissolved. Heat the hot solution to regain boiling. Measure the new boiling point and freezing point. Discuss potential applications of these principles (cooking, ice cream, melting ice, antifreeze). (NOTE: Do this demonstration only if not done by students in Experiment 21.)
4. Heat of Solution:
   Review results obtained in Experiment 9, Part II. Demonstrate the endothermic dissolving of KNO_3. Relate this result to the shape of the KNO_3 solubility curve (Figure 13.6). Ask students to predict the effect of temperature on the solubility of a compound which dissolves exothermically.

V. Answers to Questions

1. a. household ammonia: solute is a gas (NH_3), solvent is a liquid (H_2O)
   b. brass: solute is a solid (Zn), solvent is a solid (Cu)
   c. tincture of iodine: solute is a solid (I_2), solvent is a liquid (C_2H_5OH)
2. Mass percent of solute is the percent of the solution's mass which is solute. Molarity is the moles of solute in a liter of solution.
3. $\%\ \text{solute} + \%\ H_2O = 100$

   $\%\ H_2O = 100 - 42.6 = 57.4$
4. (1) Calculate the mass of solute required:

   $$\text{mass KCl} = 0.500\ \ell \times \frac{1.00\ \text{mol KCl}}{1\ \ell} \times \frac{74.6\ \text{g KCl}}{1\ \text{mol}} = 37.3\ \text{g}$$

   (2) Add 37.3 g of KCl to a 500-cm^3 volumetric flask. Add sufficient water to dissolve the KCl. Then add additional water to obtain a solution volume of exactly 500 cm^3.
5. Ionic solutes tend to be soluble in water because the polar water molecules are electrically attracted to positive and negative ions. This attraction provides some of the energy required to break down the ionic crystal.
6. The polar HCl molecules are broken into ions by the polar water molecules. The Cl_2 molecules are nonpolar and do not interact with the H_2O molecules.
7. H_2O_2 can form hydrogen bonds with water.

8. a. $NaNO_3(s) \rightarrow Na^+(aq) + NO_3^-(aq)$

   b. $KCl(s) \rightarrow K^+(aq) + Cl^-(aq)$

   c. $Br_2(l) \rightarrow Br_2(aq)$

   d. $O_2(g) \rightarrow O_2(aq)$

9. Add an additional crystal of solute to the solution. If unsaturated the solute will dissolve, and if saturated the solute will not dissolve. If the solution is supersaturated the excess solute will crystallize out of solution.
10. Unsaturated solutions: 0.10 M and 0.20 M
    Supersaturated solution: 0.25 M
11. a. No effect
    b. No effect
    c. The concentration would increase; some undissolved solid would go into solution.
    d. The concentration would decrease; some solid will crystallize out as the solution is cooled.
12. a. $AgNO_3$ b. $MgCl_2$ c. NaCl
13. a. about 32 g at 20°C b. about 62 g at 40°C
    c. about 110 g at 60°C d. about 167 g at 80°C
14. Water contains a small amount of dissolved N_2 and O_2 gases. The solubility of these gases is decreased when the water is heated and they escape as gas bubbles.
15. Density is not a colligative property. While density does vary with concentration, it is not independent of the solute.
16. Vapor pressure, (b), and freezing point, (c), are colligative properties. While color does vary with concentration, the particular color of a solution (if any), depends upon the nature of the solute. The electrical conductivity also depends upon the solute as some are electrolytes while others are nonelectrolytes.
17. (1) Conductivity: Because NaCl is an ionic solute its solution will conduct electricity, while that of molecular sucrose will not.
    (2) Freezing point: The freezing point of the NaCl solution will be lower than that of the sugar solution. Electrolytes, such as NaCl, ionize to produce more particles.
18. Electrolytes: NaCl, KNO_3 and HCl
    Nonelectrolytes: Br_2 and $C_6H_{12}O_6$
19. The vapor pressure of a solution is lower than that of the solvent (see Figure 13.10). For boiling, the solution's temperature must be increased until the solutions's vapor pressure is equal to atmospheric pressure. This requires a higher temperature than is needed to boil the pure solvent. For freezing, the temperature must be decreased until the solution's vapor pressure is equal to that of ice. That temperature is lower than the freezing point of the pure solvent.
20. a. Both solutions are colorless.
    b. Both solutions are nonconductive.
    c. The vapor pressure of the 0.2 M solution will be less than that of the 0.1 M solution.
    d. The boiling point of the 0.2 M solution will be higher than that of the

0.1 M solution.

e. The freezing point of the 0.2 M solution will be lower than that of the 0.1 M solution.

21. Rock salt is added to ice because it melts the ice by forming a solution with a freezing point below 0°C.
22. a. sucrose: 1 b. $CaCl_2$: 3 c. NaCl: 2 d. KNO_3: 2
 e. $Al(NO_3)_3$: 4

VI. Solutions to Problems

1. $$\% \text{ NaCl} = \frac{2.78 \text{ g NaCl}}{32.4 \text{ g soln}} \times 100 = 8.58$$

$$\% \text{ H}_2\text{O} = 100.00 - 8.58 = 91.42$$

2. a. $$\text{mass HCl} = 25 \text{ g soln} \times \frac{2.0}{100} = 0.50 \text{ g}$$

b. $$\text{mass H}_2\text{O} = 75.0 \text{ g soln} \times \frac{98.0}{100} = 73.5 \text{ g}$$

c. $$\text{mass soln} = 5.6 \text{ g HCl} \times \frac{100}{2.0} = 280 \text{ g}$$

3. $$\text{Molarity} = \frac{25 \text{ g NaOH}}{0.250 \ \ell} \times \frac{1 \text{ mol}}{40.0 \text{ g NaOH}} = 2.5 \text{ M}$$

4. $$\text{moles K}_2\text{CrO}_4 = 0.10 \ \ell \times \frac{0.50 \text{ mol}}{1 \ \ell} = 0.050$$

$$\text{mass K}_2\text{CrO}_4 = 0.050 \text{ mol} \times \frac{194 \text{ g K}_2\text{CrO}_4}{1 \text{ mol}} = 9.7 \text{ g}$$

5. a. $$\text{volume} = 1.00 \text{ mol Mg(NO}_3)_2 \times \frac{1 \ \ell}{0.200 \text{ mol Mg(NO}_3)_2} = 5.00 \ \ell$$

b. $$\text{volume} = 1.00 \text{ g Mg(NO}_3)_2 \times \frac{1 \text{ mol}}{148.3 \text{ g Mg(NO}_3)_2} \times \frac{1 \ \ell}{0.200 \text{ mol}}$$
$$= 3.37 \times 10^{-2} \ \ell$$

6. a. MM $AgNO_3$ = 170

Add 170 g of $AgNO_3$ to a 1 ℓ flask and add water to a solution volume of 1.0 ℓ.

b. $$\text{mass NH}_4\text{Br} = 0.50 \ \ell \times \frac{3.0 \text{ mol NH}_4\text{Br}}{1 \ \ell} \times \frac{98 \text{ g}}{\text{mol NH}_4\text{Br}} = 150 \text{ g}$$

Add 150 g of NH_4Br to a 500-cm^3 flask and add water to a solution volume of 0.50 ℓ.

7. Using Figure 13.6:
   a. The solubility of KCl at 20°C is about 34 g/100 g H_2O.
   The solubility of KCl at 80°C is about 51 g/100 g H_2O.

   b. $\text{mass } H_2O = 20 \text{ g KCl} \times \dfrac{100 \text{ g } H_2O}{51 \text{ g KCl}} = 39 \text{ g}$

   c. $\text{mass KCl} = 39 \text{ g } H_2O \times \dfrac{34 \text{ g KCl}}{100 \text{ g } H_2O} = 13 \text{ g}$

8. $\text{Molarity A} = \dfrac{20.0 \text{ g A}}{1\ \ell} \times \dfrac{\text{mol}}{80 \text{ g A}} = 0.250 \text{ M}$

   $\text{Molarity B} = \dfrac{20.0 \text{ g B}}{1\ \ell} \times \dfrac{\text{mol}}{120 \text{ g B}} = 0.167 \text{ M}$

   a. higher vapor pressure: soln B b. higher boiling point: soln A
   c. higher freezing point: soln B

9. The K_2SO_4 solution has the lowest freezing point because the solute dissociates into three ions:

$$K_2SO_4(s) \rightarrow 2\ K^+(aq) + SO_4^{2-}(aq)$$

Each unit of KBr and $AgNO_3$ breaks apart into two ions, while urea, a molecular compound, does not ionize in solution.

10. a. $\text{mass soln} = 5.02 \text{ g } CaCl_2 \times \dfrac{100 \text{ g soln}}{3.00 \text{ g } CaCl_2} = 167 \text{ g}$

    b. $\text{mass } H_2O = 167 \text{ g soln} - 5.02 \text{ g } CaCl_2 = 162 \text{ g}$

11. $\text{mass soln} = 2.00 \text{ g } NaNO_3 + 5.60 \text{ g KI} + 122 \text{ g } H_2O = 130 \text{ g (3 s.f.)}$

    $\% \ NaNO_3 = \dfrac{2.00 \text{ g } NaNO_3}{130 \text{ g soln}} \times 100 = 1.54$

    $\% \text{ KI} = \dfrac{5.60 \text{ g KI}}{130 \text{ g soln}} \times 100 = 4.31$

    $\% \ H_2O = \dfrac{122 \text{ g } H_2O}{130 \text{ g soln}} \times 100 = 93.8$

12. a. $\text{Molarity} = \dfrac{15.0 \text{ g sucrose}}{0.250\ \ell} \times \dfrac{1 \text{ mol}}{342 \text{ g sucrose}} = 0.175 \text{ M}$

    b. In evaporation, only the water is lost. As the volume is halved, the concentration would be doubled to 0.350 M.

13. a. $\text{mass } CaCl_2 = 220 \text{ g soln} \times \dfrac{6.0 \text{ g } CaCl_2}{100 \text{ g soln}} = 13 \text{ g}$

b. $\text{mass } CaCl_2 = 0.220\ \ell \times \dfrac{2.50 \text{ mol } CaCl_2}{1\ \ell} \times \dfrac{111 \text{ g } CaCl_2}{\text{mol } CaCl_2} = 61.0 \text{ g}$

c. $\text{mass } Ca(NO_3)_2 = 0.220\ \ell \times \dfrac{2.50 \text{ mol } Ca(NO_3)_2}{1\ \ell} \times \dfrac{164 \text{ g } Ca(NO_3)_2}{\text{mol } Ca(NO_3)_2}$

$= 90.2 \text{ g}$

14. a. $\text{volume} = \dfrac{1.60 \text{ mol}}{0.100 \text{ mol}/\ell} = 16.0\ \ell$

b. $\text{moles needed} = \dfrac{1.60 \text{ g } K_2CrO_4}{194 \text{ g } K_2CrO_4/\text{mol}} = 8.25 \times 10^{-3}$

$\text{volume} = \dfrac{8.25 \times 10^{-3} \text{ mol}}{0.100 \text{ mol}/\ell} = 0.0825\ \ell$

15. Using Figure 13.6:
The solubility of KNO_3 at 60°C is about 110 g/100 g H_2O.
The solubility of KNO_3 at 20°C is about 32 g/100 g H_2O.

a. $\text{mass } KNO_3 = 200 \text{ g } H_2O \times \dfrac{110 \text{ g } KNO_3}{100 \text{ g } H_2O} = 220 \text{ g (3 s.f.)}$

b. $\text{mass } KNO_3 = 200 \text{ g } H_2O \times \dfrac{32 \text{ g } KNO_3}{100 \text{ g } H_2O} = 64 \text{ g}$

$\text{mass } KNO_3 \text{ (excess)} = 220 \text{ g} - 64 \text{ g} = 156 \text{ g}$

16. The 70 g/ℓ solution ($M > 1$) has the lowest freezing point. The 60 g/ℓ solution ($M < 1$) has the highest freezing point.

17. a. 0.20 M NaCl: FP = −0.72°C
(The NaCl concentration is doubled.)

b. 0.10 M KNO_3: FP = −0.36°C
(KNO_3 produces two ions per unit, the same as NaCl.)

c. 0.10 M $CaCl_2$: FP = −0.54°C
($CaCl_2$ produces three ions per unit, or 3/2 more than NaCl or KNO_3.)

18. (b) glucose = (c) sucrose > (d) $KClO_3$ > (a) $Mg(NO_3)_2$
19. (a) 0.10 M $Al(NO_3)_3$ = (b) 0.20 M NaCl > (c) 0.10 M $CaBr_2$

20. $\text{molality} = \dfrac{15 \text{ g glycol}}{0.0600 \text{ kg } H_2O} \times \dfrac{1 \text{ mol}}{62 \text{ g glycol}} = 4.0 \text{ m}$

$\Delta T_f = 1.86°C \times m = 1.86°C\ (4.0) = 7.5°C$

$T_f = -7.50°C$

21. $$\text{molality} = \frac{\Delta T_f}{1.86°C} = \frac{5.0°C}{1.86°C} = 2.7\ m$$

$$\text{moles } CH_3OH = \frac{2.7 \text{ mol } CH_3OH}{1 \text{ kg } H_2O} \times 0.10 \text{ kg } H_2O = 0.27$$

$$\text{mass } CH_3OH = 0.27 \text{ mol } CH_3OH \times \frac{32 \text{ g } CH_3OH}{\text{mol } CH_3OH} = 8.6 \text{ g}$$

22. $$\text{molality} = \frac{\Delta T_f}{1.86°C} = \frac{0.61°C}{1.86°C} = 0.33\ m$$

$$\text{moles} = 0.0300 \text{ kg } H_2O \times \frac{0.33 \text{ mol}}{1 \text{ kg } H_2O} = 9.9 \times 10^{-3}$$

$$MM = \frac{2.00 \text{ g}}{9.9 \times 10^{-3} \text{ mol}} = 2.0 \times 10^{2} \text{ g/mole}$$

CHAPTER 14

Organic Chemistry: Hydrocarbons

I. Basic Skills

A. Qualitative Students should be able to:

1. a. Define hydrocarbons and describe the general properties of hydrocarbons.
   b. Distinguish between saturated, unsaturated, and aromatic hydrocarbons.
2. a. Define alkane, alkene, and alkyne and give the general formula of each class.
   b. Describe the structural features of each class.
3. Explain what is meant by isomerism. Give an example of isomers.
4. Explain what is meant by an addition reaction.
5. Describe the benzene molecule, including its structural features.
6. a. List the sources of hydrocarbons and the principal hydrocarbons available from each source (Tables 14.4 and 14.5).
   b. Explain fractional distillation and cracking with regard to the processing of petroleum.
   c. Explain destructive distillation with regard to the processing of coal.

B. Quantitative Students should be able to:

1. Determine the molecular formula of a saturated or unsaturated hydrocarbon from its general formula (Example 14.1).
2. Name the simpler straight chain and branched hydrocarbons (Tables 14.1 and 14.3 and Example 14.3).
3. a. Draw the structural formulas of characteristic members of any hydrocarbon class.
   b. Draw the structural formula of a hydrocarbon from its IUPAC name.
   c. Draw the structural formulas for the isomers of a particular hydrocarbon (Examples 14.2 and 14.4).
4. Identify the hydrocarbon class to which a compound may belong from its molecular formula (Example 14.5).
5. a. Describe how a particular compound can be made by an addition reaction (Example 14.6).
   b. Write equations for addition reactions (Equations 14.4–14.7).

II. Chapter Development

1. The chapter's major stress is on understanding the structural features of the different classes of hydrocarbons. Structural formulas are used throughout

and will continue to be important in Chapters 15 and 16. Structural isomerism is brought out as an important property of hydrocarbons. Geometrical isomerism (cis-trans) is not covered, except in a footnote.

2. The IUPAC nomenclature system is presented in a simplified form for the naming of alkanes. While we have no objections to the system we have minimized its use to avoid "the tail wagging the dog." A detailed study of the IUPAC system is best left for college.
3. As far as hydrocarbon reactions are concerned, primary emphasis is placed on addition reactions. These are most easy to understand and most likely to be encountered in the laboratory. Combustion reactions and heats of combustion were previously discussed in Chapter 4.
4. Cyclic hydrocarbons are briefly mentioned as minor components of petroleum. Their general formulas are not covered (C_nH_{2n} for cycloalkanes and C_nH_{2n-2} for cycloalkenes).
5. Additional material on natural gas, petroleum, and coal is presented in Chapter 27, Energy Resources. It includes the making of gaseous and liquid fuels from coal.

III. Problem Areas

1. If students mastered Lewis formulas in Chapter 11 they should have little difficulty with the organic structural formulas of this chapter. Carbon atoms always have a total of 4 bonds (4 valence electrons) while hydrogen atoms form only 1 bond (1 valence electron). All electron pair bonds are shown with a dash, and nonbonding electron pairs are omitted.
2. In drawing structural isomers students will sometimes be deceived by the planar nature of the paper on which they write. Nonexistent isomers will be discovered because they do not visualize the molecule's correct geometry. Other mistakes are caused by forgetting that atoms can rotate around a single bond. The best remedy, of course, is the use of molecular models to clarify the actual structure. Experiment 22, Isomerism in Organic Chemistry, is particularly helpful in this regard.
3. Teachers (and students) vary considerably in the importance they attach to organic nomenclature. While the IUPAC has provided a logical system, a certain amount of memorization is still required for its use. Students should be able to name branched hydrocarbons or organic halides. Point out that location numbers must be used when the same name describes more than one isomer.
4. Students may not appreciate the subtle nature of the benzene structure, particularly how it differs from an alkene. Explain that many types of tests indicate that all carbon-carbon bonds are identical. Point out how benzene and alkenes differ in their ability to undergo addition reactions.

IV. Suggested Activities

1. Hydrocarbon Structures:

Use a molecular model kit to show the structures of alkanes, alkenes, and alkynes. Point out that while atomic radii and bond length are not to scale, the shapes of the molecules (bond angles) are correct. Show how rotation is permitted around single bonds but not around double bonds. Show how structural isomerism results from a rearrangement of atoms which require the breaking of bonds (not rotation or twisting). Discuss equivalent bonding sites (CH_4 and C_2H_6) and nonequivalent sites (C_3H_8) by using a different colored ball to represent a chlorine atom.

2. Hydrocarbon Product Exhibit:
   Exhibit selected hydrocarbon products (plastics, synthetic fibers, medicinal drugs, etc.). Without going into the formula of the product, relate the product to the parent hydrocarbon molecule and to its source. Example 1: plastic bottles, polyethylene from ethylene from petroleum. Example 2: aspirin from benzene from coal or petroleum. Example 3: nylon from hexane from petroleum.

V. Answers to Questions

1. a. Carbon atoms can bond to each other to form long chains.
   b. Most hydrocarbons of a given molecular formula can exist in more than one structure. The different molecules, called isomers, have different properties.
2. a. Saturated hydrocarbons contain all single bonds between the carbon atoms.
   b. Unsaturated hydrocarbons contain one or more multiple bonds between carbon atoms.
   c. Aromatic hydrocarbons are those whose structure is related to the benzene molecule.
3. The boiling points of straight-chain hydrocarbons increase with mass because the attractive forces (Van der Waals forces) increase with mass. Hydrocarbons are insoluble in water because they are not able to break down the hydrogen bonds between water molecules.
4. In all hydrocarbon molecules, a hydrogen atom forms one bond to a carbon atom. Carbon atoms always form four bonds.
5. a. alkane: C_nH_{2n+2} b. alkene: C_nH_{2n} c. alkyne: C_nH_{2n-2}
6. Any alkane with four or more carbon atoms, such as C_4H_{10}, can form an isomer by branching from the main chain. The C_3H_8 molecule does not have sufficient carbon atoms to form a branch.
7. CH_4 methane C_5H_{12} pentane

   C_2H_6 ethane C_6H_{14} hexane

   C_3H_8 propane C_7H_{16} heptane

   C_4H_{10} butane C_8H_{18} octane
8. An alkyl group is the residue left when a hydrogen atom is removed from a

hydrocarbon molecule. Examples are: CH_3—, methyl; C_2H_5—, ethyl; C_3H_7—, propyl.

9. a. 1-methylpropane is correctly called butane.

   b. 3-methylbutane is correctly called 2-methylbutane.

   c. 2,2,2-trimethylbutane is correctly called 2,2-dimethylbutane.

   d. 2-ethylbutane is correctly called 3-methylpentane.

10. a. An alkane contains all single bonds.

    b. An alkene contains a double bond.

    c. An alkyne contains a triple bond.

11. The C_2H_4 molecule has a planar structure with bond angles of 120°.

    The C_2H_2 molecule has a linear structure with bond angles of 180°.

12.
```
H         H                  H                  H  H                       H
 \       /                   |                  |  |                       |
  C = C              H       C - H      H       C - C - H        H        C - H
 /       \            \     /|           \     /|  |               \      /|
H         H            C = C  H           C = C  H  H         H  C = C  H
                      /     \            /     \                |  /     \
                     H       H          H       H           H - C          H
                                                                |
                                                                H

  C2H4                   C3H6                   C4H8                   C4H8

          H
          |
H         C - H
 \       /|
  C = C   H
 /     \
/       \ H
         \|
H         C - H
          |
          H

   C4H8
```

13.
```
                            H                    H  H             H               H
                            |                    |  |             |               |
H - C ≡ C - H      H - C ≡ C - C - H   H - C ≡ C - C - C - H   H - C - C ≡ C - C - H
                            |                    |  |             |               |
                            H                    H  H             H               H

    C2H2                 C3H4                   C4H6                   C4H6
```

14. $6\ CH_4(g) + O_2(g) \rightarrow 2\ C_2H_2(g) + 2\ CO(g) + 10\ H_2(g)$

    $CaC_2(s) + H_2O \rightarrow C_2H_2(g) + CaO(s)$

    Acetylene is used in welding and as a reactant to produce other organic compounds.

15. $C_6H_5-CH_2-CH_3$ ethyl benzene $C_6H_5-CH_2-CH_2-CH_3$ propyl benzene

16. a.
$$\begin{array}{ccccccc} & & H & & H & & \\ & & | & & | & & \\ H & - & C & - & C & - & H \\ & & | & & | & & \\ & & H & & H & & \end{array}$$

b.
$$\begin{array}{ccccccc} & & H & & H & & \\ & & | & & | & & \\ Cl & - & C & - & C & - & Cl \\ & & | & & | & & \\ & & H & & H & & \end{array}$$

c.
$$\begin{array}{ccccccc} & & H & & H & & \\ & & | & & | & & \\ Br & - & C & - & C & - & Br \\ & & | & & | & & \\ & & H & & H & & \end{array}$$

d.
$$\begin{array}{ccccccc} & & H & & H & & \\ & & | & & | & & \\ H & - & C & - & C & - & Br \\ & & | & & | & & \\ & & H & & H & & \end{array}$$

17. a.
$$\begin{array}{ccc} H & & H \\ & C=C & \\ H & & H \end{array}$$

b.
$$\begin{array}{ccc} H & & H \\ & C=C & \\ Cl & & Cl \end{array}$$

c.
$$\begin{array}{ccc} H & & H \\ & C=C & \\ Br & & Br \end{array}$$

d.
$$\begin{array}{ccc} H & & H \\ & C=C & \\ H & & Br \end{array}$$

18. Natural gas and petroleum are found in underground deposits and are obtained by drilling. Coal is mined from both underground and surface deposits.
19. In fractional distillation crude oil is separated into portions which have specific uses. The separation is based on the different boiling points of the fractions.

20. $C_{12}H_{26}(l) \rightarrow C_6H_{14}(l) + C_6H_{12}(l)$

21. Hexane, C_6H_{14}, would be most prevalent as alkanes predominate in petroleum.
22. Destructive distillation of coal produces coke, coal gas and coal tar.

VI. Solutions to Problems

1. a. C_6H_{14} b. C_5H_{12} c. $C_{18}H_{38}$
2. a. C_5H_{12}: alkane b. C_5H_{10}: alkene c. C_5H_8: alkyne
   d. C_5H_6 cannot be any of the given classes of hydrocarbons.
   e. C_6H_{10}: alkyne
3. Structures (a), (b) and (d) are all methylbutane.

4.

```
    H  H  H  H  H  H
    |  |  |  |  |  |
 H- C -C -C -C -C -C -H
    |  |  |  |  |  |
    H  H  H  H  H  H
```

hexane

```
        H
        |
     H- C -H
        |
    H   |  H  H  H
    |   |  |  |  |
 H- C - C -C -C -C -H
    |   |  |  |  |
    H   H  H  H  H
```

2-methylpentane

```
           H
           |
        H- C -H
           |
    H  H   |  H  H
    |  |   |  |  |
 H- C -C - C -C -C -H
    |  |   |  |  |
    H  H   H  H  H
```

3-methylpentane

```
        H
        |
     H- C -H
        |
    H   |  H  H
    |   |  |  |
 H- C - C -C -C -H
    |   |  |  |
    H   |  H  H
        |
     H- C -H
        |
        H
```

2,2-dimethylbutane

```
        H
        |
     H- C -H
        |
    H   |  H  H
    |   |  |  |
 H- C - C -C -C -H
    |   |  |  |
    H   H  |  H
           |
        H- C -H
           |
           H
```

2,3-dimethylbutane

5. C_4H_6, (c), and C_5H_8, (d), show structural isomerism.
6. See answer to Problem 4 for IUPAC names of isomers of C_6H_{14}.
7. a. 2-methylhexane

```
        H
        |
     H- C -H
        |
    H   |  H  H  H  H
    |   |  |  |  |  |
 H- C - C -C -C -C -C -H
    |   |  |  |  |  |
    H   H  H  H  H  H
```

b. 2,2-dimethylpentane

```
        H
        |
     H- C -H
        |
    H   |  H  H  H
    |   |  |  |  |
 H- C - C -C -C -C -H
    |   |  |  |  |
    H   |  H  H  H
        |
     H- C -H
        |
        H
```

c. 2-methyl, 3-ethylhexane

```
        H
        |
     H- C -H
        |
    H   |  H  H  H  H
    |   |  |  |  |  |
 H- C - C -C -C -C -C -H
    |   |  |  |  |  |
    H   H  |  H  H  H
           |
        H- C -H
           |
        H- C -H
           |
           H
```

d. octane

```
    H  H  H  H  H  H  H  H
    |  |  |  |  |  |  |  |
 H- C -C -C -C -C -C -C -C -H
    |  |  |  |  |  |  |  |
    H  H  H  H  H  H  H  H
```

e. 4-ethylheptane

```
     H   H   H   H   H   H   H
     |   |   |   |   |   |   |
 H - C - C - C - C - C - C - C - H
     |   |   |   |   |   |   |
     H   H   H   |   H   H   H
                 |
             H - C - H
                 |
             H - C - H
                 |
                 H
```

8. a. React with Cl_2. b. React with H_2. c. React with HBr.
9. a. React with H_2. b. React with H_2 followed by HCl.

10. a. alkane:

```
  |   |   |   |   |   |   |
- C - C - C - C - C - C - C -
  |   |   |   |   |   |   |
```

(heptane)

b. alkene:

```
          |   |   |   |   |   |
 >C = C - C - C - C - C - C -
              |   |   |   |   |
```

(1-heptene)

c. alkyne:

```
              |   |   |   |   |
 - C ≡ C - C - C - C - C - C -
              |   |   |   |   |
```

(1-heptyne)

d. aromatic:

```
          |
 (C6H5) - C -
          |
```

(methyl benzene or toluene)

NOTE: Isomers are possible for structures a–c above.

11. a. C_4H_{10}: alkane b. C_4H_{12}: does not exist
c. C_4H_6: alkyne d. C_4H_8: alkene

12.

```
          |   |
 - C ≡ C - C - C -
          |   |
```

1-butyne

```
   |           |
 - C - C ≡ C - C -
   |           |
```

2-butyne

NOTE: Isomers of C_4H_6 are possible which are not alkynes.

13. a. C_8H_{16} b. C_8H_{14} c. C_4H_8 d. C_5H_8

14.

```
(1)      H   H   H            (2)      H   H   H
         |   |   |                     |   |   |
     H - C - C - C - H             H - C - C - C - Cl
         |   |   |                     |   |   |
         Cl  Cl  Cl                    H   Cl  Cl
```

```
(3)      H   H   H            (4)      H   Cl  H            (5)      H   H   Cl
         |   |   |                     |   |   |                     |   |   |
    Cl - C - C - C - Cl            H - C - C - C - Cl            H - C - C - C - Cl
         |   |   |                     |   |   |                     |   |   |
         H   H   Cl                    H   Cl  H                     H   H   Cl
```

15. a. 2-chlorobutane b. 2,3-dichlorobutane c. 1,1,1-trichlorobutane

d. 1,3-dichloropropane

16. a. H_2 b. Cl_2 c. HCl d. H_2O

17.

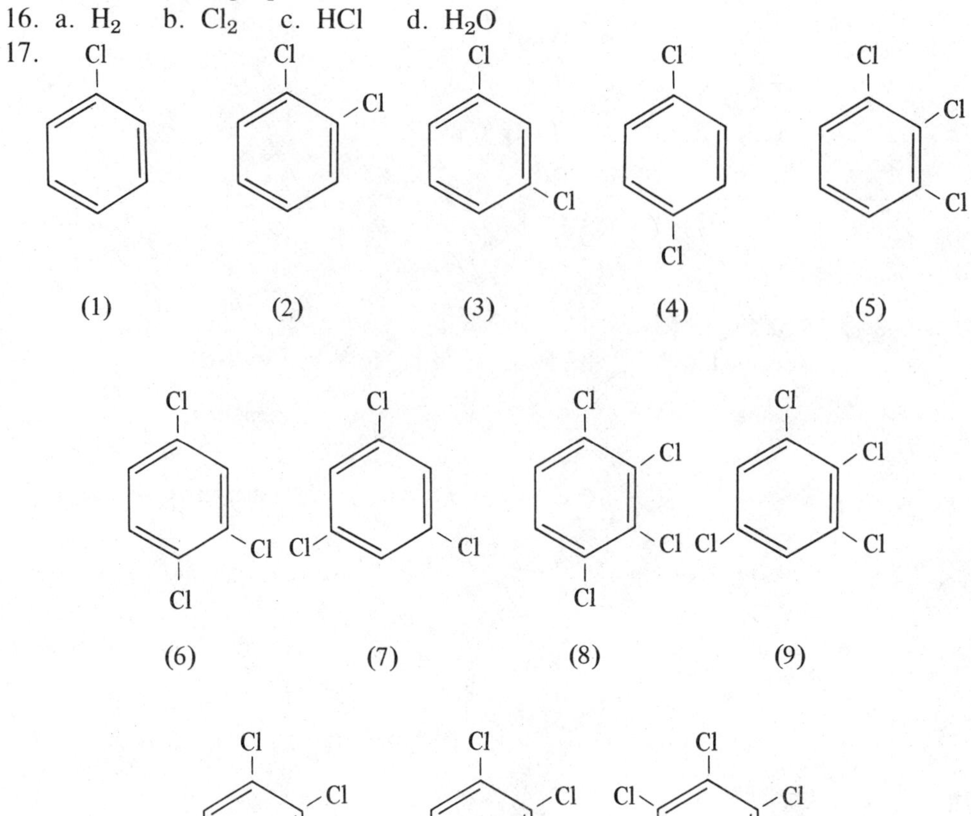

18. Octane number = 0.5 (103) + 0.5 (25) = 64

19. $0.0875 \times 92 = 8.0$ g H

$$\text{moles H} = \frac{8.0 \text{ g H}}{1.0 \text{ g H/mol}} = 8.0$$

$0.9125 \times 92 = 84$ g C

$$\text{moles C} = \frac{84 \text{ g C}}{12 \text{ g C/mol}} = 7.0$$

molecular formula = C_7H_8

Possible structural formula:

CH_3 (benzene ring structure) toluene

CHAPTER 15

Organic Chemistry: Oxygen Compounds

I. Basic Skills

A. Qualitative   Students should be able to:

1. a. Explain what is meant by a functional group.
   b. Give the functional groups for the organic oxygen compounds.
2. a. Describe the general properties of alcohols.
   b. Describe the fermentation process for the production of beverage alcohol.
   c. Explain what is meant by denatured alcohol.
3. Explain the acidic character of carboxylic acids.
4. a. Describe the composition of fats, soaps, and detergents.
   b. Distinguish between saturated and unsaturated fats.

B. Quantitative   Students should be able to:

1. a. Draw the structural formulas of characteristic members of all classes of organic oxygen compounds.
   b. Draw the structural formulas for the isomers of a particular organic oxygen compound (Examples 15.1 and 15.2).
2. Identify the class of oxygen compound from a given structural formula (Examples 15.3 and 15.4).
3. a. Identify the acid and alcohol from which an ester is made.
   b. Name an ester from its structural formula (Table 15.7).
4. Write balanced chemical equations for the preparation of:
   a. methyl alcohol (Equation 15.1)
   b. ethyl alcohol (Equation 15.2)
   c. diethyl ether (Equation 15.4)
   d. formaldehyde (Equation 15.5)
   e. acetone (Equation 15.6)
   f. acetic acid (Equation 15.10)
5. Describe general methods for the preparation of:
   a. esters (Equation 15.11)
   b. fats (Equation 15.13)
   c. soaps (Equation 15.14)

II. Chapter Development

1. The chapter surveys the more important organic oxygen compounds. Em-

phasis is placed on alcohols, acids, and esters. These classes occur again in Chapter 16 in the study of polyesters. Amides, R—$CONH_2$, are introduced in Chapter 16 as the functional group in polyamides.

2. The nomenclature of oxygen compounds is not stressed. In general, common names are used in preference to IUPAC system names.

III. Problem Areas

1. The variety of functional groups and classes will confuse some students in drawing structural formulas or in identifying them. Actually, the six classes of this chapter involve only three functional groups, individually or in combination; the carbonyl group, the hydroxyl group, and the oxy group. Again, the use of molecular models helps to distinguish these groups and helps in visualizing their geometry.
2. Students may be challenged by the drawing of structural formulas of isomeric oxygen compounds. The job is simplified if all isomers must be of the same class. If Experiment 24 is done, students will learn that alcohols are isomeric with ethers, aldehydes with ketones, and acids with esters.

IV. Suggested Activities

1. Comparison of Isomers: **Teacher Demonstration Only**
   Exhibit bottles of butyl alcohol and diethyl ether together with molecular models of the compounds. Have students determine their molecular formulas to show that they are isomeric. Compare their physical properties and relate them to hydrogen bonding; odor, solubility in water, boiling point (117°C vs 34°C), and density (0.81 vs 0.71 g/cm^3). (CAUTION: Keep ether away from any flame.) Discuss the differences in their physiological properties. Demonstrate reaction differences by adding a small piece of fresh sodium metal to 50-ml portions of each liquid. The butyl alcohol will evolve hydrogen gas showing the presence of a hydroxyl group.
2. Preparation of Sterno: **Teacher Demonstration Only**
   Make a saturated solution of calcium acetate in advance (about 40 g/100 ml). Add 2 drops of 6 M NaOH to 25 ml of the filtered solution. Add 2 drops of phenolphthalein to 100 ml of 95% ethyl alcohol in a 150-ml beaker. Add the calcium acetate solution to the alcohol and stir rapidly. The mixture will form a colored gel. Remove the gel with a spatula, place on a heat resistant pad, and light. Have students write the equation for the combustion reaction. Stop the reaction by smothering it with a larger beaker.

V. Answers to Questions

1. Hydrogen forms 1 bond, carbon forms 4 bonds, and oxygen forms 2 bonds.
2. Alcohols contain the hydroxyl group, —OH. The general formula of an alcohol is R—OH.
3. CH_3OH is more soluble in water than CH_3—$(CH_2)_5$—OH. The solubility of alcohols decreases as the hydrocarbon portion of the molecule increases in size. The hydrogen bonding of the hydroxyl group does not provide sufficient energy for the hydrocarbon portion to break into the water structure.
4. Methyl alcohol molecules form hydrogen bonds between —OH groups. The stronger attractive forces cause CH_3OH to have a higher boiling point than CH_3—O—CH_3 even though its molecular mass is less.
5. a. methyl alcohol b. ethyl alcohol c. phenol

```
     H              H   H
     |              |   |
 H - C - OH     H - C - C - OH
     |              |   |
     H              H   H
```

(c. phenol: benzene ring with OH)

6. The destructive distillation of wood produces a mixture of solid, liquid, and gaseous products. The solid is charcoal, the liquid contains methyl alcohol and other organic compounds, and the gas is mostly CO and CO_2.
7. Industrial methyl alcohol is made from CO and H_2 obtained from natural gas:

$$CO(g) + 2\ H_2(g) \rightarrow CH_3OH(l)$$

   Methyl alcohol may also be made by the destructive distillation of wood.
8. Industrial ethyl alcohol is made from ethylene obtained from petroleum:

$$C_2H_4(g) + H_2O(l) \rightarrow C_2H_5OH(l)$$

   Ethyl alcohol is also made by the fermentation of carbohydrates obtained from natural products:

$$C_6H_{12}O_6(aq) \rightarrow 2\ C_2H_5OH(aq) + 2\ CO_2(g)$$

9. Denatured ethyl alcohol contains a small amount of a compound which makes the alcohol unfit to drink. This is done so that users will not have to pay the large tax on drinkable alcohol.
10. The functional group of an ether is —O—, an oxygen atom which is singly bonded to two alkyl groups. The general formula of an ether is R—O—R′.

11. a. dimethyl ether b. diethyl ether

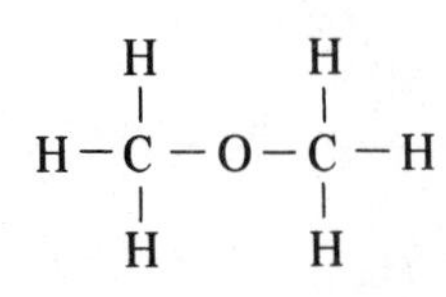

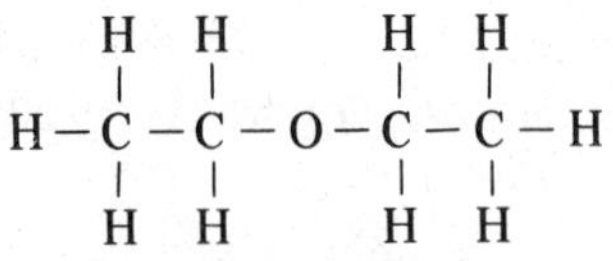

```
  H H     H H
  | |     | |
H-C-C-O-C-C-H
  | |     | |
  H H     H H
```

12.

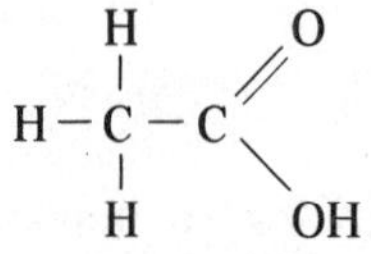

```
  H   O
  |  //
H-C-C
  |  \
  H   OH
```

The bond angles around the CH_3— group are 109°. The bond angles around the —COOH group are 120°.

13. The general formula of an aldehyde is

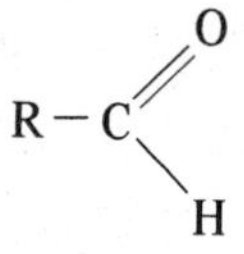

```
    O
   //
R-C
   \
    H
```

while that of a ketone is $R-\overset{\overset{\displaystyle O}{\|}}{C}-R'$. The carbonyl group in an aldehyde is on an end carbon atom. In a ketone, the carbonyl group is on an interior carbon atom.

14. a. methyl ethyl ketone b. diethyl ketone

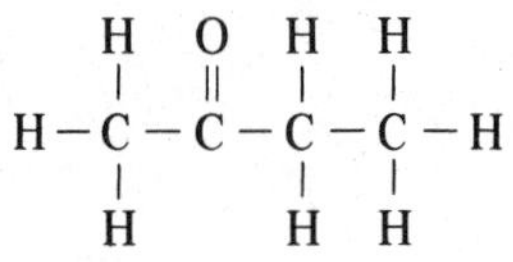

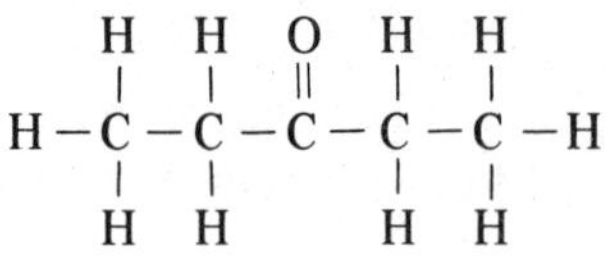

```
  H O H H
  | ‖ | |
H-C-C-C-C-H
  |   | |
  H   H H
```

```
  H H O H H
  | | ‖ | |
H-C-C-C-C-C-H
  | |   | |
  H H   H H
```

15. alcohol 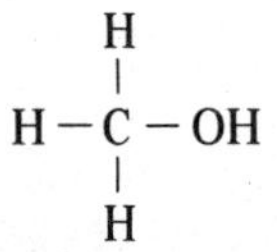 aldehyde 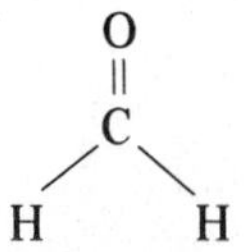 acid

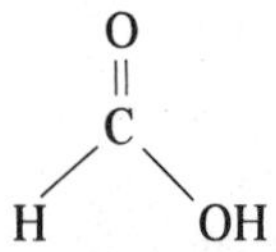

```
  H
  |
H-C-OH
  |
  H
```

```
   O
   ‖
   C
  / \
 H   H
```

```
   O
   ‖
   C
  / \
 H   OH
```

Esters and ethers require at least two carbon atoms while ketones require at least three carbon atoms.

16. Only the alcohols and acids form hydrogen bonds with like molecules. Only in these compounds is a hydrogen atom bonded to an oxygen atom.

17. a. acetic acid b. acetone

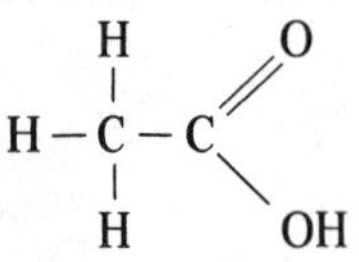

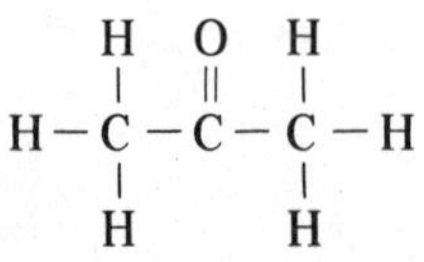

```
  H   O
  |  //
H-C-C
  |  \
  H   OH
```

```
  H O H
  | ‖ |
H-C-C-C-H
  |   |
  H   H
```

18. $CH_3\text{—}CH_2\text{—}OH(aq) + O_2(g) \qquad CH_3\text{—}COOH(aq) + H_2O(l)$

$CH_3\text{—}CHO(l) + 1/2\ O_2(g) \rightarrow CH_3\text{—}COOH(l)$

19. The general formula of an ester is $\begin{array}{c}\mathrm{O}\\ \| \\ \mathrm{R-C-O-R'}\end{array}$, where R′ may not be a hydrogen atom.

   a. In an acid, R′ is an H atom.

   b. In a ketone, the —O— group is absent.

   c. In an aldehyde, the —O—R′ group is replaced by an H atom.

20. a. methyl formate b. methyl acetate c. ethyl propionate

$$\begin{array}{ccccccccc} & & \mathrm{O} & & & & \mathrm{H} & & \\ & & \| & & & & | & & \\ \mathrm{H} & - & \mathrm{C} & - & \mathrm{O} & - & \mathrm{C} & - & \mathrm{H} \\ & & & & & & | & & \\ & & & & & & \mathrm{H} & & \end{array} \qquad \begin{array}{ccccccccccc} & & \mathrm{H} & & \mathrm{O} & & & & \mathrm{H} & & \\ & & | & & \| & & & & | & & \\ \mathrm{H} & - & \mathrm{C} & - & \mathrm{C} & - & \mathrm{O} & - & \mathrm{C} & - & \mathrm{H} \\ & & | & & & & & & | & & \\ & & \mathrm{H} & & & & & & \mathrm{H} & & \end{array} \qquad \begin{array}{ccccccccccccccc} & & \mathrm{H} & & \mathrm{H} & & \mathrm{O} & & & & \mathrm{H} & & \mathrm{H} & & \\ & & | & & | & & \| & & & & | & & | & & \\ \mathrm{H} & - & \mathrm{C} & - & \mathrm{C} & - & \mathrm{C} & - & \mathrm{O} & - & \mathrm{C} & - & \mathrm{C} & - & \mathrm{H} \\ & & | & & | & & & & & & | & & | & & \\ & & \mathrm{H} & & \mathrm{H} & & & & & & \mathrm{H} & & \mathrm{H} & & \end{array}$$

21. Fats are triesters of glycerol, $CH_2OH\text{—}CHOH\text{—}CH_2OH$.
22. Saturated fats are made from saturated acids, while unsaturated fats are made from unsaturated acids. A saturated acid contains all carbon-carbon single bonds. An unsaturated acid contains one or more carbon-carbon double bonds. An unsaturated fat is saturated by undergoing an addition reaction with hydrogen.
23. Soaps are the sodium salts of long-chain acids. They are made by reacting fats with sodium hydroxide.
24. The hydrocarbon end of a negative soap ion is soluble in organic material such as grease. The ionic end of the soap is soluble in water. This combination of properties allows the water to carry off the grease with the soap. (See Figure 15.11.) Soap causes problems in hard water because the soap ions react with Ca^{2+} and Mg^{2+} ions in the water to form an insoluble precipitate.

VI. Solutions to Problems

1.

$$\begin{array}{ccccccccccccc} & & \mathrm{H} & & \mathrm{H} & & \mathrm{H} & & \mathrm{H} & & \mathrm{H} & & \\ & & | & & | & & | & & | & & | & & \\ \mathrm{H} & - & \mathrm{C} & - & \mathrm{C} & - & \mathrm{C} & - & \mathrm{C} & - & \mathrm{C} & - & \mathrm{OH} \\ & & | & & | & & | & & | & & | & & \\ & & \mathrm{H} & & \mathrm{H} & & \mathrm{H} & & \mathrm{H} & & \mathrm{H} & & \end{array} \qquad \begin{array}{ccccccccccccc} & & \mathrm{H} & & \mathrm{OH} & & \mathrm{H} & & \mathrm{H} & & \mathrm{H} & & \\ & & | & & | & & | & & | & & | & & \\ \mathrm{H} & - & \mathrm{C} & - & \mathrm{C} & - & \mathrm{C} & - & \mathrm{C} & - & \mathrm{C} & - & \mathrm{H} \\ & & | & & | & & | & & | & & | & & \\ & & \mathrm{H} & & \mathrm{H} & & \mathrm{H} & & \mathrm{H} & & \mathrm{H} & & \end{array}$$

$$\begin{array}{ccccccccccccc} & & \mathrm{H} & & \mathrm{H} & & \mathrm{OH} & & \mathrm{H} & & \mathrm{H} & & \\ & & | & & | & & | & & | & & | & & \\ \mathrm{H} & - & \mathrm{C} & - & \mathrm{C} & - & \mathrm{C} & - & \mathrm{C} & - & \mathrm{C} & - & \mathrm{H} \\ & & | & & | & & | & & | & & | & & \\ & & \mathrm{H} & & \mathrm{H} & & \mathrm{H} & & \mathrm{H} & & \mathrm{H} & & \end{array}$$

2.

```
    H      H   H   H   H                         H   H       H   H   H
    |      |   |   |   |                         |   |       |   |   |
H - C - O - C - C - C - C - H               H - C - C - O - C - C - C - H
    |      |   |   |   |                         |   |       |   |   |
    H      H   H   H   H                         H   H       H   H   H
```

3.

```
    H   H   H       O                            H   H       O
    |   |   |      //                            |   |      //
H - C - C - C - C                            H - C - C - C
    |   |   |      \                             |   |      \
    H   H   H       OH                           H   CH3     OH
```

4. a. aldehyde b. acid c. ester d. ether e. alcohol

5. alcohol ether aldehyde

```
                                          O
                                         //
CH3-OH         CH3-O-CH3         CH3-C
                                         \
                                          H
```

ketone acid ester

```
      O              O              O
      ||             ||             ||
CH3-C-CH3       CH3-C-OH       CH3-C-O-CH3
```

6. a. alcohol b. ether c. aldehyde

```
    H   H   H               H       H   H               H   H       O
    |   |   |               |       |   |               |   |      //
H - C - C - C - OH      H - C - O - C - C - H       H - C - C - C
    |   |   |               |       |   |               |   |      \
    H   H   H               H       H   H               H   H       H
```

(or isopropyl alcohol)

d. ketone e. acid f. ester

```
    H   O   H               H   H   O                   H   O       H
    |   ||  |               |   |   ||                  |   ||      |
H - C - C - C - H       H - C - C - C - OH          H - C - C - O - C - H
    |       |               |   |                       |           |
    H       H               H   H                       H           H
```

(or ethyl formate)

7. a. b.

```
    O       H   H   H                 H   O       H   H
    ||      |   |   |                 |   ||      |   |
H - C - O - C - C - C - H         H - C - C - O - C - C - H
            |   |   |                 |           |   |
            H   H   H                 H           |   H
                                                  |
                                              H - C - H
                                                  |
                                                  H
```

8.

```
    H   H   H     O          H   H     O           H   O   H   H
    |   |   |    //          |   |    //           |   ||  |   |
H - C - C - C - C        H - C - C - C         H - C - C - C - C - H
    |   |   |    \           |   |    \            |       |   |
    H   H   H     H          H  CH3    H           H       H   H
```

9.

```
    H   H   O       H                H   O       H   H
    |   |   ||      |                |   ||      |   |
H - C - C - C - O - C - H        H - C - C - O - C - C - H
    |   |           |                |           |   |
    H   H           H                H           H   H

    O       H   H   H                O           H       H
    ||      |   |   |                ||          |       |
H - C - O - C - C - C - H        H - C - O ----- C ----- C - H
            |   |   |                            |       |
            H   H   H                            |       H
                                             H - C - H
                                                 |
                                                 H
```

10. a. alcohol: 1   b. ether: 2   c. aldehyde: 1
    d. ketone: 3   e. acid: 1   f. ester: 2

11.

```
     H   O       H   H               H   O       H   H
     |   ||      |   |               |   ||      |   |
Cl - C - C - O - C - C - H       H - C - C - O - C - C - Cl
     |           |   |               |           |   |
     H           H   H               H           H   H

    H   O       H   H
    |   ||      |   |
H - C - C - O - C - C - H
    |           |   |
    H           Cl  H
```

12. alcohol, ether, aldehyde

```
CH3 - SH      CH3 - S - CH3       H
                                   \
                                    C = S
                                   /
                                  H
```

ketone, acid, ester

```
        S               S               S
        ||              ||              ||
CH3 -   C - CH3     H - C - SH      H - C - S - CH3
```

13. methyl formate

```
    O     H
    ||    |
H - C - O - C - H
          |
          H
```

methyl acetate

```
    H   O       H
    |   ||      |
H - C - C - O - C - H
    |           |
    H           H
```

ethyl formate

```
    O       H   H
    ||      |   |
H - C - O - C - C - H
            |   |
            H   H
```

ethyl acetate

```
    H   O       H   H
    |   ||      |   |
H - C - C - O - C - C - H
    |           |   |
    H           H   H
```

14.

```
                                         O
                                         ||
a. CH3 - CH2 - OH             and    H - C - OH

                                               O
                                               ||
b. CH3 - CH2 - CH2 - OH       and      CH3 - C - OH

                                         O
                                         ||
c. CH3 - CHOH - CH3           and    CH3 - C - OH
```

15. CH_3OH (MM = 32.0)

CH_2OH—CH_2OH (MM = 62.1)

As freezing point lowering is related to the concentration of particles in the solution, 62.1 g of ethylene glycol would be required to produce the same freezing point as 32.0 g of methyl alcohol.

16. $C_6H_{12}O_6(aq) \rightarrow 2\ C_2H_5OH(aq) + 2\ CO_2(g)$

volume $C_2H_5OH = 0.10\ (1.00\ \ell) = 0.10\ \ell$

$$\text{mass } C_2H_5OH = 0.10\ \ell \times \frac{10^3\ \text{cm}^3}{1\ \ell} \times \frac{0.80\ \text{g}}{\text{cm}^3} = 8.0 \times 10^1\ \text{g}$$

MM $(C_6H_{12}O_6) = 180$ g MM $(C_2H_5OH) = 46.0$ g

$$\text{mass } C_6H_{12}O_6 = 8.0 \times 10^1\ \text{g } C_2H_5OH \times \frac{180\ \text{g } C_6H_{12}O_6}{2(46.0)\ \text{g } C_2H_5OH} = 160\ \text{g}$$

17. a. alcohol

```
    H   H
    |   |
H - C - C - OH
    |   |
    H   H
```

ether

```
    H       H
    |       |
H - C - O - C - H
    |       |
    H       H
```

b. aldehyde

```
    H   H       O
    |   |      //
H - C - C - C
    |   |      \
    H   H       H
```

ketone

```
    H   O   H
    |   ||  |
H - C - C - C - H
    |       |
    H       H
```

c. acid

```
    H   O
    |   ||
H - C - C - OH
    |
    H
```

ester

```
    O       H
    ||      |
H - C - O - C - H
            |
            H
```

18. $\text{Density} = \dfrac{46.07\ \text{g}}{\text{mol}} \times \dfrac{1\ \text{mol}}{22.4\ \ell} = 2.06\ \text{g}/\ell$

19. 90 proof = 45 vol %

volume of alcohol in whiskey = 0.45 (2.0 oz) = 0.90 oz

volume of alcohol in beer = 0.060 (12 oz) = 0.72 oz

$\text{number of beers} = \dfrac{0.90\ \text{oz}}{0.72\ \text{oz/beer}} = 1.2$

CHAPTER 16

Organic Chemistry: Polymers

I. Basic Skills

A. Qualitative Students should be able to:

1. a. Explain what is meant by a monomer and a polymer.
   b. Describe the types of monomers required to make addition polymers and condensation polymers.
   c. Distinguish between a polyester and a polyamide with respect to their monomers.
2. Describe the reaction processes in forming addition polymers and condensation polymers.
3. a. Describe the characteristics of rubber and its uses.
   b. Explain how the monomers used in rubber differ from other addition monomers.
4. a. Explain what is meant by a carbohydrate.
   b. Explain the primary function of carbohydrates and list their sources.
5. a. Explain how the structures of glucose, fructose, and sucrose are related.
   b. Explain how the structures of glucose, starch, and cellulose are related.
6. a. Explain what is meant by a protein and describe its monomer.
   b. Describe the major functions of proteins and list their sources.

B. Quantitative Students should be able to:

1. a. Relate the molecular mass of an addition polymer to the number of monomers (Example 16.1).
   b. Determine the percent composition of an addition polymer (Example 16.4).
2. Identify the monomers used to make a polymer from the polymer's structural formula (Example 16.2).
3. Draw the structural formula, knowing the monomers, of:
   a. an addition polymer (Example 16.3)
   b. a natural or synthetic rubber (Example 16.5)
   c. a polyester (Example 16.6) or polyamide (Example 16.7)
   d. a protein (Example 16.8)

II. Chapter Development

1. The chapter includes the major synthetic linear polymers. Under condensation polymers, only polyesters and polyamides are covered. Polyethers and

silicones are omitted. Two-dimensional and three-dimensional covalent networks were discussed previously in Chapter 12.

2. A simplified approach is taken in presenting addition polymerization. The chain reaction mechanism and its initiation are not necessary to the understanding of a polymer's structural formula. All addition polymers are assumed to be of head-to-tail type.
3. The polymeric carbohydrates and proteins are included as important examples of natural polymers. No attempt is made to include all amino acids. The structure of nucleic acids is left for other courses.

III. Problem Areas

1. Some difficulty may arise when having to identify the monomers from a polymer's structural formula. If the chain contains only carbon atoms, it must be an addition polymer. Except for rubber and Teflon, most addition polymers are made of monomers having the general formula $CH_2 = CHX$. The student should be able to complete the formula by identifying X. A condensation polymer will contain the ester group and be a polyester, or contain the amide group and be a polyamide. Proteins, made from amino acids, are a special type of polyamide. They can be identified by having only two carbon atoms between the amide groups and having a variable side chain off the alpha carbon atom.
2. Those who did not master ester formation in Chapter 15 will have trouble drawing the structure of a polyester. Review this reaction, pointing out the elimination of the water molecule and the forming of the ester linkage. Follow the same procedure in showing how amines and acids form amides and how this may lead to a polyamide.

IV. Suggested Activities

1. Plastics Exhibit:
   Exhibit a selection of plastic items to show a variety of applications. (You may want to have student help in this.) As much as possible, determine the chemical identity of each plastic: polyethylene, polyvinyl chloride, styrofoam, Teflon, nylon, dacron, etc. Ask students to then identify the monomer or monomers.
2. Natural Rubber: **Teacher Demonstration Only**
   Liquid latex is a colloidal suspension of natural rubber, polyisoprene. Add 10 ml of liquid latex to 40 ml of water and stir. Add about 5 ml of 1 M HCl and stir well. The rubber particles will coagulate into a rubbery mass. Form the rubber into a ball under running water and show it to the class. Bounce it on the floor or off the wall to demonstrate its properties.
3. Nylon:
   Follow the procedure in the Laboratory Manual to make nylon if Experiment 26 is not to be performed by the students.

4. Test for Starch:
   Add 1% solutions of starch, glucose, fructose, and sucrose to separate test tubes. Add 2–3 drops of iodine to each tube and note the positive test result for starch. Test a potato slice or cooked spaghetti. If time permits, add 1 ml of 6 M HCl to another tube of starch solution and heat in a boiling water bath for 30 minutes. The iodine test should be negative indicating the polymer has been hydrolyzed to small chains (dextrins) or to glucose molecules. NOTE: The reverse of this test may be used as a test for iodine.
5. Protein Variety:
   Show that the condensation of two different amino acids can form two different products (isomers). Using letters (A, B, C) for amino acids, ask students to figure the number of different arrangements using each letter once. (Answer: 6) Ask students to try to determine the relationship between the number of amino acids (used once) and the number of isomers. (Answer: n!) For 20 different amino acids, n! is 20! or 2.4×10^{18}.

V. Answers to Questions

1. A polymer is a long chain molecule in which the atoms are held together by covalent bonds. Monomers are the small molecules which are strung together to make a polymer. As a polymer is made from many monomers, its molecular mass is much higher than that of the monomer.
2. As the formula of ethylene is C_2H_4, a polymer with 2400 —CH_2— groups was made from 1200 ethylene molecules.
3. An addition polymer is made by addition reactions using monomers which contain a carbon-carbon double bond. An addition polymer has the same chemical composition as the monomer from which it is made. A condensation polymer is made by condensation reactions in which a small molecule is removed between reacting monomer units. Hence, its composition is slightly different from those of the monomers involved.
4. a. ethylene   b. vinyl chloride   c. tetrafluoroethylene

```
H       H        H       H        F       F
 \     /          \     /          \     /
  C = C            C = C            C = C
 /     \          /     \          /     \
H       H        H       Cl       F       F
```

5. a. polyethylene: kitchenware plastic, bottles, toys
   b. polyvinylchloride: floor tile, phonograph records, water pipes
   c. Teflon: pan coatings, bearings, gaskets, tubing
6. Isoprene

```
H     CH3  H      H
 \     |   |     /
  C =  C - C = C
 /               \
H                 H
```

7. Neoprene contains carbon-carbon double bonds. Dacron contains carbon-oxygen double bonds.
8. A copolymer is made from two different monomers. Dacron is a copolymer of terephthalic acid and ethylene glycol.
9. Butadiene Styrene

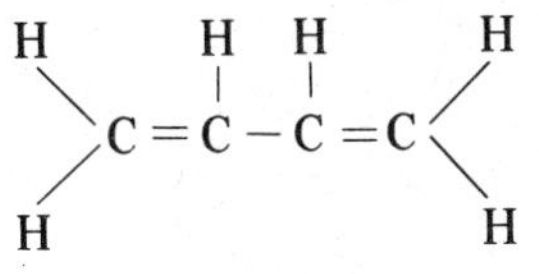

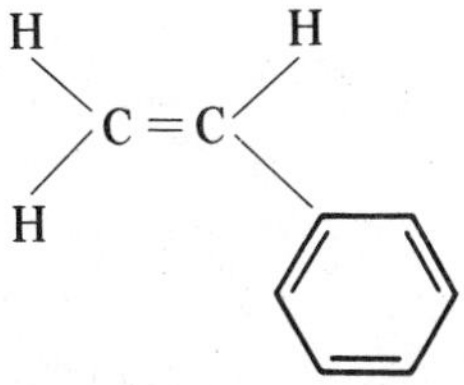

```
H      H  H      H      H            H
 \     |  |     /        \          /
  C = C - C = C           C = C
 /              \        /          \
H                H      H        (benzene ring)
```

The polymer made from butadiene gives a rubber-like material. As there is only one double bond in styrene, its polymer would not be expected to be a synthetic rubber.

10. To form a condensation polymer, each monomer must have two functional groups. Oxalic acid, ethylene glycol, and urea are bifunctional, while formic acid, methyl alcohol, and ammonia are not.
11. For making dacron:

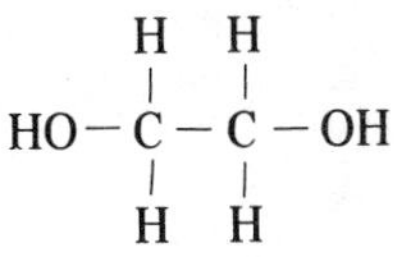

```
     H   H                      O                  O
     |   |                      ||                 ||
HO - C - C - OH     and   HO -  C - (benzene ring) - C - OH
     |   |
     H   H

ethylene glycol                 terephthalic acid
```

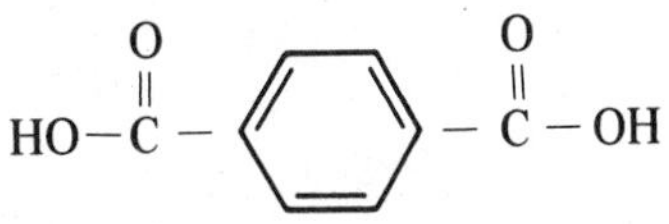

For making nylon:

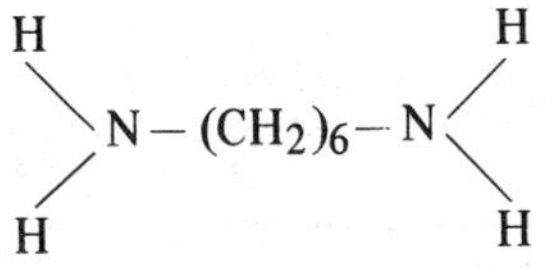

```
H                  H               O                  O
 \                /                ||                 ||
  N - (CH2)6 - N        and   HO - C - (CH2)4 - C - OH
 /                \
H                  H

hexamethylenediamine                 adipic acid
```

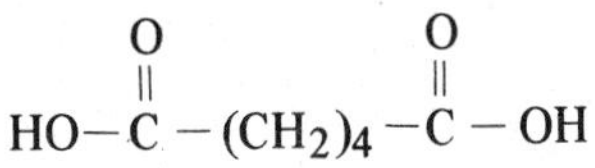

12.

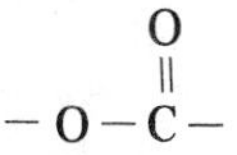

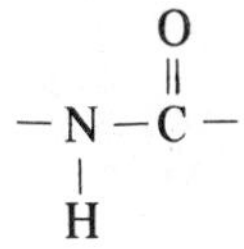

```
       O               O
       ||              ||
 - O - C -       - N - C -
                   |
                   H

ester group      amide group
```

13. Glucose and fructose are isomers; both have a molecular formula of $C_6H_{12}O_6$. While starch is a polymer of glucose, its repeating formula is $C_6H_{10}O_5$ due to the loss of an H_2O molecule in the condensation reaction.

14. A compound with the formula $C_5H_{10}O_5$ is most likely to be (d) a carbohydrate. (Ribose is a $C_5H_{10}O_5$ carbohydrate.)

15. The smallest molecular mass is (d) water. The largest molecular mass is (b) starch, a polymer of glucose.
16. An alpha amino acid has the amine group on the carbon atom adjacent to the carboxyl group. In the general formula, the variable side chain is designated as R:

```
        H       O
        |     //
  H—N—C—C
     |   |    \
     H   R     OH
```

17. a. methionine

```
        H  O
        |  ||
  H—N—C—C—OH
     |   |
     H  CH2—CH2—S—CH3
```

b. aspartic acid

```
        H  O
        |  ||
  H—N—C—C—OH
     |   |
     H  CH2—COOH
```

c. phenylalanine

```
        H  O
        |  ||
  H—N—C—C—OH
     |   |
     H   |
        CH2—C6H5 (benzene ring)
```

18.

```
        H  O                 H  O                        H  O       H  O
        |  ||                |  ||                       |  ||      |  ||
  H—N—C—C—OH  +  H—N—C—C—OH  →  H—N—C—C—N—C—C—OH
     |   |              |   |                      |   |     |   |
     H  CH3             H   H                      H  CH3    H   H
                                                   \_______/ \_______/
   alanine            glycine                      alanine    glycine
                                                    unit       unit
```

19. There are at least 20 different monomers (amino acids) used in making protein polymers. As proteins may contain $10^2 - 10^5$ monomers, the number of different arrangements of monomers is extremely large.

VI. Solutions to Problems

1. a. $$\frac{\text{mass of polymer}}{\text{mass of —CF}_2\text{—}} = \frac{1.0 \times 10^5}{50} = 2.0 \times 10^3 \ \frac{\text{—CF}_2\text{— groups}}{\text{polymer}}$$

b. $$\text{C}_2\text{F}_4 \text{ molecules} = \frac{2.0 \times 10^3 \text{ —CF}_2\text{— groups}}{2 \text{ —CF}_2\text{— groups/molecule}} = 1.0 \times 10^3$$

c. 2.0×10^3 —CF_2— groups $\simeq 2.0 \times 10^3$ C atoms

2.

```
H         H
 \       /
  C = C          acrylonitrile
 /       \
H         C ≡ N
```

3. a. polypropylene

```
 H    H     H    H     H    H     H    H
 |    |     |    |     |    |     |    |
-C  - C  -  C -  C  -  C -  C  -  C -  C-
 |    |     |    |     |    |     |    |
 H   CH3    H   CH3    H   CH3    H   CH3
```

b. polytetrafluoroethylene

```
 F   F   F   F   F   F   F   F
 |   |   |   |   |   |   |   |
-C - C - C - C - C - C - C - C -
 |   |   |   |   |   |   |   |
 F   F   F   F   F   F   F   F
```

c. polystyrene

```
 H     H     H     H     H     H     H     H
 |     |     |     |     |     |     |     |
-C  -  C  -  C  -  C  -  C  -  C  -  C  -  C -
 |     |     |     |     |     |     |     |
 H    C6H5   H    C6H5   H    C6H5   H    C6H5
```

4. a. For C_2H_4:

$$\% \ C = \frac{24.02 \text{ g C}}{28.05 \text{ g } C_2H_4} \times 100 = 85.63$$

$$\% \ H = 100.00 - 85.63 = 14.37$$

b. The composition of polyethylene, $(\text{—}CH_2\text{—})_x$, is the same as ethylene, C_2H_4, as both have the simplest formula CH_2.

c. The composition of polypropylene,

```
   H  H
   |  |
( -C -C - )x
   |  |
   H CH3
```

is also the same as ethylene as its simplest formula is also CH_2.

5. Polybutadiene

```
 H   H   H   H   H   H   H   H   H   H   H   H
 |   |   |   |   |   |   |   |   |   |   |   |
-C - C = C - C - C - C = C - C - C - C = C - C -
 |           |   |           |   |           |
 H           H   H           H   H           H
```

6. The monomer pairs in (a) and (c) could react to form a polyester. To form a polyester, the acid and alcohol must both be bifunctional. The acid in (b), CH_3COOH, has only one functional group.

7. a.

```
  O  H  O        H  H       O  H  O        H  H
  ‖  |  ‖        |  |       ‖  |  ‖        |  |
 -C -C -C -O -C -C -O -C -C -C -O -C -C -O-
     |           |  |          |           |  |
     H           H  H          H           H  H
```

b.

```
  O  H  O        H  CH3     O  H  O        H  CH3
  ‖  |  ‖        |  |       ‖  |  ‖        |  |
 -C -C -C -O -C -C -O -C -C -C -O -C -C -O-
     |           |  |          |           |  |
     H           H  H          H           H  H
```

8.

```
  O  H  O      O       O  H  O      O
  ‖  |  ‖      ‖       ‖  |  ‖      ‖
 -C -C -C -N -C -N -C -C -C -N -C -N-
     |     |      |       |     |      |
     H     H      H       H     H      H
```

9.

```
       H  O       H  O       H  O
       |  ‖       |  ‖       |  ‖
 H -N -C -C -N -C -C -N -C -C -OH
    |  |       |  |       |  |
    H  CH3     H  H       H  CH2 -C6H5
```

NOTE: An isomer of this molecule is obtained if the student uses the amine end of alanine, instead of the carboxyl end, in the first condensation.

10.

```
       H  O                      H  O
       |  ‖                      |  ‖
 H -N -C -C -OH       and  H -N -C -C -OH
    |  |                      |  |
    H  CH2-OH                 H  CH2-COOH

       serine                    aspartic acid
```

11.

```
 CH3       H
    \     /
     C = C
    /     \
 H         Cl
```

12.

```
 H      CH3 CH3     H
  \      |   |     /
   C  =  C - C  = C
  /                \
 H                  H
```

13.

```
      O              O                  O
      ‖              ‖                  ‖
 HO -C -(CH2)5 -C -OH     and    H2N -C -NH2
```

14.

```
  H  CH3   H  H  Cl  H  H  H  CH3   H  H  Cl  H  H
  |   |    |  |   |  |  |  |   |    |  |   |  |  |
 -C - C = C - C - C - C = C - C - C - C = C - C - C - C = C - C -
  |        |  |   |          |  |        |  |   |           |
  H        H  H   H          H  H        H  H   H           H
```

15.

```
      O  CH3  O                        H
      ‖   |   ‖                        |
HO - C - C - C - OH     and    H2N - C - NH2
          |                            |
          H                            H
```

16. $\dfrac{\text{mass of polymer}}{\text{mass of } C_2H_4} = \dfrac{5.6 \times 10^4}{28} = 2.0 \times 10^3 \; C_2H_4 \dfrac{\text{molecules}}{\text{polymer}}$

17. $\text{mass of polymer} = 6.0 \times 10^{26} \text{ molecules} \times \dfrac{62.5 \text{ g } C_2H_3Cl}{6.02 \times 10^{23} \text{ molecules}}$

$= 6.2 \times 10^4 \text{ g}$

18. An addition polymer, such as PVC, has the same mass composition as its monomer.

MM of C_2H_3Cl = 62.49

$\% \text{ C} = \dfrac{24.02 \text{ g C}}{62.49 \text{ g } C_2H_3Cl} \times 100 = 38.44$

$\% \text{ Cl} = \dfrac{35.45 \text{ g Cl}}{62.49 \text{ g } C_2H_3Cl} \times 100 = 56.73$

$\% \text{ H} = 100.00 - 38.44 - 56.73 = 4.83$

19.

```
                O         O                 O         O
                ‖         ‖                 ‖         ‖
 -O        O - C-[C6H4]-C - O         O - C-[C6H4]-C-
    \[C6H4]/                  \[C6H4]/
```

20. a. $C_9H_{11}NO_2$: phenylalanine

b. MM = 9(12.01) + 11(1.01) + 14.01 + 2(16.00) = 165.11

c. $\% \text{ C} = \dfrac{108.1 \text{ g C}}{165.1 \text{ g PA}} \times 100 = 65.48$

$\% \text{ N} = \dfrac{14.01 \text{ g N}}{165.1 \text{ g PA}} \times 100 = 8.48$

$\% \text{ O} = \dfrac{32.00 \text{ g O}}{165.1 \text{ g PA}} \times 100 = 19.38$

$\% \text{ H} = 100.00 - 65.48 - 8.48 - 19.38 = 6.66$

21\. a. glycine: $C_2H_5NO_2$ MM = 75.07

b. serine: $C_3H_7NO_3$ MM = 105.10

c. glycine + serine → dimer + H_2O

MM of dimer = 75.07 + 105.10 − 18.02 = 162.15

22\.

```
       H  O      H  O      H  O      H  O
       |  ||     |  ||     |  ||     |  ||
H-N - C - C - N - C - C - N - C - C - N - C - C - OH
  |   |       |   |       |   |       |   |
  H  CH2      H  CH3      H   H       H  CH2
      |                                   |
      OH                                 COOH
```

NOTE: The student may obtain an isomer of this polymer by using the amine end of serine in the first condensation.

23\.

```
H      H         H
 \     |        /
  C = C - C = C
 /        |     \
H         H      H
```

MM = 4(12.01) + 6(1.01) = 54.10

butadiene (C_4H_6)

```
H       H
 \     /
  C = C
 /     \
H
```

MM = 8(12.01) + 8(1.01) = 104.16

styrene (C_8H_8)

The mole ratio of butadiene to styrene in SBR is 3:1.

$$\frac{\text{mass styrene}}{\text{mass butadiene}} = \frac{104.16}{3(54.10)} = \frac{0.6418 \text{ g styrene}}{\text{g butadiene}}$$

24\. The repeating unit of nylon is:

```
[-C - (CH2)4 - C - N - (CH2)6 - N -]
  ||           ||  |            |
  O            O   H            H
```

The formula of the repeating unit is $C_{12}H_{22}N_2O_2$ and its molecular mass is 226.3.

$$\% \text{ C} = \frac{144.1 \text{ g C}}{226.3 \text{ g nylon}} \times 100 = 63.68 \qquad \% \text{ N} = \frac{28.01 \text{ g N}}{226.3 \text{ g nylon}} \times 100 = 12.38$$

$$\% \text{ O} = \frac{32.00 \text{ g O}}{226.3 \text{ g nylon}} \times 100 = 14.14$$

$\%\ H = 100.00 - 63.68 - 12.38 - 14.14 = 9.80$

25. glycine: $C_2H_5NO_2$ MM = 75.07

alanine: $C_3H_7NO_2$ MM = 89.10

In the making of a protein, n − 1 condensation reactions take place for every n monomers involved; therefore, only n − 1 water molecules are eliminated. For the given reaction:

30 (glycine) + 60 (alanine) → protein + 89 H_2O

MM of protein = 30(75.07) + 60(89.10) − 89(18.02)

MM of protein = 2252 + 5346 − 1604 = 5994

CHAPTER 17

Rate of Reaction

I. Basic Skills

A. Qualitative Students should be able to:

1. Explain the meaning of reaction rate and give the units in which it is expressed.
2. Explain chemical reactions in terms of molecular collisions.
3. Explain how the concentration of the reactants affects the reaction rate in terms of molecular collisions.
4. a. Explain what is meant by a single step reaction.
   b. Explain what is meant by a multi-step reaction and a reaction mechanism.
5. Explain what is meant by a rate equation and a rate constant.
6. Describe the effect of surface area on a reaction rate.
7. Describe the general effect of a temperature change upon the rate of a reaction.
8. Explain what is meant by activation energy and how it affects the rate of a reaction.
9. a. Describe the kinetic energy distribution of molecules in a gas sample.
   b. Explain the effect of temperature upon reaction rate in terms of activation energy.
10. a. Explain what is meant by a catalyst.
    b. Describe how a catalyst affects the rate of a reaction in terms of activation energy and mechanism.
    c. Explain what is meant by an inhibitor and the methods by which an inhibitor works.

B. Quantitative Students should be able to:

1. Convert reaction rates to different time units (Example 17.1).
2. Determine the reaction rate for a particular time interval (Example 17.2).
3. Write the rate equation for a single step reaction.
4. Determine the effect of concentration changes on the rates of a single step reaction (Example 17.3).
5. Write the overall equation for a multi-step reaction given the equations of the individual steps.
6. Draw energy diagrams of reactions which show ΔH and the activation energy (Example 17.4).

II. Chapter Development

1. The rate equation for a single step reaction is developed and used to predict the effect of concentration changes. The rate constant, k, is not evaluated. Rate equations for multi-step reactions are discussed in an optional section.
2. The concept of activation energy is presented as the minimum energy colliding molecules must have in order to react. No mention is made of the effect of orientation on collisions and rate. No attempt is made to describe the geometry of the activated complex.
3. Methods of determining reaction rates experimentally are not discussed. The reaction must produce a change in a measurable property which is related to the concentration of one of the species in the reaction.

III. Problem Areas

1. Students may confuse the reaction rate over a time interval (an average rate) with an instantaneous reaction rate. If so, try comparing the instantaneous rates in a drag race to the average rate for the complete race. All calculations in the text involve average rates. Use Figure 17.1 and Table 17.1 to show how rates are calculated and how they vary with time.
2. It is tempting to explain the effect of temperature upon rate by relating it to the increased velocity of the molecules and the resulting increase in collision rate. The actual size of this effect can be shown to be quite small. The relationship between molecular velocity and the absolute temperature is $v = kT^{1/2}$ The effect of a temperature change from T_1 to T_2 on the velocity is then:

$$v_2 = v_1 \left[\frac{T_2}{T_1}\right]^{1/2}$$

For a 10°C change (from 25°C to 35°C), the new velocity, v_2, will be equal to

$$v_1 \left[\frac{308\ \text{K}}{298\ \text{K}}\right]^{1/2}$$

or 1.02 v_1. The 2% increase in velocity cannot account for the large increase in rate. The concept of activation energy must be used.
3. The concept of activation energy can be explained as a threshold energy. The system must have this minimum energy in order for a reaction to take place. Analogous examples: (1) A falling beaker or glass will not break until dropped from some minimum height; (2) An electron is not removed from an atom until the ionization energy is absorbed.

IV. Suggested Activities

Factors That Affect Rate: **Teacher Demonstrations Only**
Temperature: Compare the rate at which calcium metal reacts in a beaker of cold water and a beaker of hot water. Add phenolphthalein to each beaker before the reaction.
Concentration: Compare the rate of evolution of hydrogen gas when zinc is added to 1 M HCl and to 6 M HCl.
Catalyst: Add MnO_2 to hydrogen peroxide to catalyze its decomposition to water and oxygen. (Raw hamburger or blood is also effective here due to the presence of the enzyme catalase.)
Nature of the Reactants: Add a magnesium strip to water and compare with the reaction of calcium metal above. (Mg reacts slowly if the water is heated.) Add copper wire to 6 M HCl and compare with the reaction of zinc above.

V. Answers to Questions

1. The rate of a reaction is the change in concentration of a substance with respect to the change in time. Mathematically, reaction rate is expressed as:

$$\text{Rate} = \frac{\Delta \text{ conc. X}}{\Delta \text{ time}} = \frac{(\text{conc. X})_2 - (\text{conc. X})_1}{\text{time}_2 - \text{time}_1}$$

   To determine a reaction rate, the concentration of any reactant or product in the reaction system is measured over an interval of time.
2. a. miles/hour or km/hour b. inches/year or cm/year
   c. pounds/week or kg/week d. inches/sec or cm/sec
3. a. Measure the flight distance and the flight time.
   b. Measure the child's height annually.

c. Measure the person's mass weekly.
d. Carefully!

4. a. conc. CO decreases b. conc. NO increases c. rate decreases
5. The curve would show a decreasing concentration of CO as time progresses. Using data in Table 17.1, the concentration of CO decreases from 1.00 M to 0.20 M during a time interval of 40 minutes.
6. In the gaseous state, molecules become closer together when the concentration is increased and collisions will occur more frequently. The reaction rate is directly related to the collision rate.
7. Rate = k(conc. NO)(conc. O_3)
8. $I_2 \rightarrow 2\ I$

   $I + H_2 \rightarrow H_2I$

   $H_2I + I \rightarrow 2\ HI$

   $H_2 + I_2 \rightarrow 2\ HI$ Overall reaction

9. $2\ O_3 \rightarrow 3\ O_2$ Overall reaction

   $-(O_3 \rightarrow O_2 + O)$ First step

   $O_3 + O \rightarrow 2\ O_2$

10. Coal dust has a very large surface area compared to lump coal. It can react at an explosive rate.
11. A freezer keeps meat at a lower temperature than a refrigerator and retards the rate at which it spoils.
12. The increased temperature of a fever will cause body reactions to take place at a faster rate.
13. Most of the increase in reaction rate caused by an increase in temperature can be attributed to:
    (d) the increase in the fraction of high energy molecules. (See Figure 17.8.)
14. a. 20° to 30°C: rate is doubled
    b. 20° to 10°C: rate is halved
    c. 20° to 40°C: rate is quadrupled (increased 4 times)
15. If a reaction has a low activation energy, a large fraction of colliding molecules will have the necessary amount of energy required for a reaction to take place.
16. A catalyst *decreases* E_a. At the same time, the catalyst *has no effect upon* ΔH.
17. An inhibitor is a substance which decreases the rate of a reaction. An inhibitor works in one or more of the following ways:
    a. It prevents the reactants from coming in contact.
    b. It reacts preferentially with one of the reactants.
    c. It destroys a catalyst or makes it ineffective.

VI. Solutions to Problems

1. $\text{Rate} = \dfrac{0.50 \text{ mol}}{\ell \cdot \text{min}} \times \dfrac{60 \text{ min}}{\text{hr}} = \dfrac{30 \text{ mol}}{\ell \cdot \text{hr}}$

2. $\text{Rate} = \dfrac{(0.26 - 0.10) \text{ mol}/\ell}{15 \text{ min}} = \dfrac{1.1 \times 10^{-2} \text{ mol}}{\ell \cdot \text{min}}$

3. a. $\text{Rate} = \dfrac{(0.75 - 0.67) \text{ mol}/\ell}{(30 - 20) \text{ min}} = \dfrac{8 \times 10^{-3} \text{ mol}}{\ell \cdot \text{min}}$

   b. $\text{Rate} = \dfrac{(0.80 - 0.75) \text{ mol}/\ell}{(40 - 30) \text{ min}} = \dfrac{5 \times 10^{-3} \text{ mol}}{\ell \cdot \text{min}}$

4. Rate = k (conc. NO)(conc. O_3)

   a. As the concentration of NO is tripled, $\dfrac{(0.30 \text{ M})}{(0.10 \text{ M})} = 3$, the reaction rate is tripled.

   b. As the new concentration of O_3 is one-fifth, $\dfrac{(0.020 \text{ M})}{(0.10 \text{ M})} = 0.20$, the reaction rate will be one-fifth (20%) of its original value.

5.

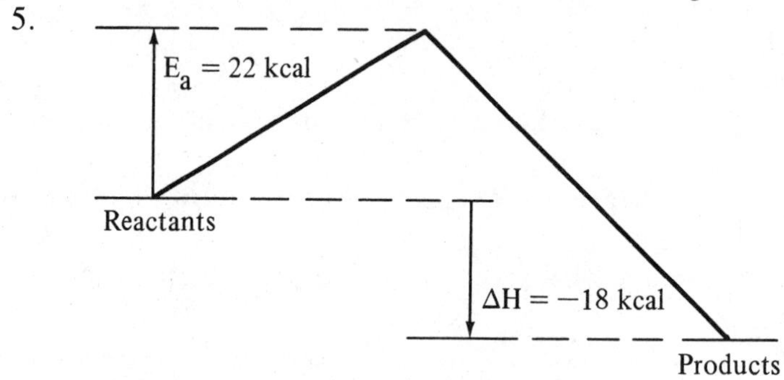

6.

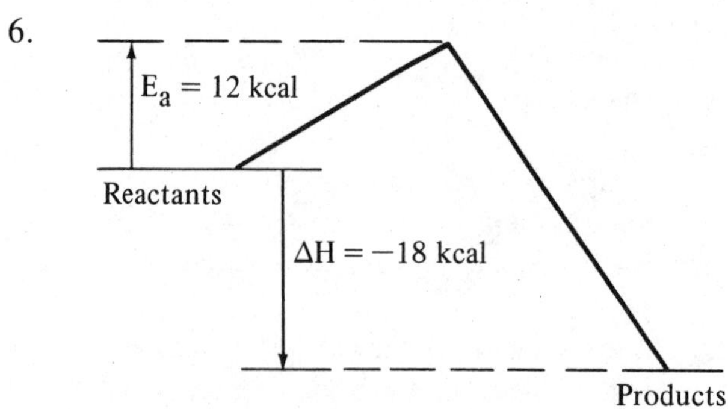

7. $\text{Rate} = \dfrac{0.50 \text{ mol}}{\ell \cdot \text{sec}} \times \dfrac{60 \text{ sec}}{\text{min}} = \dfrac{30 \text{ mol}}{\ell \cdot \text{min}}$

8. Rate = k(conc. NO)(conc. O_3)

If the concentration of NO is doubled (× 2) and the concentration of O_3 is cut in half (× 1/2), the net effect is that the reaction rate is unchanged.

9. a. $\text{Rate} = \dfrac{(0.75 - 0.50)\ \text{mol}/\ell}{(2 - 1)\ \text{hr}} = \dfrac{0.25\ \text{mol}}{\ell \cdot \text{hr}}$

b. $\text{Rate} = \dfrac{(0.50 - 0)\ \text{mol}/\ell}{(1 - 0)\ \text{hr}} = \dfrac{0.50\ \text{mol}}{\ell \cdot \text{hr}}$

c. $\text{Rate} = \dfrac{(0.88 - 0.75)\ \text{mol}/\ell}{(3 - 2)\ \text{hr}} = \dfrac{0.13\ \text{mol}}{\ell \cdot \text{hr}}$

10.

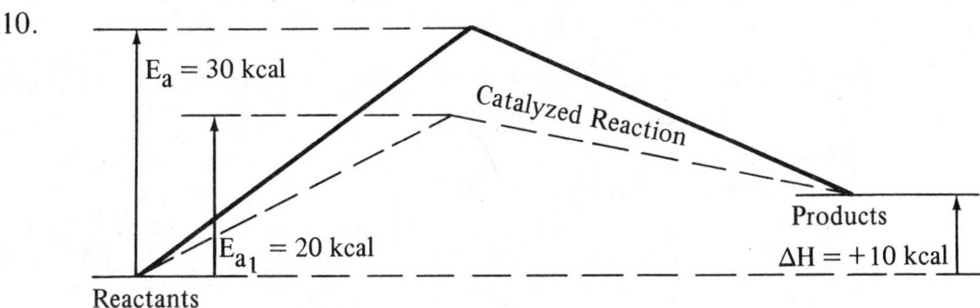

11. $\Delta H = -10$ kcal and $E_a = +20$ kcal

12. The rate of the reaction at t = 10s can be estimated as the average of the rates between 0 and 10 seconds and 10 and 20 seconds.

$\text{Rate 1} = \dfrac{(.004 - 0)\ \text{mol}/\ell}{(10 - 0)\ \text{sec}} = \dfrac{4 \times 10^{-4}\ \text{mol}}{\ell \cdot \text{sec}}$

$\text{Rate 2} = \dfrac{(.006 - .004)\ \text{mol}/\ell}{(20 - 10)\ \text{sec}} = \dfrac{2 \times 10^{-4}\ \text{mol}}{\ell \cdot \text{sec}}$

$\text{Rate (avg)} = \dfrac{4 \times 10^{-4} + 2 \times 10^{-4}}{2} = \dfrac{3 \times 10^{-4}\ \text{mol}}{\ell \cdot \text{sec}}$

13. $\text{Rate (HI)} = \dfrac{(0.24 - 0.42)\ \text{mol}/\ell}{20\ \text{min}} = \dfrac{-9.0 \times 10^{-3}\ \text{mol}}{\ell \cdot \text{min}}$

$\text{Rate } (H_2) = \dfrac{-\text{Rate (HI)}}{2} = \dfrac{-(-9.0 \times 10^{-3}\ \text{mol}/\ell \cdot \text{min})}{2}$

$= \dfrac{4.5 \times 10^{-3}\ \text{mol}}{\ell \cdot \text{min}}$

14. Compare rates in two runs in which one variable (conc. X or conc. Y) remains constant. For example, in Runs 1 and 2, the conc. X remains constant while the conc. Y is doubled and the rate remains constant. This means that rate is independent upon conc. Y; that is, it is zero order with respect to Y.
In Runs 1 and 3, the conc. Y remains constant while the conc. X is doubled and the rate is quadrupled (× 4). This means that the rate is proportional to the square of conc. X and that it is second order with respect to X. The overall rate equation is: Rate = k(conc. X)^2.

CHAPTER 18

Chemical Equilibrium

I. Basic Skills

A. Qualitative Students should be able to:

1. Explain chemical equilibrium with regard to reaction rates, the concentrations of reactants and products, and the direction from which it is approached.
2. Describe the equilibrium constant, K, and explain how it is (or is not) affected by a change in concentration, volume (pressure), or temperature.
3. Predict qualitatively the extent of reaction based on the equilibrium constant and identify the predominant species present (Example 18.5).
4. Explain Le Chatelier's Principle as it applies to chemical systems.
5. Using Le Chatelier's Principle, predict the direction that a reaction will shift when its equilibrium is disturbed by:
   a. changing concentration of a reactant or product (Example 18.6).
   b. expansion or compression (Example 18.7).
   c. changing the temperature (Example 18.8).
6. a. Explain the effect of pressure, temperature, and catalyst on the yield of a reaction (Example 18.9).
   b. Select the optimum conditions for a high product yield and explain how the rate is affected by these conditions.

B. Quantitative Students should be able to:

1. Write the equilibrium expression for a chemical system involving solids, liquids, or gases (Examples 18.2 and 18.4).
2. Solve an equilibrium expression for K or an unknown concentration when other values are known (Examples 18.1 and 18.3).
3. Analyze experimental data to determine if a reaction is at equilibrium.

II. Chapter Development

1. The chapter is limited to gaseous equilibrium systems. Solution equilibria will appear later in the study of acid-base equilibria, Chapter 19; solubility equilibria, Chapter 20; and oxidation-reduction equilibria, Chapter 23.
2. The equilibrium expression is derived empirically. The $PCl_5 - PCl_3 - Cl_2$ system is used to demonstrate the constancy of the expression.
3. Students are only required to solve the equilibrium expression when one term is unknown. The complex algebra which can come out of these equations is left for college chemistry.

4. The effect on K of changing the coefficients in an equation (multiplying by 2) is not covered.

III. Problem Areas

1. The concept of an equilibrium state is often difficult for students to accept. Most reactions that they have seen have been of the type which go to completion. Seldom are they faced with a reaction where there is a standoff between reactants and products. Give examples of reversible reactions. Explain that each reaction is driven by a force and that these forces oppose each other. If the forces are somewhat equal, then an equilibrium position (a compromise) will be reached. This is quite similar to what may happen in a balanced tug-of-war.
2. Students may misuse Le Chatelier's Principle and incorrectly predict a system's response to a disturbance. Make the point that the system attempts to "fight back." If a substance is added, the system tries to remove it; if a substance is removed, the system tries to replace it. It is similar to trying to maintain a minimum of $200 in a checking account. A fee is charged if the balance drops below $200 and potential interest is lost when the balance exceeds $200. Temperature changes can be related to maintaining temperature equilibrium in a house. If the outside temperature drops, you fight back by turning the heat source up (exothermic reaction). If the outside temperature increases, you fight back by turning on the air conditioner (endothermic reaction).
3. Students may not understand how the yield and rate of a reaction are interrelated. The maximum yield is determined by the equilibrium position as measured by the equilibrium constant, K. For practical purposes, the rate, which is independent of K, must also be considered. For exothermic reactions, the value of K increases as the temperature is decreased. The optimum conditions require a compromise between yield and rate. It is reactions of this type which particularly benefit from the use of a catalyst.

IV. Suggested Activities

1. $NO_2 - N_2O_4$ Equilibrium: **Teacher Demonstration Only**
   Demonstrate the effect of temperature on the equilibrium position of the $NO_2 - N_2O_4$ system. If available, use sealed tubes of the gas mixture. If not, tubes of the gas mixture may be made by the reaction of concentrated nitric acid on copper. Place one tube in an ice water bath and the other in a boiling water bath. The former will appear light yellow (N_2O_4) while the other will appear reddish brown (NO_2). This is in agreement with LeChatelier's Principle and the ΔH of the reaction.

   $$N_2O_4(g) \rightleftarrows 2\ NO_2(g) \qquad \Delta H = +13.9\ \text{kcal}$$

   Ask students to predict the effect of a pressure change.

V. Answers to Questions

1. a. At the beginning of the experiment, the rate of the reverse reaction is zero as no chlorine atoms are present. In general, the rate of the forward reaction will be greater until equilibrium is reached.
   b. At equilibrium, the rates of the forward and reverse reactions are equal.
2. a. No b. Yes c. No
   The concentration ratio changes temporarily when a system at equilibrium is disturbed. When the new equilibrium position is reached, only a change in temperature produces a new value of K.
3. a. 0.60 mol PCl_3 b. 0.60 mol Cl_2
4. a. 0.036 mol I_2 b. 0.072 mol HI
5. Equations (b), (d), and (e); when N_2 is a reactant, (a) and (c), the N_2 term appears in the denominator.
6. a. Equation (e) b. Equations (a), (b), (c), and (d)
7. Equation (c); in the other cases, H_2O is present as a liquid.
8. A K of 10^{20} means that: (c) the product is formed in high yield.
9. For a K of 1, you would expect to find: (c) about the same amount of products as reactants.
10. For a K of 10^{-7}, you would expect to find: (d) mostly CO and H_2O.
11. Le Chatelier's Principle states that when a system at equilibrium is disturbed, the system shifts so as to partially counteract the change. It is used in chemistry to predict how a chemical system at equilibrium will react when it is disturbed. (See pages 457–465.)
12. A chemical system at equilibrium can be disturbed by:
    a. adding or removing a reactant or product.
    b. expanding or compressing a gaseous system.
    c. changing the temperature.
13. Adding a reactant increases the concentration of the reactant. To reduce the reactant's concentration, the forward reaction takes place.
14. The endothermic reaction is favored because it consumes heat and tends to counteract the increase in temperature.
15. A system at equilibrium can only partially counteract a disturbance. It will return to a new equilibrium state which is between the original equilibrium state and the temporary nonequilibrium disturbed state.
16. The equilibrium position can be shifted by: (e) none of these. As the number of gas moles does not change, the equilibrium position is not affected by compression or expansion. As $\Delta H = 0$, the temperature also has no effect on the equilibrium.
17. $K = \frac{[PCl_3][Cl_2]}{[PCl_5]} = 0.050$

$$\text{For Exp 1: } K = \frac{(0.22)(0.22)}{(0.98)} = 0.049$$

$$\text{For Exp 2: } K = \frac{(0.18)(0.18)}{(0.62)} = 0.052$$

For both experiments, K = 0.050 within an uncertainty of ± 0.002.

18. The Haber process for making ammonia is:

$$N_2(g) + 3\ H_2(g) \rightleftarrows 2\ NH_3(g); \quad \Delta H = -22\ \text{kcal}$$

a. A pressure decrease would cause a shift in favor of more moles of gas causing a decrease in the yield of NH_3. The rate would also decrease due to the decrease in the concentration of reactant gases.
b. A temperature decrease would favor the exothermic reaction causing an increase in the yield of NH_3. The rate would decrease as temperature is decreased.
c. The yield of NH_3 is not changed by the use of a catalyst. Without a catalyst, the rate would decrease.

VI. Solutions to Problems

1. a. $K = \dfrac{[SO_3]^2}{[SO_2]^2\,[O_2]}$ b. $K = \dfrac{[HBr]^2}{[H_2][Br_2]}$ c. $K = \dfrac{[N_2][H_2]^2}{[N_2H_4]}$

2. a. $K = [H_2O]$ b. $K = \dfrac{[CO_2][H_2]}{[CO]}$ c. $K = \dfrac{[H_2O]^3}{[H_2]^3}$

3. $K = \dfrac{[HI]^2}{[H_2][I_2]} = 60$

$[HI]^2 = K\,[H_2][I_2]$

$[HI]^2 = 60(0.10)^2 = 0.60$

$[HI] = 0.77\ \text{mol}/\ell$

4. $K = \dfrac{[HI]^2}{[H_2][I_2]} = 60$

$[I_2] = \dfrac{[HI]^2}{60[H_2]}$ $[I_2] = \dfrac{(0.60)^2}{60(0.10)} = 0.060\ \text{mol}/\ell$

5. $K = \dfrac{[NH_3]^2}{[N_2][H_2]^3}$

$K = \dfrac{(0.10)^2}{(0.10)(0.20)^3} = 12$

6. The better yield of NH_3 would be obtained at 200°C where K is 650. The yield of products increases as the value of K increases.

7. The reaction would proceed in the net direction which would consume an added substance or replace a substance which is removed.
   a. forward b. reverse c. forward d. reverse
8. a. compress b. expand c. no effect
9. As the reaction is exothermic, a temperature increase would decrease the yield and a temperature decrease would increase the yield.
10. a. An increase in pressure would increase the rate and decrease the yield of products.
    b. An increase in temperature would increase both the rate and the yield of products.
    c. A catalyst would increase the rate but have no effect upon the yield.
11. $K = [H_2O]^2$

    $K = (0.16)^2 = 2.6 \times 10^{-2}$

12. $$K = \frac{[SO_2]^2[O_2]}{[SO_3]^2}$$

    $$K = \frac{(0.10)^2(0.10)}{(0.10)^2} = 0.10$$

13. $$K = \frac{[CO_2]^2}{[CO]^2[O_2]} = 2.0 \times 10^5$$

    $$[O_2] = \frac{[CO_2]^2}{K[CO]^2} = \frac{(1)^2}{2.0 \times 10^5(1)^2} = 5 \times 10^{-6} \text{ mol}/\ell$$

14. The small K of 1×10^{-12} indicates that the equilibrium position strongly favors the reactants. The concentration of Cl atoms will be very small compared to that of Cl_2 molecules.
15. A decrease in K from 2.0×10^5 to 200 would be accompanied by a decrease in the yield of CO_2.
16. The large K of 3×10^9 indicates that the equilibrium position strongly favors the products. The concentration of I_2 molecules will be very large compared to that of I atoms.
17. Adding H_2 will shift the equilibrium toward the production of more HI.
18. A volume increase shifts the equilibrium position in the direction of the larger number of moles of gas.
    a. forward b. reverse c. reverse d. no effect
19. As the reaction is endothermic, an increase in temperature would increase the value of K.
20. The reaction rate could be increased by an increase in temperature, an increase in pressure or by the use of a catalyst. The yield of NO_2 could be increased by an increase in temperature or a decrease in pressure.
21. The yield of NO_2 could be increased by a decrease in temperature, an increase in pressure or by an increase in the concentration of O_2.
22. For the given reaction, an increase in pressure would move the equilibrium position to the left (toward the reactants). An increase in temperature would move it to the right (toward the products).

23. The reaction rate could be increased by an increase in temperature, an increase in pressure or by the use of a catalyst. The yield of SO_3 would be decreased by an increase in temperature and increased by an increase in pressure. The yield is not affected by the use of a catalyst.

24. a. $\dfrac{[PCl_3][Cl_2]}{[PCl_5]} = \dfrac{(0.50)(0.10)}{1.00} = 0.050$

As K = 0.050, the reaction is at equilibrium.

b. $\dfrac{[PCl_3][Cl_2]}{[PCl_5]} = \dfrac{(0.25)(0.25)}{1.00} = 0.062$

The reaction is proceeding toward the reactants as $0.062 > K$.

c. $\dfrac{[PCl_3][Cl_2]}{[PCl_5]} = \dfrac{(0.050)(0.10)}{(0.10)} = 0.050$

As K = 0.050, the reaction is at equilibrium.

25. a. $\dfrac{[HI]^2}{[H_2][I_2]} = \dfrac{(6.0)^2}{(0.60)(1.00)} = 60$

As K = 60, the reaction is at equilibrium.

b. $\dfrac{[HI]^2}{[H_2][I_2]} = \dfrac{(60.0)^2}{(1.0)(1.0)} = 3600$

As 3600 is $> K$, the reaction is proceeding toward the reactants.

c. $\dfrac{[HI]^2}{[H_2][I_2]} = \dfrac{(1.00)^2}{(0.10)(0.17)} = 59$

As K = 60, the reaction is essentially at equilibrium.

26. a. $[PCl_5] = \dfrac{0.10 \text{ mol}}{4.0\ \ell} = 0.025 \text{ mol}/\ell$

b. As 0.10 mol of PCl_5 reacted, 0.10 mol of PCl_3 and Cl_2 was formed.

$[PCl_3] = [Cl_2] = \dfrac{0.10 \text{ mol}}{4.0\ \ell} = 0.025 \text{ mol}/\ell$

c. $K = \dfrac{[PCl_3][Cl_2]}{[PCl_5]} = \dfrac{(0.025)^2}{0.025} = 0.025$

27. a. $\dfrac{V_2}{V_1} = \dfrac{T_2}{T_1}$ $T_1 = 273 \text{ K}$ $T_2 = 850°C + 273 = 1123 \text{ K}$

$$V_2 = V_1 \times \frac{T_2}{T_1} = 22.4\ \ell \times \frac{1123\ K}{273\ K} = 92.1\ \ell$$

b. $[CO_2] = \dfrac{1\ mol}{92.1\ \ell} = 1.09 \times 10^{-2}\ mol/\ell$

c. $K = [CO_2] = 1.09 \times 10^{-2}$

28. Compression causes an equilibrium shift toward the fewer moles of gas. In this case, the reverse reaction would occur so as to produce more CO and H_2O.
29. The reaction of a heated metal oxide with H_2 to produce the metal and H_2O would give the desired expression. Two examples are:

$$CuO(s) + H_2(g) \rightleftarrows Cu(s) + H_2O(g)$$

$$SnO(s) + H_2(g) \rightleftarrows Sn(s) + H_2O(g)$$

CHAPTER 19

Acids and Bases

I. Basic Skills

A. Qualitative Students should be able to:

1. a. Define an acid and a base.
   b. Describe the general properties of acidic and basic solutions.
2. Explain what is meant by the pH of a solution.
3. Classify solutions as acidic, basic, or neutral according to their pH or $[H^+]$ (Example 19.3).
4. a. Explain what is meant by strong and weak acids and strong and weak bases.
   b. Name the more common strong acids and bases.
5. a. Explain what is meant by an ionization constant.
   b. Compare the relative strengths of weak acids by using their ionization constants (Example 19.7).
6. Explain what is meant by a neutralization reaction.

B. Quantitative Students should be able to:

1. Write the chemical equation and the equilibrium expression for the ionization of water.
2. Determine the $[H^+]$ or $[OH^-]$ in a water solution when the other concentration is known (Example 19.1).
3. Calculate the pH of a solution from the $[H^+]$, or the $[H^+]$ from the pH (Example 19.2).
4. a. Write equations for the ionization of a strong acid (Example 19.4) or a strong base (Example 19.8).
   b. Calculate the $[H^+]$ of a strong acid (Example 19.4) or the $[OH^-]$ of a strong base (Example 19.8) knowing the molarity of the solution.
5. a. Write the equation for the ionization of a weak acid (Example 19.5).
   b. Write the equilibrium expression for the ionization of a weak acid.
   c. Calculate the $[H^+]$ of a weak acid knowing its molarity and percent ionization (Example 19.5).
   d. Calculate the ionization constant, K_a, of a weak acid given its molarity and the $[H^+]$ (Example 19.6).
6. Write equations for:
   a. the successive ionizations of an acid which contains more than one ionizable hydrogen atom,
   b. the reaction of a weak base with water,
   c. the neutralization reactions of acids and bases (Example 19.9).

II. Chapter Development

1. In general, acid-base principles are presented from the viewpoint of the Arrhenius Theory. This is believed to be the most practical and useful approach for high school students. The Brønsted-Lowry Theory is introduced in an optional section.
2. The hydronium ion is introduced but not used in equations except in applications of the Brønsted-Lowry Theory. Equations are simplified by using the H^+ formula. In general, pH to $[H^+]$ conversions are limited to integral pH values.
3. In general, equilibrium discussions and problems are restricted to weak acids and K_a. Ammonia is the only molecular weak base considered. The hydrolysis of basic anions and acidic cations is discussed in an optional section.
4. The use of acids to dissolve precipitates and the use of bases to form complex ions is discussed in Chapter 20, Precipitation Reactions.
5. Acid-base titrations are covered in Chapter 21, Quantitative Analysis.
6. Buffering and the common ion effect with respect to weak acids are not discussed. The common ion effect is covered in Chapter 20 with respect to precipitation reactions.

III. Problem Areas

1. The use of exponents and logarithms in this chapter presents some difficulties. The use of an inexpensive calculator can solve most of these problems. Students can get help in the use of exponents from the Review of Mathematics in the Appendix.
2. Equilibrium problems are kept at the same level of difficulty as originally presented in Chapter 18. The major change is the need to determine the concentration of the un-ionized weak acid. Using the percent ionization, or equivalent information, the student must subtract the amount ionized from the total to find the amount which remains.

IV. Suggested Activities

1. Strong Acids and Weak Acids:
   Demonstrate differences between strong acids (HCl, HNO_3) and weak acids (CH_3COOH, H_2CO_3) of the same molarity.
   a. Compare electrical conductivity of each solution.
   b. Compare the rate of reaction with zinc or magnesium.
   c. Using pH paper, compare the pH values.
2. Dehydration Action of H_2SO_4: **Teacher Demonstration Only**
   Add about 10 ml of concentrated H_2SO_4 to 4 g of sucrose in a 150 ml beaker. The sucrose is dehydrated in an exothermic reaction leaving a residue of

carbon. CAUTION: Do not handle the product as it may contain unreacted H_2SO_4.

3. Neutralization Reactions:
   a. Add exactly 50 ml of 1 M HCl to 50 ml of 1 M NaOH. (Check strength of solutions in advance.) Measure pH of the resulting solution and show that it is neutral. Demonstrate the electrical conductivity to show the presence of ions.
   b. Add exactly 50 ml of 0.01 M H_2SO_4 to 50 ml of 0.01 M $Ba(OH)_2$. (Check strength of solutions in advance.) A white precipitate of $BaSO_4$ forms. Stir well and show the low conductivity due to the absence of ions.
4. pH of Common Products:
   Using pH paper (or pH meter if available) measure the pH of a variety of common products. Have students predict in advance whether each solution is acidic, basic, or neutral. Suggestions are tap water, local river, lake, or sea water, carbonated beverages, beer, baking powder (soln.), fruit juices, milk of magnesia, and vinegar.

V. Answers to Questions

1. The solution is acidic because it turns blue litmus red. It will react with (a) Zn and (c) OH^- and will not react with (b) H^+ and (d) Mg^{2+}

$$2\ H^+(aq) + Zn(s) \rightarrow H_2(g) + Zn^{2+}(aq)$$

$$H^+(aq) + OH^-(aq) \rightarrow H_2O$$

2. The solution is basic because it turns red litmus blue. It will react with (b) H^+ and (d) Mg^{2+} and will not react with (a) Zn and (c) OH^-.

$$OH^-(aq) + H^+(aq) \rightarrow H_2O$$

$$2\ OH^-(aq) + Mg^{2+}(aq) \rightarrow Mg(OH)_2(s)$$

3. a. neutral b. acidic c. neutral d. acidic e. basic

4. Solutions (a) pH = 9, (c) $[H^+] = 3 \times 10^{-12}$ M, and (d) $[OH^-] = 2 \times 10^{-2}$ M are basic.

5. For solution A, pH = 2: $[H^+] > [OH^-]$ (10^{-2} M vs 10^{-12} M)

   For solution B, pH = 7: $[H^+] = [OH^-] = 10^{-7}$ M

   For solution C, pH = 9: $[OH^-] > [H^+]$ (10^{-5} M vs 10^{-9} M)

6. a. vinegar, acidic (CH_3COOH) b. lye, basic (NaOH)
   c. grapefruit, acidic (citric acid) d. ammonia, basic (NH_3)

7. a. It is more correct to state that strong acids ionize completely in water. It is possible to have a dilute solution of a strong acid.
   b. Only concentrated sulfuric acid reacts violently with water.
   c. Strong bases also react with clothing.
   d. Strong acids ionize to produce H^+ ions in solution.
8. The common strong acids are HCl, HBr, HI, HNO_3, $HClO_4$, and H_2SO_4.
9. a. HCl, strong b. HF, weak c. HNO_2, weak
   d. HNO_3, strong e. HCOOH, weak
10. a. sodium phosphate, Na_3PO_4 b. sodium carbonate, Na_2CO_3
    c. sodium monohydrogen phosphate, Na_2HPO_4
    d. sodium hydrogen carbonate, $NaHCO_3$
    e. sodium dihydrogen phosphate, NaH_2PO_4
11. $H_2CO_3(aq) \rightleftarrows H^+(aq) + HCO_3^-(aq)$

$$K_a = \frac{[H^+][HCO_3^-]}{[H_2CO_3]}$$

12. If $K_a = \dfrac{[H^+][HCOO^-]}{[HCOOH]}$, then the chemical equation is:

$$HCOOH(aq) \rightleftarrows H^+(aq) + HCOO^-(aq)$$

13. The ionization constant of the strong acid HNO_3 would be: (d) a very large number.
14. Recall that HCl is a strong acid, completely ionized. The strength of a weak acid is related to its ionization constant, K_a. Using the K_a data in Table 19.2, the acids can be ranked in order of decreasing strength: HCl > CH_3COOH > HCN.
15. The strength of an acid decreases as K_a decreases. The three strongest acids in Table 19.2 are H_3PO_4, HF, and HCOOH.
16. a. NaOH, strong base b. $HClO_4$, strong acid
    c. HClO, weak acid d. HNO_2, weak acid
    e. $Ca(OH)_2$, strong base f. NH_3, weak base
17. If one mole of a substance added to one liter of water produces a $[OH^-]$ of 0.004 M, then the substance is a weak base. Only (a) NH_3 of the listed substances is a weak base.
18. Only a strong acid such as (b) HCl could produce a solution of pH = 0.
19. In order of increasing pH of 1 M solutions:

$$HBr = HNO_3 < CH_3COOH < NH_3 < NaOH = KOH$$

20. In general, an ammonium salt is formed from ammonia by the addition of the appropriate acid:
    a. NH_4Cl: HCl b. $(NH_4)_2SO_4$: H_2SO_4 c. NH_4NO_3: HNO_3

VI. Solutions to Problems

1. $K = [H^+][OH^-] = 1.0 \times 10^{-14}$

a. $[OH^-] = \dfrac{1.0 \times 10^{-14}}{1 \times 10^{-6}} = 1 \times 10^{-8}$ M

b. $[OH^-] = \dfrac{1.0 \times 10^{-14}}{2 \times 10^{-4}} = 5 \times 10^{-11}$ M

c. $[OH^-] = \dfrac{1.0 \times 10^{-14}}{4.0 \times 10^{-12}} = 2.5 \times 10^{-3}$ M

2. $K = [H^+][OH^-] = 1.0 \times 10^{-14}$

a. $[H^+] = \dfrac{1.0 \times 10^{-14}}{1 \times 10^{-4}} = 1 \times 10^{-10}$ M

b. $[H^+] = \dfrac{1.0 \times 10^{-14}}{5 \times 10^{-10}} = 2 \times 10^{-5}$ M

c. $[H^+] = \dfrac{1.0 \times 10^{-14}}{3.6 \times 10^{-7}} = 2.8 \times 10^{-8}$ M

3. $pH = -\log [H^+]$

a. $pH = -\log (10^{-8}) = 8$

b. $pH = -\log (10^{-1}) = 1$

c. $pH = -\log (10^{-3}) = 3$

4. $\log [H^+] = -pH$

a. $\log [H^+] = -5$ $\qquad [H^+] = 1 \times 10^{-5}$ M

b. $\log [H^+] = -9$ $\qquad [H^+] = 1 \times 10^{-9}$ M

c. $\log [H^+] = 1$ $\qquad [H^+] = 10$ M

5. $HBr(aq) \rightarrow H^+(aq) + Br^-(aq)$

HBr is a strong acid and is completely ionized in water. For this reason, the $[H^+]$ is 0.20 M, the same as the molarity of the acid.

6. $HNO_2(aq) \rightleftarrows H^+(aq) + NO_2^-(aq)$

$$[H^+] = \frac{\%\ \text{ionized}}{100}\ (\text{conc. } HNO_2)$$

$[H^+] = 0.05\ (0.20\ M) = 0.01$ M

7. $HB(aq) \rightleftarrows H^+(aq) + B^-(aq)$

   a. $[H^+] = 0.040\ M$ b. $[B^-] = 0.040\ M$

   c. $[HB] = (\text{conc. } HB) - [H^+]$

   $[HB] = 0.40\ M - 0.040\ M = 0.36\ M$

   d. $K_a = \dfrac{[H^+][B^-]}{[HB]}$

   $K_a = \dfrac{(0.040)^2}{0.36} = 4.4 \times 10^{-3}$

8. $Sr(OH)_2(s) \rightarrow Sr^{2+}(aq) + 2\ OH^-(aq)$

   $0.20\ \text{mol } Sr(OH)_2 \hat{=} 0.40\ \text{mol } OH^-$

   $[OH^-] = \dfrac{0.40\ \text{mol}}{4.0\ \ell} = 0.10\ M$

9. Each reaction involves the neutralization of a strong acid by a strong base. Reactions of this type are written showing only the ions which undergo a change:

   $H^+(aq) + OH^-(aq) \rightarrow H_2O$

10. As the reactions all involve a strong acid, they are written as:

    $NH_3(aq) + H^+(aq) \rightarrow NH_4^+(aq)$

11. a. $CH_3COOH(aq) + OH^-(aq) \rightarrow CH_3COO^-(aq) + H_2O$

    b. $HNO_2(aq) + OH^-(aq) \rightarrow NO_2^-(aq) + H_2O$

    c. $H_2CO_3(aq) + OH^-(aq) \rightarrow HCO_3^-(aq) + H_2O$

    $HCO_3^-(aq) + OH^-(aq) \rightarrow CO_3^{2-}(aq) + H_2O$

12. $[H^+] = (\text{conc. } HNO_3) = 0.01\ M$

    $pH = -\log [H^+] = -\log (10^{-2}) = 2$

    $[OH^-] = \dfrac{1.0 \times 10^{-14}}{1 \times 10^{-2}} = 1 \times 10^{-12}\ M$

13. $[OH^-] = (\text{conc. } KOH) = \dfrac{0.30\ \text{mol}}{3.0\ \ell} = 0.10\ M$

    $[H^+] = \dfrac{1.0 \times 10^{-14}}{0.10} = 1.0 \times 10^{-13}\ M$

$pH = -\log [H^+] = -\log (1.0 \times 10^{-13}) = 13.0$

14. $K = [H^+][OH^-] = 1.0 \times 10^{-14}$

a. $[OH^-] = \dfrac{1.0 \times 10^{-14}}{1 \times 10^{-5}} = 1 \times 10^{-9}$ M

b. $[OH^-] = \dfrac{1.0 \times 10^{-14}}{1 \times 10^{-9}} = 1 \times 10^{-5}$ M

c. $[OH^-] = \dfrac{1.0 \times 10^{-14}}{10} = 1 \times 10^{-15}$ M

15. $[H^+] = \dfrac{1.0 \times 10^{-14}}{1 \times 10^{-12}} = 1 \times 10^{-2}$ M

$pH = -\log [H^+] = -\log (10^{-2}) = 2$

16. a. $[H^+] = (\text{conc. HCl}) = 0.1$ M

$pH = -\log [H^+] = -\log (10^{-1}) = 1$

b. $[OH^-] = (\text{conc. NaOH}) = 0.1$ M

$[H^+] = \dfrac{1.0 \times 10^{-14}}{0.1} = 1 \times 10^{-13}$ M

$pH = -\log [H^+] = -\log (10^{-13}) = 13$

17. a. $CH_3COOH(aq) \rightleftarrows H^+(aq) + CH_3COO^-(aq)$

$K_a = \dfrac{[H^+][CH_3COO^-]}{[CH_3COOH]} = 1.8 \times 10^{-5}$

b. $[H^+] = \dfrac{K_a\ [CH_3COOH]}{[CH_3COO^-]}$

$[H^+] = \dfrac{1.8 \times 10^{-5}\ (0.10)}{1.0} = 1.8 \times 10^{-6}$ M

18. $[H^+] = \dfrac{\%\ \text{ionized (conc. HB)}}{100}$

$[H^+] = \dfrac{0.10\ (0.10\ \text{mol})}{(0.500\ \ell)} = 2.0 \times 10^{-2}$ M

$[HB] = (\text{conc. HB}) - [H^+]$

$[HB] = 0.20\ M - 0.020\ M = 0.18\ M$

$$K_a = \frac{[H^+][B^-]}{[HB]}$$

$$K_a = \frac{(2.0 \times 10^{-2})^2}{0.18} = 2.2 \times 10^{-3}\ M$$

19. $[H^+][OH^-] = 1.0 \times 10^{-12}$ and $[H^+] = [OH^-]$

$[H^+]^2 = 1.0 \times 10^{-12}$

$[H^+] = 1.0 \times 10^{-6}\ M$

$pH = -\log [H^+] = -\log (1.0 \times 10^{-6}) = 6.0$

20. a. $NH_3(aq) + H_2O \rightleftarrows NH_4^+(aq) + OH^-(aq)$

b. $NH_3(aq) + H^+(aq) \rightarrow NH_4^+(aq)$

c. $HF(aq) + OH^-(aq) \rightarrow F^-(aq) + H_2O$

21. a. $HCl(aq) \rightarrow H^+(aq) + Cl^-(aq)$

b. $HCOOH(aq) \rightleftarrows H^+(aq) + HCOO^-(aq)$

c. $RbOH(s) \rightarrow Rb^+(aq) + OH^-(aq)$

d. $Ca(OH)_2(s) \rightarrow Ca^{2+}(aq) + 2\ OH^-(aq)$

22. $H_2CO_3(aq) \rightleftarrows H^+(aq) + HCO_3^-(aq)$

a. $$K_a = \frac{[H^+][HCO_3^-]}{[H_2CO_3]}$$

$$[H^+] = \frac{K_a\ [H_2CO_3]}{[HCO_3^-]}$$

$$[H^+] = \frac{4.2 \times 10^{-7}\ (0.20)}{0.10} = 8.4 \times 10^{-7}\ M$$

b. $$[H^+] = \frac{4.2 \times 10^{-7}\ (2.0 \times 10^{-6})}{1.0 \times 10^{-5}} = 8.4 \times 10^{-8}\ M$$

23. a. $HClO(aq) + OH^-(aq) \rightarrow ClO^-(aq) + H_2O$

b. $CH_3CH_2COOH(aq) + OH^-(aq) \rightarrow CH_3CH_2COO^-(aq) + H_2O$

c. $HCO_3^-(aq) + OH^-(aq) \rightarrow CO_3^{2-}(aq) + H_2O$

24. a. $pH = -\log [H^+] = -\log (2 \times 10^{-4}) = 3.7$

b. $[H^+] = \dfrac{1.0 \times 10^{-14}}{2.5 \times 10^{-6}} = 4.0 \times 10^{-9}$

$pH = -\log (4.0 \times 10^{-9}) = 8.40$

25. $\log [H^+] = -pH = -2.1$

$[H^+] = 8 \times 10^{-3}$ M

$[HF] = (\text{conc. HF}) - [H^+]$

$[HF] = 0.10\text{ M} - 8 \times 10^{-3}\text{ M} = 0.09\text{ M}$

$$K_a = \frac{[H^+][F^-]}{[HF]}$$

$$K_a = \frac{(8 \times 10^{-3})^2}{0.09} = 7 \times 10^{-4}$$

26. a. $F^-(aq) + H_2O \rightleftarrows HF(aq) + OH^-(aq)$

b. $CH_3COO^-(aq) + H_2O \rightleftarrows CH_3COOH(aq) + OH^-(aq)$

c. $NO_2^-(aq) + H_2O \rightleftarrows HNO_2(aq) + OH^-(aq)$

27. a. $HF(aq) + H_2O \rightleftarrows F^-(aq) + H_3O^+(aq)$

acid base base acid

b. $F^-(aq) + H_2O \rightleftarrows HF(aq) + OH^-(aq)$

base acid acid base

c. $HCO_3^-(aq) + HCOOH(aq) \rightleftarrows H_2CO_3(aq) + HCOO^-(aq)$

base acid acid base

CHAPTER 20

Precipitation Reactions

I. Basic Skills

A. Qualitative   Students should be able to:

1. Explain what is meant by a precipitation reaction.
2. Explain what is meant by a compound being soluble or insoluble.
3. Explain the common ion effect and how it affects the solubility of the ionic compound.
4. Describe the solubility product constant and what it tells about the relative solubility of a compound.
5. Describe what is meant by a complexing agent and a complex ion.

B. Quantitative   Students should be able to:

1. Use solubility rules (Table 20.1) to predict the relative solubility of a compound (Example 20.1).
2. Determine the formulas of the possible precipitates when ionic solutions are mixed (Example 20.2).
3. a. Write the chemical equation showing the solubility equilibrium of a saturated solution.
   b. Write the equilibrium expression using K_{sp} (Example 20.6).
4. Given the value of K_{sp} for a compound:
   a. calculate the concentration of one ion in a saturated solution, when the concentration of the other ion is known (Examples 20.4 and 20.7a).
   b. calculate the concentrations of both ions in a saturated solution when their concentrations are equal (Examples 20.5 and 20.7b).
5. Write the chemical equations for the:
   a. dissolving of insoluble hydroxides or carbonates by a strong acid (Example 20.8).
   b. dissolving of insoluble compounds by a complexing agent (Table 20.3 and Example 20.9).

II. Chapter Development

1. The chapter lays important groundwork for Quantitative Analysis (Chapter 21) and Qualitative Analysis (Chapter 22). Precipitation reactions are used in the gravimetric techniques of analysis. Precipitation reactions and the dissolving of precipitates are techniques used to separate and identify ions.
2. K_{sp} calculations are limited to symmetrical electrolytes (AgCl, $CaCO_3$). The use of K_{sp} to predict whether or not a precipitate will form is covered in an

optional section.

3. Dashes are used to represent the covalent bonds between the complexing agent and the metal ion in a complex ion. No attempt is made to give a detailed description of the bonding in complex ions or their resulting geometry.

III. Problem Areas

1. The use of net ionic equations is important in writing correct equations and equilibrium expressions in this chapter. Some students hate to leave anything out of an equation. Try to convince this group that nonprecipitating ions are unchanged in the reaction and should be omitted from the equation. Above all, combinations of ions which are soluble should not be written in formula form (NaCl, $CuSO_4$) implying that they form precipitates.
2. Students tend to believe that the ion concentrations for symmetrical electrolytes are always equal. For example, that in a solution in which AgCl is precipitated the $[Ag^+] = [Cl^-]$. Only when a single compound is dissolved in pure water will the ion concentrations be equal. A real effort to make this point has been made by providing examples and problems in the text which cover it. K_{sp} expressions are inverse relationships between ion concentrations. Students have seen these before in $PV = k$ and $K = [H^+] \times [OH^-]$.

IV. Suggested Activities

1. Precipitation and the Common Ion Effect:
   a. Dilute 15 ml portions of 2 M H_2SO_4 and 2 M $CaCl_2$ to 50 ml in separate graduated cylinders. Pour the solutions together in a conical graduate. A precipitate of $CaSO_4$ forms slowly. The concentration product of 9×10^{-2} (after dilution) exceeds the K_{sp} of 2.4×10^{-5}. The common ion effect may be shown by adding a small piece of chalk ($CaCO_3$) and noting the increased precipitation.
   b. Repeat the demonstration with 50 ml portions of the undiluted solutions. A precipitate is formed instantly and the graduate may be carefully inverted to show its solidity. The concentration product is 1.
2. Complex Ion Formation:
   Add about 20 ml of 0.5 M $CuSO_4$ solution to a beaker and point out the characteristic color of the Cu^{2+} ion in water solution. Add a small amount of dilute NH_3 and precipitate blue $Cu(OH)_2$. Add additional NH_3 until the precipitate dissolves and the dark blue $Cu(NH_3)_4{}^{2+}$ ion is formed. Add dilute H_2SO_4 until the complex ion is destroyed and the original solution color returns. Have students write equations for all reactions.

3. Le Chatelier's Principle: **Teacher Demonstration Only**
Exhibit test tubes of yellow K_2CrO_4 and orange $K_2Cr_2O_7$ solutions. Add 2–3 drops of $Ba(NO_3)_2$ solution to show the lower solubility of $BaCrO_4$. Put the equation on the board showing the equilibrium between the chromate and dichromate ions:

$$2\ CrO_4^{2-}(aq) + 2\ H^+(aq) \rightleftarrows Cr_2O_7^{2-}(aq) + H_2O$$

Ask students to suggest a method of dissolving the $BaCrO_4$ precipitate in the first tube (add H^+) and forming a $BaCrO_4$ precipitate in the second tube (add OH^-). Explain results in terms of LeChatelier's Principle. (Note: This demonstration is also an extension of Experiment 28.)

V. Answers to Questions

1. When $NiCl_2$ and KOH solutions are mixed, the two possible precipitates are $Ni(OH)_2$ and KCl. Using the solubility rules in Table 20.1, the precipitate is found to be $Ni(OH)_2$. The green color is also characteristic of nickel compounds.
2. a. KBr: K^+ and Br^- b. $Pb(NO_3)_2$: Pb^{2+} and NO_3^-
c. $(NH_4)_2CO_3$: NH_4^+ and CO_3^{2-}
d. $CuCl_2$: Cu^{2+} and Cl^-
e. $Al_2(SO_4)_3$: Al^{3+} and SO_4^{2-}
3. The soluble compounds are (a) $NaNO_3$ and (d) CaS.
4. The soluble sulfides are those of the Group I metals, the Group 2 metals, and $(NH_4)_2S$. The Group 1 sulfides are Li_2S, Na_2S, K_2S, Rb_2S, and Cs_2S; the Group 2 sulfides are BeS, MgS, CaS, SrS, and BaS.
5. All hydroxides except those of the Group 1 metals and $Sr(OH)_2$ and $Ba(OH)_2$ of Group 2 are insoluble. Examples are $Mg(OH)_2$, $Al(OH)_3$, $Ni(OH)_2$, $Cu(OH)_2$, and $Zn(OH)_2$.
6.

	NO_3^-	Cl^-	SO_4^{2-}	CO_3^{2-}	OH^-	S^{2-}
Pb^{2+}	S	I	I	I	I	I
Ca^{2+}	S	S	I	I	I	S
Ag^+	S	I	I	I	I	I

"S": soluble "I": insoluble

7. $CaSO_4$ may be precipitated from a saturated solution by adding:
a. any solution which contains the SO_4^{2-} ion such as H_2SO_4.
b. any solution which contains the Ca^{2+} ion such as $Ca(NO_3)_2$.
8. Silver iodide is less soluble in a silver nitrate solution than in water because of the "common ion" effect. The Ag^+ from the $AgNO_3$ solution reduces the amount of AgI which can dissolve by shifting the equilibrium to cause AgI to precipitate from solution.

9. K_{sp} of $Cu(OH)_2 = [Cu^{2+}][OH^-]^2$; Answer (b).
10. The increase in solubility, as temperature is increased, indicates an increase in the value for K_{sp}. Answer (c), $K_{sp} = 1.7 \times 10^{-8}$, shows an increase compared to the K_{sp} of 1.7×10^{-10} at 25°C.
11. As the product of the ion concentrations equals a constant (the K_{sp}), the $[Cl^-]$ must be (a) half as large when the $[Ag^+]$ is doubled.
12. The equations, $PV = k$, $K = [H^+][OH^-]$ and $K_{sp} = [Ag^+][Cl^-]$, all have the same mathematical form of xy = a constant. A graph of each would show that the variables are inversely related to each other (a hyperbola).
13. Each formula unit of $CaCO_3$ produces one Ca^{2+} ion and one CO_3^{2-} ion when dissolved:

$$CaCO_3(s) \rightarrow Ca^{2+}(aq) + CO_3^{2-}(aq)$$

14. Compounds (a), $CaSO_4$, and (d), AgI, will produce solutions in which the concentrations of the positive and negative ions are equal. In (b), there will be two K^+ ions for each SO_4^{2-}. In (c), there will be 2 OH^- ions for every Ba^{2+}.
15. Using Table 20.2:
    a. $CaSO_4$, $K_{sp} = 3 \times 10^{-5}$, is most soluble.

    b. HgS, $K_{sp} = 1 \times 10^{-52}$, is least soluble.
16. Compounds (b), $BaCO_3$, and (d), $Mg(OH)_2$, dissolve in strong acid but not in water. The reactions are:

$$BaCO_3(s) + 2\ H^+(aq) \rightarrow Ba^{2+}(aq) + CO_2(g) + H_2O$$

$$Mg(OH)_2(s) + 2\ H^+(aq) \rightarrow Mg^{2+}(aq) + 2\ H_2O$$

17. When a strong acid is added to $CuCO_3$, the Cu^{2+} ion and the H^+ ion compete for the CO_3^{2-} ion. The H^+ ion wins.
18. When ammonia is added to $Cu(OH)_2(s)$, there is a competition for the Cu^{2+} ion between the OH^- ion and the NH_3 molecule. (The NH_3 wins.)
19. $Zn(OH)_2$ is soluble in 6 M HCl, NaOH, and NH_3. The reactions are:

$$Zn(OH)_2(s) + 2\ H^+(aq) \rightarrow Zn^{2+}(aq) + 2\ H_2O$$

$$Zn(OH)_2(s) + 2\ OH^-(aq) \rightarrow Zn(OH)_4^{2-}(aq)$$

$$Zn(OH)_2(s) + 4\ NH_3(aq) \rightarrow Zn(NH_3)_4^{2+}(aq) + 2\ OH^-(aq)$$

20. In general, a complex ion is one which contains more than one atom. In a more restricted sense, a complex ion consists of a metal ion which is covalently bonded to neutral molecules and/or negative ions. Examples of complex ions are $Ag(NH_3)_2^+$ and $Zn(OH)_4^{2-}$. The complexing agent is the species which reacts with the metal ion to form the complex ion. The complexing agents for the complex ions shown above are NH_3 and the OH^- ion.

VI. Solutions to Problems

1. a. $AgNO_3$: Ag^+ and NO_3^- Na_2S: Na^+ and S^{2-}

   Possible precipitates are Ag_2S and $NaNO_3$.

   b. $CuSO_4$: Cu^{2+} and SO_4^{2-} $BaCl_2$: Ba^{2+} and Cl^-

   Possible precipitates are $CuCl_2$ and $BaSO_4$.

   c. $NiCl_2$: Ni^{2+} and Cl^- $Ca(OH)_2$: Ca^{2+} and OH^-

   Possible precipitates are $Ni(OH)_2$ and $CaCl_2$.

   d. $Cu(NO_3)_2$: Cu^{2+} and NO_3^- $ZnCl_2$: Zn^{2+} and Cl^-

   Possible precipitates are $CuCl_2$ and $Zn(NO_3)_2$

2. a. $2\ Ag^+(aq) + S^{2-}(aq) \rightarrow Ag_2S(s)$

   b. $Ba^{2+}(aq) + SO_4^{2-}(aq) \rightarrow BaSO_4(s)$

   c. $Ni^{2+}(aq) + 2\ OH^-(aq) \rightarrow Ni(OH)_2(s)$

   d. No reaction

3. $K_{sp} = [Ag^+][Br^-] = 1 \times 10^{-13}$

   a. $[Ag^+] = \dfrac{K_{sp}}{[Br^-]} = \dfrac{1 \times 10^{-13}}{5 \times 10^{-4}} = 2 \times 10^{-10}\ M$

   b. $[Br^-] = \dfrac{K_{sp}}{[Ag^+]} = \dfrac{1 \times 10^{-13}}{0.10} = 1 \times 10^{-12}\ M$

4. a. $K_{sp} = [Ag^+][I^-] = 1 \times 10^{-16}$

   $[Ag^+] = [I^-] = \text{solubility of AgI} = s$

   $s^2 = 1 \times 10^{-16}$

   $s = 1 \times 10^{-8}\ M$

b. $K_{sp} = [Mn^{2+}][S^{2-}] = 1 \times 10^{-13}$

$[Mn^{2+}] = [S^{2-}] = \text{solubility of MnS} = s$

$s^2 = 1 \times 10^{-13} = 10 \times 10^{-14}$

$s = 3 \times 10^{-7}$ M

5. a. $K_{sp} = [Pb^{2+}][Cl^-]^2$

b. $K_{sp} = [Cr^{3+}][OH^-]^3$

c. $K_{sp} = [As^{3+}]^2[S^{2-}]^3$

d. $K_{sp} = [Ba^{2+}][CrO_4^{2-}]$

6. a. $Zn(OH)_2(s) + 2\ H^+(aq) \rightarrow Zn^{2+}(aq) + 2\ H_2O$

b. $AgOH(s) + H^+(aq) \rightarrow Ag^+(aq) + H_2O$

c. $Al(OH)_3(s) + 3\ H^+(aq) \rightarrow Al^{3+}(aq) + 3\ H_2O$

7. $BaCO_3(s) + 2\ H^+(aq) \rightarrow Ba^{2+}(aq) + CO_2(g) + H_2O$

8. a. $Cd(OH)_2(s) + 4\ NH_3(aq) \rightarrow Cd(NH_3)_4^{2+}(aq) + 2\ OH^-(aq)$

b. $CdCO_3(s) + 4\ NH_3(aq) \rightarrow Cd(NH_3)_4^{2+}(aq) + CO_3^{2-}(aq)$

c. $AgBr(s) + 2\ NH_3(aq) \rightarrow Ag(NH_3)_2^+(aq) + Br^-(aq)$

9. $Cr(OH)_3(s) + 3\ OH^-(aq) \rightarrow Cr(OH)_6^{3-}(aq)$

10. $K_{sp} = [Ba^{2+}][SO_4^{2-}]$

$K_{sp} = (5.0 \times 10^{-4})(3.0 \times 10^{-6}) = 1.5 \times 10^{-9}$

11. a. $Ba^{2+}(aq) + SO_4^{2-}(aq) \rightarrow BaSO_4(s)$

b. $Fe^{2+}(aq) + 2\ OH^-(aq) \rightarrow Fe(OH)_2(s)$

c. $Cr^{3+}(aq) + 3\ OH^-(aq) \rightarrow Cr(OH)_3(s)$

d. $Ag^+(aq) + Cl^-(aq) \rightarrow AgCl(s)$

e. No reaction

12. $Fe(OH)_2(s) + 2\ H^+(aq) \rightarrow Fe^{2+}(aq) + 2\ H_2O$

$Cr(OH)_3(s) + 3\ H^+(aq) \rightarrow Cr^{3+}(aq) + 3\ H_2O$

13. $AgCl(s) + 2\ NH_3(aq) \rightarrow Ag(NH_3)_2^+(aq) + Cl^-(aq)$

14. $Cr(OH)_3(s) + 3\ OH^-(aq) \rightarrow Cr(OH)_6^{3-}(aq)$

15. $K_{sp} = [Mg^{2+}][OH^-]^2$

$K_{sp} = (1 \times 10^{-3})(1 \times 10^{-4})^2 = 1 \times 10^{-11}$

16. Iron (III) sulfide: Fe_2S_3

$K_{sp} = [Fe^{3+}]^2[S^{2-}]^3$

17. a. $K_{sp} = [Cd^{2+}][S^{2-}] = 1 \times 10^{-27}$

$[Cd^{2+}] = [S^{2-}] = \text{solubility} = s$

$s^2 = 1 \times 10^{-27} = 10 \times 10^{-28}$

$s = 3 \times 10^{-14}$ M

b. $K_{sp} = [Zn^{2+}][S^{2-}] = 1 \times 10^{-23}$

$[Zn^{2+}] = [S^{2-}] = \text{solubility} = s$

$s^2 = 1 \times 10^{-23}$ or 10×10^{-24}

$s = 3 \times 10^{-12}$ M

c. $K_{sp} = [Ca^{2+}][SO_4^{2-}] = 3 \times 10^{-5}$

$[Ca^{2+}] = [SO_4^{2-}] = \text{solubility} = s$

$s^2 = 3 \times 10^{-5}$ or 30×10^{-6}

$s = 5 \times 10^{-3}$ M

18. $K_{sp} = [Pb^{2+}][S^{2-}] = 1 \times 10^{-28}$

$$[Pb^{2+}] = \frac{K_{sp}}{[S^{2-}]} = \frac{1 \times 10^{-28}}{1 \times 10^{-4}} = 1 \times 10^{-24}\ M$$

19. $K_{sp} = [Pb^{2+}][CrO_4^{2-}] = 2 \times 10^{-14}$

$$[CrO_4^{2-}] = \frac{K_{sp}}{[Pb^{2+}]} = \frac{2 \times 10^{-14}}{5 \times 10^{-7}} = 4 \times 10^{-8}\ M$$

20. $K_{sp} = [Hg^{2+}][S^{2-}] = 1 \times 10^{-52}$

$[Hg^{2+}] = [S^{2-}] = \text{solubility} = s$

$s^2 = 1 \times 10^{-52}$

$s = 1 \times 10^{-26}$ M

$$\text{solubility } (g/\ell) = \frac{1 \times 10^{-26} \text{ mol}}{\ell} \times \frac{232 \text{ g HgS}}{\text{mol}} = 2 \times 10^{-24} \text{ g}/\ell$$

21. a. $CoCO_3(s) + 2\ H^+(aq) \rightarrow Co^{2+}(aq) + CO_2(g) + H_2O$

b. $Al_2(CO_3)_3(s) + 6\ H^+(aq) \rightarrow 2\ Al^{3+}(aq) + 3\ CO_2(g) + 3\ H_2O$

22. a. $$V_{(seawater)} = 1.00 \text{ mol Mg} \times \frac{1\ \ell}{0.0520 \text{ mol Mg}} = 19.2\ \ell$$

b. $$V_{(seawater)} = 1.00 \times 10^3 \text{ g Mg} \times \frac{1 \text{ mol Mg}}{24.3 \text{ g Mg}} \times \frac{1\ \ell}{0.0520 \text{ mol Mg}} = 791\ \ell$$

23. $PbCl_2(s) \rightleftarrows Pb^{2+}(aq) + 2\ Cl^-(aq)$

$K_{sp} = [Pb^{2+}][Cl^-]^2$

As solubility = moles $PbCl_2$ which dissolve in 1 liter:

$[Pb^{2+}] = s$ and $[Cl^-] = 2s$

Then, $K_{sp} = (s)(2s)^2 = 4s^3$

24. $K_{sp} = [Ca^{2+}][SO_4^{2-}] = 3 \times 10^{-5}$

(conc. Ca^{2+})(conc. SO_4^{2-}) $= (2 \times 10^{-3})(1 \times 10^{-2}) = 2 \times 10^{-5}$

As $2 \times 10^{-5} < K_{sp}$ of 3×10^{-5}, a precipitate of $CaSO_4$ will not form.

25. $K_{sp} = [Ag^+][Br^-] = 5 \times 10^{-13}$

$[Ag^+] = [Br^-] = \text{solubility} = s$

$s^2 = 5 \times 10^{-13}$ or 50×10^{-14}

$s = 7 \times 10^{-7}$ M

$$\text{mass AgBr} = 0.1\ \ell \times \frac{7 \times 10^{-7} \text{ mol AgBr}}{1\ \ell} \times \frac{188 \text{ g AgBr}}{1 \text{ mol AgBr}} = 1 \times 10^{-5} \text{ g}$$

CHAPTER 21

Quantitative Analysis

I. Basic Skills

A. Qualitative Students should be able to:

1. Explain what is meant by quantitative analysis.
2. Describe the technique of gravimetric analysis and how it is used to determine the composition of a sample.
3. a. Describe the technique of volumetric analysis.
   b. Explain how an acid-base titration is performed.
   c. Explain the necessary requirements for the volumetric analysis of a reaction.
4. Explain what is meant by instrumental analysis and list some of its advantages.
5. Explain how a colorimeter is used to determine the concentration of a colored species in solution.

B. Quantitative Students should be able to:

1. Determine the percentage composition of a sample using gravimetric data (Examples 21.1 and 21.2).
2. Determine the molarity of an acid or base using volumetric data (Example 21.3).
3. Determine the percentage composition of a sample using gravimetric and volumetric data (Examples 21.4 and 21.5).
4. a. Determine the absorbance constant graphically from colorimetric data.
   b. Determine the concentration of a solute using colorimetric data (Example 21.6).
5. Determine by calculation how to dilute solutions to a particular concentration (Review Chapter 13).

II. Chapter Development

1. For the most part, the chapter deals with analytical applications of principles which have been covered previously. Precipitation reactions and acid-base reactions are the basis of the gravimetric and volumetric techniques discussed. Considerable recall from Chapter 13 is required on the use of molarity in calculations.
2. Under instrumental methods, only visible spectroscopy is discussed. Nonvisible techniques (IR, UV, and NMR) are not included. Students are not likely to have access to these instruments.

3. The chemistry of indicators is not discussed. For this reason, students cannot be expected to know how to choose the proper indicator for a particular acid-base titration. Indicators are covered in the lab in Experiments 28 and 29.
4. The normality system of expressing concentrations in acid-base calculations is not used.

III. Problem Areas

1. Many of the problem solutions of this chapter involve going through a series of steps. This bothers those students who want to dive in and solve everything in a single step. Work carefully on the board (or overhead) to show how a solution is organized in steps to go from the starting information to the final answer.
2. In gravimetric or volumetric problems have students always begin by writing the balanced chemical equation for the reaction which takes place. Otherwise, they do not know the molar reaction ratio or they assume that it is 1:1.

IV. Suggested Activities

1. Titrations:
   Demonstrate an acid-base titration showing the correct technique of reading and using a buret. Demonstrate an end-point using phenolphthalein indicator. Discuss why the pH of a weak acid-strong base or strong acid-weak base neutralization is not 7. Explain how this affects the choice of an indicator.
2. Colorimetry:
   a. Demonstrate the effect of concentration and depth on the color of solutions. Show equal depths of 1 M and 0.5 M $K_2Cr_2O_7$ solutions in flat-bottom vials. (Any colored solution, including tea, may be used.) Have students compare the intensities of the colors by looking down through the solutions. Pour out exactly half of the more concentrated solution and have students compare colors again. Develop the idea that the color of a solution is related to its depth (path length of light) as well as to its concentration. The technique of adjusting depth until colors appear to be the same will be used in Experiment 39.
   b. If available, demonstrate the use of a colorimeter. Explain the importance of using light of a particular wavelength.

V. Answers to Questions

1. a. volumetric analysis b. gravimetric analysis
   c. instrumental analysis d. volumetric analysis

2. The HNO_3 is added first to dissolve the coin metals. The HCl is added next to precipitate the Ag^+ ion as AgCl.
3. The percentage of silver in the coin will be:
   a. low b. low c. high
4. The sulfate product must have a low solubility. Suitable products would be (b) $BaSO_4$ and (c) $PbSO_4$.
5. a. To determine the % Cl^- in a solid:
   Dissolve a weighed amount of the compound in water. Add $AgNO_3$ solution until precipitation is complete. To find the mass of chlorine, note that 143.32 g AgCl $\triangleq$ 35.45 g Cl.
   b. To determine the % C in an organic compound:
   Burn a weighed sample of the compound in excess oxygen. The products will be CO_2 and H_2O. Collect and weigh the CO_2 produced in a CO_2 absorber. To find the mass of carbon, note that 44.01 g CO_2 $\triangleq$ 12.01 g C.
6. An acid-base titration is a volumetric analysis method used to determine the concentration of an acidic or basic solution or to determine the percentage of an acid or base in a solid mixture. A solution of an acid or base of known concentration is added to the opposite compound until neutralization occurs. An indicator is used to give a visual indication of when the reaction is complete.
7. As NaOH solution is added to an HCl solution, the concentration of H^+ ions decreases and the pH increases.
8. Any water in the buret would decrease the concentration of the NaOH solution. (See Question 10c.)
9. Phenolphthalein is the indicator of choice when titrating acids with a strong base. A pink color will appear where the base contacts the acid as the end point is approached. The color will disappear when the solution is swirled and will remain at the end point.
10. a. Less HCl will be present, so less NaOH will be used in the titration. This will make the concentration of NaOH appear to be too high.
    b. The student will not obtain any results as an end point cannot be reached without an indicator.
    c. The NaOH solution is diluted by water in the buret. As more NaOH solution will be required to neutralize the acid, the calculated NaOH concentration will be low.
    d. If the end point is passed by adding too much NaOH, the calculated concentration of NaOH will be low.
11. Statements (a) and (d) are true. Statements (b) and (c) are false because 1 mole of $Ca(OH)_2$ is equivalent to 2 moles of OH^- ions which will neutralize 2 moles of H^+ ions.
12. Using Table 21.1 to match wavelength with color:

    a. 350 nm, colorless (UV) b. 500 nm, green

    c. 600 nm, orange d. 750 nm, colorless (IR)
13. The light absorbed at 400–450 nm is violet while that absorbed at 580–700 nm is orange-red. The color of a solution is determined by the light which is transmitted, blue, green, and yellow. The mixture of these colors will appear green to the eye of an observer.

14. a. Calculate the moles of CrO_4^{2-} ion required in 30 cm^3 of the dilute solution:

$$\text{moles } CrO_4^{2-} = 0.0300\ \ell \times 1.00 \times 10^{-4}\ \frac{\text{mol } CrO_4^{2-}}{1\ \ell} = 3.00 \times 10^{-6}$$

b. Calculate the volume of the concentrated solution required:

$$\text{volume} = 3.00 \times 10^{-6}\ \text{mol } CrO_4^{2-} \times \frac{1\ \ell}{3.00 \times 10^{-4}\ \text{mol } CrO_4^{2-}}$$
$$= 1.00 \times 10^{-2}\ \ell \text{ or } 10.0\ cm^3$$

c. Measure out 10.0 cm^3 of the concentrated solution and add water to a solution volume of 30.0 cm^3.

15. The concentration, c, of a colored solution is directly proportional to the absorbance, A. Equations (a), $c = k_1 A$, and (c), $A = k_3 c$, are in agreement with this principle. Equations (b) and (d) would require that A and c be inversely proportional.
16. If the wavelength is changed, the value of the constant, k, in the equation $c = kA$ would change.
17. A CrO_4^{2-} ion solution of about 0.1 M should be diluted to about 10^{-4} M. One cm of the original solution when diluted to a volume of 1000 cm^3 would have a concentration of 1/1000 of its original value. The absorbance of the dilute solution can then be measured with a colorimeter or spectrophotometer. Its concentration is calculated by using the equation $c = kA$. The original solution will have a concentration 1000 times that which is determined by the instrument.

VI. Solutions to Problems

1. 1 mole AgCl $\triangleq$ 1 mole Ag

143.32 g AgCl $\triangleq$ 107.87 g Ag

$$\text{mass Ag} = 1.271\text{ g AgCl} \times \frac{107.87\text{ g Ag}}{143.32\text{ g AgCl}} = 0.9566\text{ g}$$

$$\%\text{ Ag} = \frac{0.9566\text{ g Ag}}{1.063\text{ g sample}} \times 100 = 89.99\%$$

2. 1 mole AgCl $\triangleq$ 1 mole KCl

143.32 g AgCl $\triangleq$ 74.55 g KCl

$$\text{mass KCl} = 1.250\text{ g AgCl} \times \frac{74.55\text{ g KCl}}{143.32\text{ g AgCl}} = 0.6502\text{ g}$$

$$\%\text{ KCl} = \frac{0.6502\text{ g KCl}}{1.000\text{ g sample}} \times 100 = 65.02\%$$

3. moles HCl(A) = moles KOH(B)

$$V_A M_A = V_B M_B$$

$$M_B = \frac{V_A M_A}{V_B} = 0.04360\ \ell \times \frac{0.1050\ \text{mol}/\ell}{0.02840\ \ell}$$

$$M_B = 0.1612\ \text{mol}/\ell$$

4. moles NaOH(B) = $V_B \times M_B$

$$\text{moles NaOH} = 0.0400\ \ell \times 0.150\ \frac{\text{mol}}{\ell} = 6.00 \times 10^{-3}$$

$$\text{moles } C_6H_5COOH(A) = \text{moles NaOH(B)} = 6.00 \times 10^{-3}$$

$$\text{mass } C_6H_5COOH = 6.00 \times 10^{-3}\ \text{mol A} \times \frac{122\ \text{g A}}{1\ \text{mol A}} = 0.732\ \text{g}$$

$$\%\ C_6H_5COOH = \frac{0.732\ \text{g A}}{1.202\ \text{g sample}} \times 100 = 60.9$$

5. moles Ag^+ = $V(AgNO_3) \times M(AgNO_3)$

$$\text{moles } Ag^+ = 0.0262\ \ell \times 0.200\ \frac{\text{mol}}{\ell} = 5.24 \times 10^{-3}$$

$$\text{moles } Cl^- = \text{moles } Ag^+ = 5.24 \times 10^{-3}$$

$$1\ \text{mole } Cl^- \triangleq 1\ \text{mole } NH_4Cl \triangleq 53.5\ \text{g } NH_4Cl$$

$$\text{g } NH_4Cl = 5.24 \times 10^{-3}\ \text{mol } Cl^- \times \frac{53.5\ \text{g } NH_4Cl}{1\ \text{mol } Cl^-} = 0.280\ \text{g}$$

$$\%\ NH_4Cl = \frac{0.280\ \text{g } NH_4Cl}{1.542\ \text{g sample}} \times 100 = 18.2$$

6. c = kA

$$k = \frac{c}{A} = \frac{1.00 \times 10^{-4}\ M}{1.12} = 8.93 \times 10^{-5}\ M$$

The concentration of an unknown CrO_4^{2-} solution is:

$$c = kA = 8.93 \times 10^{-5}\ M \times 0.42 = 3.8 \times 10^{-5}\ M$$

7. As Zn is soluble in HCl and Cu is insoluble, the residue of the reaction is Cu.

$$\% \text{ Cu} = \frac{0.414 \text{ g Cu}}{0.642 \text{ g alloy}} \times 100 = 64.5$$

$$\% \text{ Zn} = 100.0 - 64.5 = 35.5$$

8. 1 mol AgBr ≏ 1 mol KBr

   187.8 g AgBr ≏ 119.0 g KBr

   $$\text{mass KBr} = 1.261 \text{ g AgBr} \times \frac{119.0 \text{ g KBr}}{187.8 \text{ g AgBr}} = 0.7990 \text{ g}$$

   $$\% \text{ KBr} = \frac{0.7990 \text{ g KBr}}{1.602 \text{ g sample}} \times 100 = 49.88$$

9. moles $Ag^+ = V_{AgNO_3} \times M_{AgNO_3}$

   $$\text{moles } Ag^+ = 0.0318 \ \ell \times \frac{0.200 \text{ mol}}{\ell} = 6.36 \times 10^{-3}$$

   moles Ag^+ ≏ moles I^- ≏ moles KI $= 6.36 \times 10^{-3}$

   $$\text{mass KI} = 6.36 \times 10^{-3} \text{ mol KI} \times \frac{166 \text{ g KI}}{1 \text{ mol KI}} = 1.06 \text{ g}$$

   $$\% \text{ KI} = \frac{1.06 \text{ g KI}}{1.528 \text{ g sample}} \times 100 = 69.4$$

10. 1 mole $BaSO_4$ ≏ 1 mole $MgSO_4 \cdot 7\ H_2O$ (Epsom salts)

    233.4 g $BaSO_4$ ≏ 246.5 g $MgSO_4 \cdot 7\ H_2O$

    $$\text{mass } BaSO_4 = 1.000 \text{ g Es} \times \frac{233.4 \text{ g } BaSO_4}{246.5 \text{ g Es}} = 0.9469$$

11. 1 mole $MgSO_4$ ≏ 1 mole $MgSO_4 \cdot 7\ H_2O$ (Epsom salts)

    120.4 g $MgSO_4$ ≏ 246.5 g $MgSO_4 \cdot 7\ H_2O$

    $$\text{mass } MgSO_4 = 1.444 \text{ g Es} \times \frac{120.4 \text{ g } MgSO_4}{246.5 \text{ g Es}} = 0.7053 \text{ g}$$

12. moles HCl $= V_A \times M_A$

    $$\text{moles HCl} = 0.01545 \ \ell \times \frac{0.1000 \text{ mol}}{\ell} = 1.545 \times 10^{-3}$$

    1 mole $Ca(OH)_2$ ≏ 2 mol HCl

$$\text{moles Ca(OH)}_2 = 1.545 \times 10^{-3} \text{ mol HCl} \times \frac{1 \text{ mol Ca(OH)}_2}{2 \text{ mol HCl}}$$
$$= 7.725 \times 10^{-4}$$

$$\text{M of Ca(OH)}_2 = \frac{7.725 \times 10^{-4} \text{ mol}}{0.02000\ \ell} = 0.03863 \text{ mol}/\ell$$

13. $\text{moles HCl} = V_A \times M_A$

$$\text{moles HCl} = 0.0260\ \ell \times \frac{0.250 \text{ mol}}{\ell} = 6.50 \times 10^{-3}$$

$1 \text{ mole HCl} \simeq 1 \text{ mole NaHCO}_3$

$1 \text{ mole HCl} \simeq 84.0 \text{ g NaHCO}_3$

$$\text{mass NaHCO}_3 = 6.50 \times 10^{-3} \text{ mol HCl} \times \frac{84.0 \text{ g NaHCO}_3}{1 \text{ mol HCl}} = 0.546 \text{ g}$$

$$\% \text{ NaHCO}_3 = \frac{0.546 \text{ g NaHCO}_3}{1.24 \text{ g sample}} \times 100 = 44.0$$

14. $1 \text{ mole HCl(A)} \simeq 1 \text{ mole NH}_3\text{(B)}$

$V_AM_A = V_BM_B$

$$M_B = \frac{V_AM_A}{V_B} = 0.03162\ \ell \times \frac{0.1098 \text{ mol}/\ell}{0.02416\ \ell} = 0.1437 \text{ mol}/\ell$$

15.

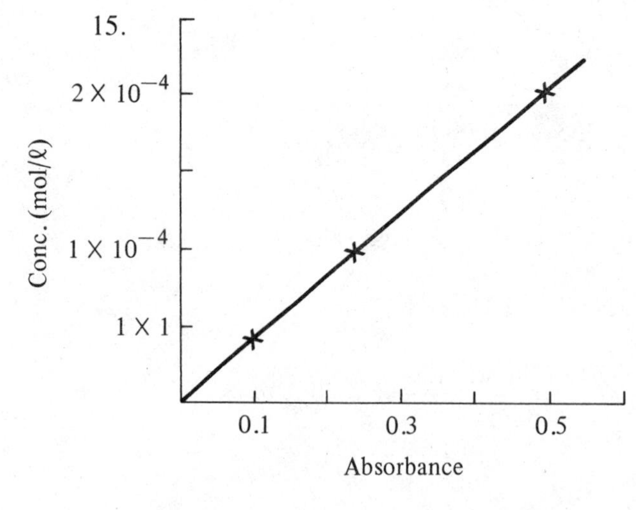

$$\text{slope} = k = \frac{\Delta c}{\Delta A}$$

$$k = \frac{(2.00 \times 10^{-4} - 0.40 \times 10^{-4})}{(0.50 - 0.10)}$$

$$k = \frac{1.60 \times 10^{-4}}{0.40}$$

$$k = 4.0 \times 10^{-4}$$

16. $k = 4.0 \times 10^{-4}$ M (See Problem 15 graph.)

$c = kA$

$c = 4.0 \times 10^{-4} \text{ M } (0.62) = 2.5 \times 10^{-4} \text{ M}$

17. $k = \frac{c}{A} = \frac{1.00 \times 10^{-4}\ M}{0.40} = 2.5 \times 10^{-4}\ M$

c (dilute soln) $= kA = 2.5 \times 10^{-4}\ M\ (1.00) = 2.5 \times 10^{-4}\ M$

c (original soln) $= \frac{100.0\ cm^3}{2.0\ cm^3}(2.5 \times 10^{-4}\ M) = 1.2 \times 10^{-2}\ M$

18. moles $Pb^{2+} = V_{Pb(NO_3)_2} \times M_{Pb(NO_3)_2}$

moles $Pb^{2+} = 0.0500\ \ell \times \frac{0.100\ mol}{\ell} = 5.00 \times 10^{-3}$

$1\ mol\ Pb^{2+} \triangleq 2\ mol\ Cl^-$

$3\ mol\ Cl^- \triangleq 1\ mol\ AlCl_3 \triangleq 133.5\ g\ AlCl_3$

$$\text{mass } AlCl_3 = 5.00 \times 10^{-3}\ mol\ Pb^{2+} \times \frac{2\ mol\ Cl^-}{1\ mol\ Pb^{2+}} \times \frac{133.5\ g\ AlCl_3}{3\ mol\ Cl^-}$$
$$= 0.445\ g$$

$$\%\ AlCl_3 = \frac{0.445\ g\ AlCl_3}{1.400\ g\ sample} \times 100 = 31.8$$

19. moles $Cl^- \triangleq$ moles $AgCl \triangleq 143.32\ g\ AgCl$

FM NaCl = 58.44 FM KCl = 74.55

Let X = g of NaCl, then 1.000 − X = g of KCl

$$X\left(\frac{143.32\ g\ AgCl}{58.44\ g\ NaCl}\right) + (1.000 - X)\left(\frac{143.32\ g\ AgCl}{74.55\ g\ KCl}\right) = 2.100\ g\ AgCl$$

X = 0.336 or 33.6% NaCl

CHAPTER 22

Qualitative Analysis

I. Basic Skills

A. Qualitative Students should be able to:

1. Define qualitative analysis and explain in general terms how it is carried out.
2. Describe how a solubility table is constructed and how it is used (Table 22.1).
3. Explain what is meant by a flow chart.
4. Explain how a general unknown is analyzed by separation into groups (Table 22.2).

B. Quantitative Students should be able to:

1. a. Use solubility data to predict the ion pairs which are insoluble (Table 22.1).
   b. Use solubility data to predict the insoluble compounds which can be dissolved by acids or complexing agents (Table 22.1).
2. Write balanced chemical equations for the forming of precipitates and the dissolving of precipitates (Examples 22.1–22.3).
3. Use solubility data to analyze an unknown containing:
   a. a single positive ion (Examples 22.4–22.6).
   b. more than one different positive ion (Examples 22.7–22.8).
4. Use the data obtained from the standard procedure for a general unknown to analyze for the positive ions of Group I (Example 22.9).
5. Draw a flow chart showing how a mixture of ions can be separated.

II. Chapter Development

1. A representative mixture of positive ions (Groups 1, 2, 3, and transition elements) is used to show the general approach to qualitative analysis. Separation schemes are developed based on chemical properties of the ions. The needed principles on the forming and dissolving of precipitates were covered previously in Chapter 20.
2. A general qualitative analysis scheme is introduced to show how a large number of ions can be detected. It is similar to most schemes used in college chemistry. For our purposes, only the reactions of the Group I ions (Ag^+, Hg_2^{2+}, and Pb^{2+}) are given in detail. A Group I analysis is carried out in Experiment 35.

III. Problem Areas

1. Problems develop when students must consider the possible presence of more than one positive ion. The need to separate ions individually must be emphasized. Have students draw flow charts which show the negative ions added and the resulting separation of the positive ions.
2. The writing of net ionic equations is required again in this chapter. While there are many examples for students to learn from, you may have to bear down on this point once again. The reaction of Equation 22.23 (Hg_2Cl_2 and NH_3) is an oxidation-reduction reaction (Chapter 23). Do not expect students to recall a reaction of this difficulty.

IV. Suggested Activities

1. Flame Tests:
   If the technique of flame testing was not shown during Chapter 8, demonstrate flame tests of Na^+ and K^+ solutions. (Students will use this technique in the analysis of Group 2 ions in Experiment 34.) Explain that the emitted light is caused by electronic transitions whose ΔE is in the visible part of the spectrum.
2. Separation of Ions:
   a. Make solutions of the ions in Table 22.1. Have students develop a separation scheme for two or three of the ions in the table. When a suitable scheme is agreed upon, carry out the reactions and make separations using a centrifuge. (Choose three-ion combinations carefully as some are not readily separated.)
   b. Have students construct a solubility table similar to Table 22.1 for a different set of ions. (Use the Solubility Rules given in Table 20.1 of Chapter 20.) Have students develop a separation scheme for two or three of the ions in the table.

V. Answers to Questions

1. a. NO_3^- b. Cl^- c. OH^- and S^{2-} d. S^{2-} e. all
2. a. Ag^+ b. Cu^{2+} and Ni^{2+} c. Na^+ and K^+
3. a. Soluble in 6 M H^+: $Cu(OH)_2$, $Zn(OH)_2$, $Ni(OH)_2$, $Al(OH)_3$, and $Mg(OH)_2$
   b. Insoluble in 6 M OH^-: $Cu(OH)_2$, $Ni(OH)_2$, and $Mg(OH)_2$
   c. Insoluble in 6 M NH_3: $Al(OH)_3$, $Mg(OH)_2$
   d. Insoluble in either 6 M OH^- or 6 M NH_3: $Mg(OH)_2$
4. a. ZnS: 6 M H^+ b. $Zn(OH)_2$: 6 M H^+, 6 M OH^- and 6 M NH_3
   c. $Ni(OH)_2$: 6 M H^+ and 6 M NH_3 d. CuS: insoluble in all
5. The complexing agents are (b) NH_3 and (c) OH^-.
6. When an NaOH solution is added dropwise to an Al^{3+} solution, the first reaction is the formation of a white $Al(OH)_3$ precipitate. When excess NaOH is

added, the precipitate dissolves by forming the $Al(OH)_4^-$ ion.

7. a. When NH_3 is added to a Mg^{2+} ion solution, a precipitate of $Mg(OH)_2$ forms.

   $$Mg^{2+}(aq) + 2\ NH_3(aq) + 2\ H_2O \rightarrow Mg(OH)_2(s) + 2\ NH_4^+$$

   b. The OH^- ions are produced by the reaction of NH_3 with water:

   $$NH_3(aq) + H_2O \rightleftarrows NH_4^+(aq) + OH^-(aq)$$

8. When NH_3 is added to a Zn^{2+} ion solution, a precipitate of $Zn(OH)_2$ forms:

   $$Zn^{2+}(aq) + 2\ NH_3(aq) + 2\ H_2O \rightarrow Zn(OH)_2(s) + 2\ NH_4^+\ (aq)$$

   The OH^- ions are produced by the reaction of NH_3 with water as shown in Question 7b. With excess NH_3, the precipitate dissolves by forming a complex ion:

   $$Zn(OH)_2(s) + 4\ NH_3(aq) \rightarrow Zn(NH_3)_4^{2+}(aq) + 2\ OH^-(aq)$$

9. a. If KCl is added, instead of HCl, the later test for the K^+ ion will not be valid.
   b. If NaOH is added, instead of NH_3, the $Al(OH)_3$ precipitate may dissolve by forming $Al(OH)_4^-$.
   c. The K^+ ion cannot be precipitated because potassium compounds are soluble.
10. Look at the solution; if the solution is blue, Cu^{2+} ions are present. Add a few drops of 6 M HCl; a white precipitate (AgCl) indicates the presence of Ag^+ ions.

11. $Zn(OH)_4^{2-}(aq) + 4\ H^+(aq) \rightarrow Zn^{2+}(aq) + 4\ H_2O$

12. Add an acid to remove the NH_3 from the complex ion:

    $$Cu(NH_3)_4^{2+}(aq) + 4\ H^+(aq) \rightarrow Cu^{2+}(aq) + NH_4^+(aq)$$

13. The copper and zinc metal in the alloy are converted to Cu^{2+} and Zn^{2+} ions by the nitric acid. The Cu^{2+} ion is in Group II and the Zn^{2+} ion is in Group III.
14. a. Groups I, II and III contain transition metal ions.
    b. Groups I and II contain mercury ions.
    c. Group IV contains the ions of the Group 1 and Group 2 metals of the Periodic Table.

15. a. $Cu^{2+}(aq) + H_2S(aq) \rightarrow CuS(s) + 2\ H^+(aq)$

    b. $2\ Bi^{3+}(aq) + 3\ H_2S(aq) \rightarrow Bi_2S_3(s) + 6\ H^+(aq)$

    c. $Mn^{2+}(aq) + H_2S(aq) \rightarrow MnS(s) + 2\ H^+(aq)$

d. $2\ Sb^{3+}(aq) + 3\ H_2S(aq) \rightarrow Sb_2S_3(s) + 6\ H^+(aq)$

16. a. $Cr^{3+}(aq) + 3\ OH^-(aq) \rightarrow Cr(OH)_3(s)$

    b. $Al^{3+}(aq) + 3\ OH^-(aq) \rightarrow Al(OH)_3(s)$

17. If the H_2S is not adjusted to a low pH, the S^{2-} ion concentration will be large enough to precipitate the sulfides of Group III.
18. A separate sample of the unknown is tested for the NH_4^+ ion by adding a OH^- ion solution. When the solution is warmed, NH_3 gas is given off which can be detected by moistened red litmus paper:

$$NH_4^+(aq) + OH^-(aq) \rightarrow NH_3(g) + H_2O$$

19. The Group I ions all form insoluble chlorides.
20. a. none b. $PbCl_2$ c. $AgCl$
21. a. $Hg_2Cl_2(s) + 2\ NH_3(aq) \rightarrow HgNH_2Cl(s) + Hg(l) + NH_4^+(aq) + Cl^-(aq)$

    b. $Pb^{2+}(aq) + CrO_4^{2-}(aq) \rightarrow PbCrO_4(s)$

VI. Solutions to Problems

1. a. $Zn^{2+}(aq) + 2\ OH^-(aq) \rightarrow Zn(OH)_2(s)$

   $Zn^{2+}(aq) + S^{2-}(aq) \rightarrow ZnS(s)$

   b. $Zn(OH)_2(s) + 2\ H^+(aq) \rightarrow Zn^{2+}(aq) + 2\ H_2O$

   $ZnS(s) + 2\ H^+(aq) \rightarrow Zn^{2+}(aq) + H_2S(g)$

2. a. $Mg^{2+}(aq) + 2\ OH^-(aq) \rightarrow Mg(OH)_2(s)$

   b. $Zn^{2+}(aq) + 2\ OH^-(aq) \rightarrow Zn(OH)_2(s)$

   With excess OH^-:

   $Zn(OH)_2(s) + 2\ OH^-(aq) \rightarrow Zn(OH)_4^{2-}(aq)$

3. $Cu^{2+}(aq) + 2\ NH_3(aq) + 2\ H_2O \rightarrow Cu(OH)_2(s) + 2\ NH_4^+(aq)$

   $Cu(OH)_2(s) + 4\ NH_3(aq) \rightarrow Cu(NH_3)_4^{2+}(aq) + 2\ OH^-(aq)$

4. A red precipitate with dimethylglyoxime proves the presence of the Ni^{2+} ion.
5. A Table 22.1 hydroxide which is soluble in both 6 M NH_3 and 6 M NaOH must be $Zn(OH)_2$. The positive ion of the original solution is Zn^{2+}.
6. a. A Table 22.1 ion which does not form a precipitate with Cl^- or OH^- solutions must be Na^+ or K^+.

b. The presence of Na^+ or K^+ can be shown by a flame test: Na^+, yellow; K^+, violet.

7. A precipitate when HCl is added indicates the presence of Ag^+.
No precipitate when NH_3 is added indicates the absence of Al^{3+}.
The K^+ ion is undetermined as no flame test has been made.

8.

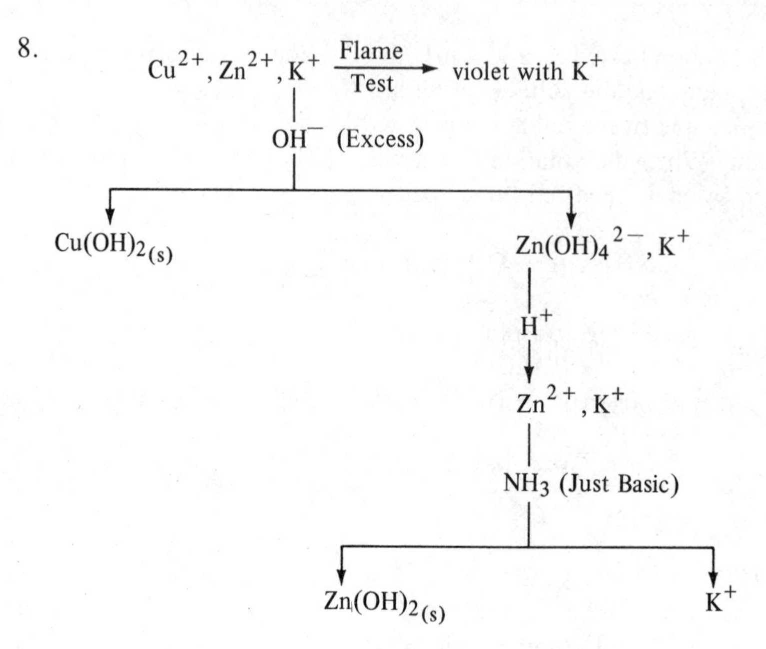

9. A yellow precipitate when CrO_4^{2-} is added to the solution obtained from the hot water test indicates the presence of the Pb^{2+} ion. The complete dissolving of the residue in NH_3 indicates the presence of the Ag^+ ion and the absence of the Hg_2^{2+} ion.

10. a. $Ni(OH)_2(s) + 2\ H^+(aq) \rightarrow Ni^{2+}(aq) + 2\ H_2O$

b. $ZnS(s) + 2\ H^+(aq) \rightarrow Zn^{2+}(aq) + H_2S(g)$

c. $CuS(s) + H^+(aq) \rightarrow$ no reaction

d. $AgCl(s) + H^+(aq) \rightarrow$ no reaction

11. a. $Zn(OH)_2(s) + 2\ OH^-(aq) \rightarrow Zn(OH)_4^{2-}(aq)$

b. $Al(OH)_3(s) + OH^-(aq) \rightarrow Al(OH)_4^-(aq)$

c. $Ni(OH)_2(s) + OH^-(aq) \rightarrow$ no reaction

12. a. $Ni(OH)_2(s) + 6\ NH_3(aq) \rightarrow Ni(NH_3)_6^{2+}(aq) + 2\ OH^-(aq)$

b. $CuS(s) + NH_3(aq) \rightarrow$ no reaction

c. $AgCl(s) + 2\ NH_3(aq) \rightarrow Ag(NH_3)_2^+(aq) + Cl^-(aq)$

13. a. Cu^{2+} (blue) or Ni^{2+} (green)

    b. If one can distinguish a blue precipitate, $Cu(OH)_2$, from a green precipitate, $Ni(OH)_2$, no further test is necessary. If not, the solution will give a red precipitate with dimethylglyoxime if Ni^{2+} is present.

14.

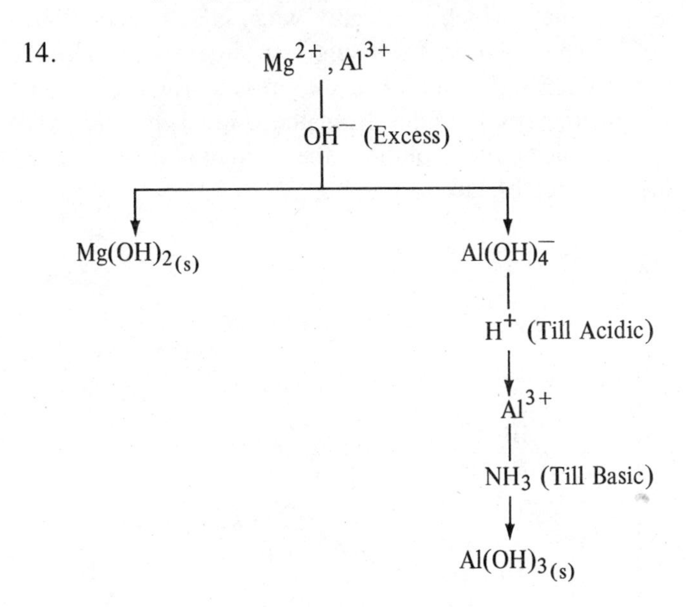

15.

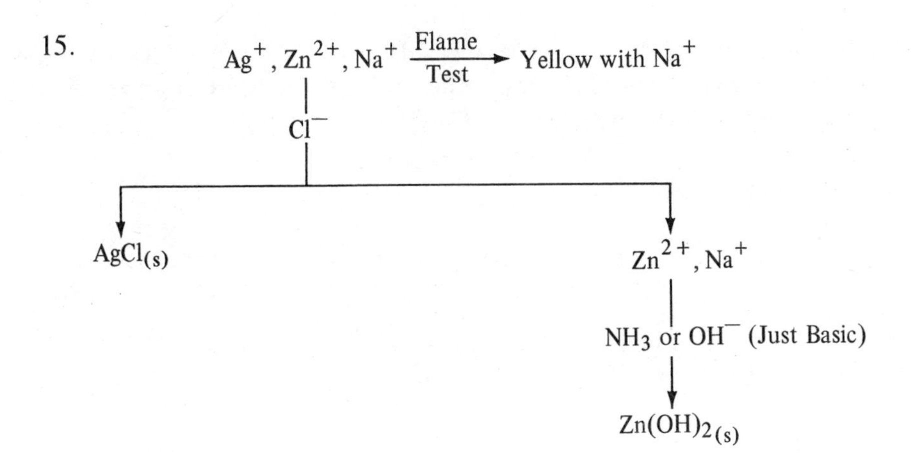

16. The precipitate formed is a hydroxide precipitate. The dissolving of the precipitate indicates the presence of Cu^{2+}, as $Cu(NH_3)_4^{2+}$. It also indicates the absence of Al^{3+} as $Al(OH)_3$ is not soluble in NH_3.
17. The addition of NH_3 to the solution produced a precipitate consisting of one or more hydroxides of Cu^{2+}, Zn^{2+}, and Mg^{2+}. The partial dissolving of the

precipitate when excess NH_3 is added confirms the presence of Mg^{2+}, as $Mg(OH)_2$ is insoluble in NH_3. The solution may contain Cu^{2+} or Zn^{2+} or both. Note color of solution—if blue, Cu^{2+} is present. To test for Zn^{2+}, add excess OH^-, separate any $Cu(OH)_2$, and then add H_2S to precipitate ZnS.

18. A Group I chloride precipitate which completely dissolves in hot water indicates the presence of Pb^{2+} and the absence of Ag^+ and Hg_2^{2+}.
19. A Group I chloride precipitate which turns grey when NH_3 is added indicates the presence of Hg_2^{2+}. The solution from this test should be acidified with HNO_3. A white precipitate, AgCl, indicates the presence of Ag^+. The Pb^{2+} ion is tested by treating the solid residue from the original reaction with hot water. When CrO_4^{2-} is added to the solution, the formation of a $PbCrO_4$ precipitate indicates the presence of Pb^{2+}.

20. $H_2S(aq) \rightleftarrows 2\,H^+(aq) + S^{2-}(aq)$

$$K = \frac{[H^+]^2[S^{2-}]}{[H_2S]}$$

a. $$[S^{2-}] = \frac{K\,[H_2S]}{[H^+]^2} = \frac{1 \times 10^{-22}\,(0.1)}{(0.3)^2} = 1 \times 10^{-22}\ M$$

b. $$[S^{2-}] = \frac{K\,[H_2S]}{[H^+]^2} = \frac{1 \times 10^{-22}\,(0.1)}{(10^{-9})^2} = 1 \times 10^{-5}\ M$$

21. a. H^+ b. HCl c. H^+ d. H^+ and NH_3
22. a. $AlCl_3$ and $Al(OH)_3$: Add H_2O and $AlCl_3$ will dissolve.
    b. $Al(OH)_3$ and $Zn(OH)_2$: Add NH_3 and $Zn(OH)_2$ dissolves by forming $Zn(NH_3)_4^{2+}$.
    c. $MgCl_2$ and $ZnCl_2$: Add excess NaOH and $ZnCl_2$ dissolves by forming $Zn(OH)_4^{2-}$ while the Mg^{2+} ion forms an insoluble $Mg(OH)_2$ precipitate.
    d. CuS and $Cu(OH)_2$: Add H^+ and $Cu(OH)_2$ will dissolve.

23. a. $2\,Ag^+(aq) + 2\,OH^-(aq) \rightarrow Ag_2O(s) + H_2O$

    b. $Al^{3+}(aq) + 3\,S^{2-}(aq) + 3\,H_2O \rightarrow Al(OH)_3(s) + 3\,HS^-(aq)$

    c. $Mg^{2+}(aq) + 2\,S^{2-}(aq) + 2\,H_2O \rightarrow Mg(OH)_2(s) + 2\,HS^-(aq)$

CHAPTER 23

Oxidation-Reduction Reactions

I. Basic Skills

A. Qualitative Students should be able to:

1. Explain what is meant by an oxidation-reduction reaction and the individual processes of oxidation and reduction.
2. a. Explain what is meant by oxidation number.
   b. Explain the relationship between oxidation and reduction and the change in an element's oxidation number.
3. Explain what is meant by an oxidizing agent and a reducing agent.
4. Explain what is meant by a half-equation and how the half-equation method is used to balance equations.
5. Locate on the Periodic Table the elements which commonly have multiple oxidation numbers.
6. Explain how sulfuric acid and nitric acid can act as oxidizing agents.

B. Quantitative Students should be able to:

1. Assign oxidation numbers to the elements in a compound or ion using the oxidation number rules (Example 23.1).
2. a. Determine from oxidation numbers the elements which are being oxidized or reduced in a reaction (Example 23.2).
   b. Determine the number of electrons transferred in an oxidation or reduction reaction.
3. a. Write oxidation and reduction half-equations.
   b. Balance oxidation-reduction equations using the half-equation method (Example 23.3).
4. Determine whether a substance can act as an oxidizing agent or reducing agent (Example 23.5).
5. Write Lewis structures of oxygen acids.

II. Chapter Development

1. A simple set of rules is given for the assignment of oxidation numbers. An example is given showing how oxidation numbers are related to the structural formula of a compound and the electronegativities of the elements in the compound.
2. Oxidation numbers are mostly used to identify the elements being oxidized and reduced and the substances which act as oxidizing agents and reducing agents. Some important compounds of chlorine, sulfur, and nitrogen are discussed from the viewpoint of their oxidation numbers.

3. The half-equation method is used in balancing oxidation-reduction equations. Emphasis is placed on reactions which take place in acidic solution.
4. The application of oxidation-reduction principles to electrochemical cells is covered in Chapter 24. The use of oxidation-reduction reactions to extract metals from their ores is covered in Chapter 25.

III. Problem Areas

1. The terms oxidation and reduction and oxidizing agent and reducing agent are often intermixed. Students must learn the basic definition of oxidation (loss of electrons) so that the other definitions which follow will be correct.
2. The main difficulty in writing half-equations is the balancing of charge. For this reason, it helps to use e^- as the symbol for the electron and its charge. Cross-multiplication is then used to eliminate electrons when adding the half-equations.

IV. Suggested Activities

1. An Oxidation-Reduction Reaction: **Teacher Demonstration Only**
   Demonstrate a redox reaction which changes color. Add 50 ml of 0.002 M $KMnO_4$ and 1 ml of dilute H_2SO_4 to an Erlenmeyer flask. Add a pinch of $MnSO_4 \cdot 2\ H_2O$ which acts as a catalyst. Add 5 ml of 0.05 M $H_2C_2O_4$ (oxalic acid) and swirl the flask. Give students the products (Mn^{2+}, CO_2, and H_2O) and have them balance the reaction equation using the half-equation method.

   $$2\ MnO_4^-(aq) + 5\ H_2C_2O_4(aq) + 6\ H^+(aq) \rightarrow 2\ Mn^{2+}(aq) + 10\ CO_2(g) + 8\ H_2O$$

   NOTE: Many other redox reactions may be used in a similar manner.
2. Oxidation Numbers of a Transition Metal: **Teacher Demonstration Only**
   Demonstrate the multiple oxidation numbers of vanadium by reacting a vanadium solution with a $KMnO_4$ solution. Prepare a $KMnO_4$ solution (20 g/ℓ) and place in a buret. Prepare a V^{2+} ion solution by dissolving 10 g V_2O_5 and 25 g Zn in 450 ml of 1 M H_2SO_4. When the solution turns violet (V^{2+}), acidify with 50 ml of 3 M H_2SO_4. (If V^{2+} solution is prepared too far in advance, it will muddy due to partial oxidation.) Add 10 ml of the V^{2+} solution to each of four large test tubes and place in a rack. Then demonstrate as follows:
   a. Add about 2.5 ml of $KMnO_4$ solution to the second tube to produce V^{3+} (forest green).
   b. Add about 5.0 ml $KMnO_4$ solution to the third tube and produce V^{4+} (sky blue). The actual ion is VO^{2+}.
   c. Add about 8.0 ml of $KMnO_4$ solution to the fourth tube to produce V^{5+} (lemon yellow). The actual ion is VO^{2+}.
   NOTE: The $KMnO_4$ volumes may need slight adjustments.

3. The Ostwald Process for Making Nitric Acid: **Teacher Demonstration Only** Beginning with nitric oxide, NO, make nitric acid, HNO_3. (Wide-mouth bottles of NO should be made in advance. The gas can be collected by water displacement by mixing 30 g $FeSO_4 \cdot 7\ H_2O$ and 15 g $NaNO_2$ at room temperature. The evolution of gas can be increased by adding water to the mixture. Leave about 50 ml of water in the gas bottle and seal it under water with a rubber stopper.) When demonstrating, unstopper the bottle of NO briefly allowing air in. Brown NO_2 gas will form. Shake the bottle and unstopper a second time. Shake the mixture again, unstopper, and test the solution with litmus paper. Review the chemical reactions. (See page 586 of the text.)

V. Answers to Questions

1. $2\ Ca(s) + O_2(g) \rightarrow 2\ CaO(s)$

   a. Ca: oxidized b. O_2: reduced
   c. O_2: oxidizing agent d. Ca: reducing agent

2. $Fe^{2+}(aq) + Ag^{+}(aq) \rightarrow Fe^{3+}(aq) + Ag(s)$

   a. Fe^{2+}: oxidized b. Ag^{+}: reduced
   c. Ag^{+}: oxidizing agent d. Fe^{2+}: reducing agent

3. $2\ Al(s) + 3\ F_2(g) \rightarrow 2\ AlF_3(s)$

   a. $Al \rightarrow Al^{3+} + 3e^{-}$: 2 Al atoms lose 6 electrons

   b. $F_2 + 2e^{-} \rightarrow 2\ F^{-}$: 3 F_2 molecules gain 6 electrons

   c. F_2: oxidizing agent d. Al: reducing agent

4. A reducing agent: (a) loses electrons, (c) increases its oxidation number, and (f) is oxidized.
5. An oxidizing agent: (b) gains electrons, (d) decreases its oxidation number, and (e) is reduced.
6. a. Na; +1 b. Mg; +2 c. Ba; +2 d. K; +1
   e. H; +1 f. O; −2
7. a. Ca^{2+}; +2 b. F^{-}; −1 c. Fe^{3+}; +3 d. S^{2-}; −2
8. The oxygen atoms in compounds have an oxidation number of −2. The oxidation number of the element O_2 is zero. The use of O_2 in place of H_2O would involve oxygen in the oxidation-reduction and upset the electron balance.
9. The Co^{2+} ion can act as a reducing agent (be oxidized to Co^{3+}) or an oxidizing agent (be reduced to Co). The Co^{3+} ion can only act as an oxidizing agent (be reduced to Co^{2+} or Co).
10. cobalt (II) chloride: $CoCl_2$ cobalt (III) chloride: $CoCl_3$
    cobalt (II) sulfate: $CoSO_4$ cobalt (III) sulfate: $Co_2(SO_4)_3$
11. The MnO_4^- ion is: (c) a strong oxidizing agent.
12. Reducing agents: Cl^-, ClO^-, ClO_2^-, ClO_3^-
    Oxidizing agents: ClO_4^-, ClO_3^-, ClO_2^-, ClO^-

13. 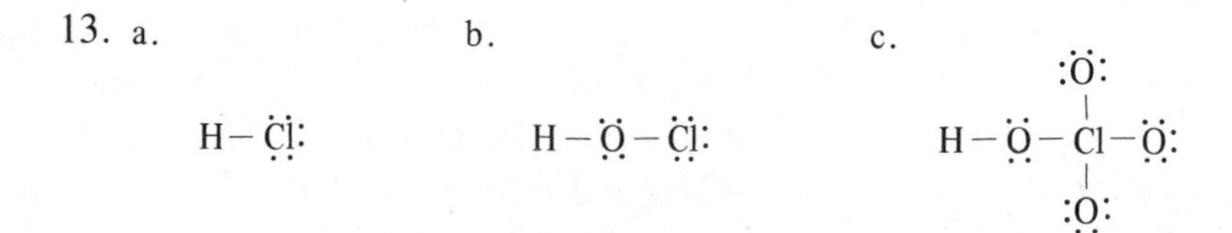

14. Bubble chlorine gas through water:

$$Cl_2(g) + H_2O \rightarrow HOCl(aq) + H^+(aq) + Cl^-(aq)$$

15. $Cl_2(g) + 2\ OH^-(aq) \rightarrow ClO^-(aq) + Cl^-(aq) + H_2O$

16. a. H–S–H b. H–O–S–O–H (with :O: above S) c. H–O–S–O–H (with :O: above and below S)

17. H_2S can be prepared in the laboratory by adding a strong acid to a sulfide ion solution or to a solid metal sulfide.

18. a. $H_2S(aq) \rightleftarrows 2\ H^+(aq) + S^{2-}(aq)$

    b. $2\ H_2S(aq) + O_2(g) \rightarrow 2\ S(s) + 2\ H_2O$

    c. $H_2S(aq) + Cu^{2+}(aq) \rightarrow CuS(s) + 2\ H^+(aq)$

19. H_2SO_4 can act as: (a) an acid, (c) an oxidizing agent, and (e) a dehydrating agent.

20. a. base b. reducing agent c. complexing agent d. base

21. Nitric acid, HNO_3, is produced from NH_3, O_2 and H_2O.
    In the first step, NH_3 is oxidized to nitric oxide, NO:

$$4\ NH_3(s) + 5\ O_2(g) \rightarrow 4\ NO(g) + 6\ H_2O$$

In the second step, NO is oxidized to nitrogen dioxide, NO_2:

$$2\ NO(g) + O_2(g) \rightarrow 2\ NO_2(g)$$

In the last step, part of the NO_2 is oxidized to HNO_3 while part is reduced back to NO:

$$3\ NO_2(g) + H_2O \rightarrow 2\ HNO_3(aq) + NO(g)$$

22. Sulfuric: H_2SO_4, sulfurous: H_2SO_3, nitric: HNO_3, nitrous: HNO_2, perchloric: $HClO_4$, and others

VI. Solutions to Problems

1. a. CrO_3: +6 b. Cr_2O_3: +3 c. Na_2CrO_4: +6 d. $Cr_2O_7^{2-}$: +6
2. a. PH_3: P (−3), H (+1) b. SiO_2: Si (+4), O (−2)
   c. Na_2MnO_4: Na (+1), Mn (+6), O (−2)
   d. HPO_4^{2-}: H (+1), P (+5), O (−2)

3. a. $NH_3 \rightarrow N_2$ b. $N_2 \rightarrow NO$ c. $NO_2 \rightarrow NO$ d. $NO \rightarrow NO_3^-$

N: $-3 \rightarrow 0$ N: $0 \rightarrow +2$ N: $+4 \rightarrow +2$ N: $+2 \rightarrow +5$

oxidation oxidation reduction oxidation

4. a. $CrO_4^{2-} \rightarrow Cr^{3+}$ b. $Cr_2O_7^{2-} \rightarrow CrO_4^{2-}$ c. $Fe^{2+} \rightarrow Fe^{3+}$

Cr: $+6 \rightarrow +3$ Cr: $+6 \rightarrow +6$ oxidation

reduction neither

d. $Al^{2+} \rightarrow Al(OH)_3$ e. $NH_3 \rightarrow NH_4^+$ f. $Al \rightarrow Al_2O_3$

Al: $+3 \rightarrow +3$ N: $-3 \rightarrow -3$ Al: $0 \rightarrow +3$

neither neither oxidation

5. a. $2\ NH_3(aq) \rightarrow N_2(g) + 6\ H^+(aq) + 6e^-$

b. $N_2(g) + 2\ H_2O \rightarrow 2\ NO(g) + 4\ H^+(aq) + 4e^-$

c. $NO_2(g) + 2\ H^+(aq) + 2e^- \rightarrow NO(g) + H_2O$

d. $NO(g) + 2\ H_2O \rightarrow NO_3^-(aq) + 4\ H^+(aq) + 3e^-$

6. $Ag(s) \rightarrow Ag^+(aq) + e^-$

$NO_3^-(aq) + 2\ H^+(aq) + e^- \rightarrow NO_2(g) + H_2O$

$Ag(s) + NO_3^-(aq) + 2\ H^+(aq) \rightarrow Ag^+(aq) + NO_2(g) + H_2O$

7. $3(2\ I^-(aq) \rightarrow I_2(s) + 2e^-)$

$2\ (MnO_4^-(aq) + 4\ H^+(aq) + 3e^- \rightarrow MnO_2(s) + 2\ H_2O)$

$6I^-(aq) + 2\ MnO_4^-(aq) + 8\ H^+(aq) \rightarrow 3\ I_2(s) + 2\ MnO_2(s) + 4\ H_2O$

8. $CuS(s) \rightarrow Cu^{2+}(aq) + S(s) + 2e^-$

$2\ (NO_3^-(aq) + 2\ H^+(aq) + e^- \rightarrow NO_2(g) + H_2O)$

$CuS(s) + 2\ NO_3^-(aq) + 4\ H^+(aq) \rightarrow Cu^{2+}(aq) + S(s) + 2\ NO_2(g) + 2\ H_2O$

9. $2\ CrO_4^{2-}(aq) + 6\ Cl^-(aq) + 16\ H^+(aq) \rightarrow 3\ Cl_2(g) + 2\ Cr^{3+}(aq) + 8\ H_2O$

a. $\text{moles } Cl_2 = 1.20 \text{ mol } Cl^- \times \frac{1 \text{ mol } Cl_2}{2 \text{ mol } Cl^-} = 0.600$

b. $\text{moles } Cr^{3+} = 1.20 \text{ mol } Cl^- \times \dfrac{1 \text{ mol } Cr^{3+}}{3 \text{ mol } Cl^-} = 0.400$

10. $Ag(s) + NO_3^-(aq) + 2\,H^+(aq) \rightarrow Ag^+(aq) + NO_2(g) + H_2O$

a. $\text{moles } NO_3^- = 1.00 \text{ g Ag} \times \dfrac{1 \text{ mol } NO_3^-}{108 \text{ g Ag}} = 9.26 \times 10^{-3}$

b. $\text{mass } NO_2 = 1.00 \text{ g Ag} \times \dfrac{1 \text{ mol } NO_2}{107.9 \text{ g Ag}} \times \dfrac{46.0 \text{ g } NO_2}{1 \text{ mol } NO_2} = 0.426 \text{ g}$

11. a. $Cu(s) \rightarrow Cu^{2+}(aq) + 2e^-$

b. $2\,NO_3^-(aq) + 12\,H^+(aq) + 10e^- \rightarrow N_2(g) + 6\,H_2O$

c. $SO_2(g) + 4\,H^+(aq) + 4e^- \rightarrow S(g) + 2\,H_2O$

d. $Al(s) \rightarrow Al^{3+}(aq) + 3e^-$

12. a. $5(Cu(s) \rightarrow Cu^{2+}(aq) + 2e^-)$

$2\,NO_3^-(aq) + 12\,H^+(aq) + 10e^- \rightarrow N_2(g) + 6\,H_2O$

$5\,Cu(s) + 2\,NO_3^-(aq) + 12\,H^+(aq) \rightarrow 5\,Cu^{2+}(aq) + N_2(g) + 6\,H_2O$

b. $2(Cu(s) \rightarrow Cu^{2+}(aq) + 2e^-)$

$SO_2(g) + 4\,H^+(aq) + 4e^- \rightarrow S(s) + 2\,H_2O$

$2\,Cu(s) + SO_2(g) + 4\,H^+(aq) \rightarrow 2\,Cu^{2+}(aq) + S(s) + 2\,H_2O$

c. $10(Al(s) \rightarrow Al^{3+}(aq) + 3e^-)$

$3(2\,NO_3^-(aq) + 12\,H^+(aq) + 10e^- \rightarrow N_2(g) + 6\,H_2O)$

$10\,Al(s) + 6\,NO_3(aq) + 36\,H^+(aq) \rightarrow 10\,Al^{3+}(aq) + 3\,N_2(g) + 18\,H_2O$

d. $4\,(Al(s) \rightarrow Al^{3+}(aq) + 3e^-)$

$3(SO_2(g) + 4\,H^+(aq) + 4e^- \rightarrow S(s) + 2\,H_2O)$

$4\,Al(s) + 3\,SO_2(g) + 12\,H^+(aq) \rightarrow 4\,Al^{3+}(aq) + 3\,S(s) + 6\,H_2O$

13. $5(H_2S(aq) + 2\,H_2O \rightarrow SO_2(g) + 6\,H^+(aq) + 6e^-)$

$3(2\,NO_3^-(aq) + 12\,H^+(aq) + 10e^- \rightarrow N_2(g) + 6\,H_2O)$

$5\,H_2S(aq) + 6\,NO_3^-(aq) + 6\,H^+(aq) \rightarrow 5\,SO_2(g) + 3\,N_2(g) + 8\,H_2O$

14. $2\ NH_3(g) + ClO^-(aq) \rightarrow N_2H_4(g) + Cl^-(aq) + H_2O$

a. $\text{mass } NH_3 = 1.00 \times 10^3 \text{ g } N_2H_4 \times \frac{2(17.0) \text{ g } NH_3}{32.0 \text{ g } N_2H_4} = 1.06 \times 10^3 \text{ g}$

b. $\text{mass } ClO^- = 1.00 \times 10^3 \text{ g } N_2H_4 \times \frac{51.5 \text{ g } ClO^-}{32.0 \text{ g } N_2H_4} = 1.61 \times 10^3 \text{ g}$

c. $\text{mass NaClO} = 1.61 \times 10^3 \text{ g } ClO^- \times \frac{74.5 \text{ g NaClO}}{51.5 \text{ g } ClO^-} = 2.33 \times 10^3 \text{ g}$

15. $Cl_2(g) + H_2O \rightleftarrows HClO(aq) + H^+(aq) + Cl^-(aq)$

a. $\text{moles } Cl_2 = 1.00 \times 10^3 \text{ g HClO} \times \frac{1 \text{ mol HClO}}{52.5 \text{ g HClO}} \times \frac{1 \text{ mol } Cl_2}{1 \text{ mol HClO}} = 19.0$

b. $\text{mass HClO} = 1.00 \times 10^3 \text{ g } Cl_2 \times \frac{52.5 \text{ g HClO}}{70.9 \text{ g } Cl_2} = 740 \text{ g}$

16. a. Au: 0, H^+: +1, NO_3^-: N(+5), and O (−2), $AuCl_4^-$: Au (+3) and Cl (−1)

NO_2: N (+4) and O (−2), H_2O: H (+1) and O (−2)

b. $Au(s) + 4\ Cl^-(aq) \rightarrow AuCl_4^-(aq) + 3e^-$ (oxidation)

$NO_3^-(aq) + 2\ H^+(aq) + e^- \rightarrow NO_2(g) + H_2O$ (reduction)

c. Multiplying the reduction half-equation by 3 gives:

$Au(s) + 4\ Cl^-(aq) + 3\ NO_3^-(aq) + 6\ H^+(aq) \rightarrow AuCl_4^-(aq) + 3\ NO_2(g) + 3\ H_2O$

17. a. $Sn(s) + 2\ H_2O \rightarrow SnO_2(s) + 4\ H^+(aq) + 4e^-$

$4(NO_3^-(aq) + 2\ H^+(aq) + e^- \rightarrow NO_2(g) + H_2O)$

$Sn(s) + 4\ NO_3^-(aq) + 4\ H^+(aq) \rightarrow SnO_2(s) + 4\ NO_2(g) + 2\ H_2O$

b. $\text{moles } NO_3^- = 1.00 \text{ g Sn} \times \frac{1 \text{ mol Sn}}{119 \text{ g Sn}} \times \frac{4 \text{ mol } NO_3^-}{1 \text{ mol Sn}} = 3.36 \times 10^{-2}$

18. $2\ H_2S(aq) + O_2(g) \rightarrow 2\ S(s) + 2\ H_2O$

$\text{Volume } O_2(\text{STP}) = 10.0 \text{ g } H_2S \times \frac{22.4\ \ell\ O_2}{2(34.0) \text{ g } H_2S} = 3.29\ \ell$

19. a. CO_2: +4 b. CO_3^-: +4 c. CH_3COOH: 0
NOTE: In (c), the H:O ratio is 2:1. When this occurs, the oxidation number of C will be zero.

20. The following ions can act as both oxidizing and reducing agents:

a. $Cr^{3+} \rightarrow Cr^{6+}$ or Cr^0 d. $Sn^{2+} \rightarrow Sn^{4+}$ or Sn^0

e. $Pb^{2+} \rightarrow Pb^{4+}$ or Pb^0

21. $3(ClO_3^-(aq) + H_2O \rightarrow ClO_4^-(aq) + 2\ H^+(aq) + 2e^-)$

$ClO_3^-(aq) + 6\ H^+(aq) + 6e^- \rightarrow Cl^-(aq) + 3\ H_2O$

$4\ ClO_3^-(aq) \rightarrow 3\ ClO_4^-(aq) + Cl^-(aq)$

22. $Bi_2S_3(s) \rightarrow 2\ Bi^{3+}(aq) + 3\ S(s) + 6e^-$

$6(NO_3^-(aq) + 2\ H^+(aq) + e^- \rightarrow NO_2(g) + H_2O)$

$Bi_2S_3(s) + 6\ NO_3^-(aq) + 12\ H^+(aq) \rightarrow 2\ Bi^{3+}(aq) + 3\ S(s) + 6\ NO_2(g) + 6\ H_2O$

23. $2\ N_2O_4(g) + (CH_3)_2N_2H_2(g) \rightarrow 3\ N_2(g) + 4\ H_2O(g) + 2\ CO_2(g)$

$$\text{Volume } N_2O_4 = 2.40\ \ell\ (CH_3)_2N_2H_2 \times \frac{2 \text{ vol } N_2O_4}{1 \text{ vol } (CH_3)_2N_2H_2} = 4.80\ \ell$$

If the pressure of the $(CH_3)_2N_2H_2$ is increased to 10 atm it will contain 10 times as many moles and will require 10 times the amount, or 48.0 ℓ, of N_2O_4.

24. $SO_3(g) + H_2O \rightarrow H_2SO_4(aq)$

$$\text{Volume } SO_3(\text{STP}) = 18 \text{ mol } H_2SO_4 \times \frac{1 \text{ mol } SO_3}{1 \text{ mol } H_2SO_4} \times \frac{22.4\ \ell}{1 \text{ mol } SO_3}$$
$$= 4.0 \times 10^2\ \ell$$

CHAPTER 24

Electrochemistry

I. Basic Skills

A. Qualitative Students should be able to:

1. Describe a voltaic cell and explain how it operates.
2. Explain what is meant by anode and cathode and anion and cation.
3. Describe the operation of commercial dry cells and the lead storage battery.
4. a. Explain what is meant by a cell voltage and half-cell voltage.
   b. Explain how half-cell voltages are calculated.
   c. List the factors which affect the cell voltage.
   d. List the necessary conditions for obtaining a standard voltage.
5. Describe the relationship between the cell voltage and the spontaneity of the reaction.
6. Describe an electrolytic cell and explain how it differs from a voltaic cell.

B. Quantitative Students should be able to:

1. a. Draw a diagram of a voltaic cell and label the anode and cathode (Example 24.1).
   b. Indicate the direction of electron flow and ion flow in a voltaic cell (Example 24.1).
2. Write equations for the anode and cathode reactions of a voltaic cell (Example 24.2).
3. Determine the standard voltages of half-cells and the total cell voltage, E° (Example 24.3).
4. a. Classify a substance as a possible oxidizing agent or reducing agent (Example 24.4).
   b. Order substances according to their relative strengths as oxidizing agents or reducing agents (Example 24.4).
5. Determine whether a given redox reaction will occur spontaneously (Example 24.5).

II. Chapter Development

1. Voltaic cells are used to introduce the terminology and the operation of electrochemical cells. Fuel cells are not discussed.
2. The term "voltage" is used in place of "potential." Both oxidation and reduction voltages are given in the table of Standard Half-Cell Voltages. The effect of concentration upon voltage is discussed in an optional section.

3. An empirical approach is used to describe what happens in the electrolysis of water solutions. (For example, that H_2 is produced at the cathode in place of Na when a NaCl solution is electrolyzed.) A strict thermodynamic approach (using E° values) will often lead to incorrect predictions as kinetics usually dictates which electrode reaction takes place.
4. While students should understand the relationship between the amount of electrons and the amount of cell products, no attempt is made to extend this to the Faraday unit involving coulombs.
5. The electrolysis of molten salts (NaCl and Al_2O_3) is discussed in Chapter 25, Metals and Their Ores.

III. Problem Areas

1. The problem of assigning electrode polarity (+ or −) is mentioned but not stressed. It is sufficient that students understand the direction of electron flow and ion flow.
2. Some students may have difficulty in describing ion flow in a cell. One approach is to show that the negative ions must move in the same direction (clockwise or counterclockwise) as negative electrons.
3. Caution students that the use of E° values to predict reaction spontaneity only applies to reactions at standard conditions. Standard conditions very rarely exist. Students should be able to predict the effects of concentration changes on voltage by using LeChatelier's Principle.

IV. Suggested Activities

1. Relative Strengths of Oxidizing Agents:
   Discuss the general oxidation reaction: $A + B^+ \rightarrow A^+ + B$. If the reaction takes place, then B^+ is a stronger oxidizing agent than A^+; if no reaction takes place, then A^+ is stronger than B^+. Then demonstrate the following reactions:

   $Zn + Cu^{2+}$, $Cu + Zn^{2+}$(NR), $Zn + Pb^{2+}$, $Pb + Zn^{2+}$(NR), $Zn + H^+$, $Pb + H^+$, $Cu + H^+$(NR), and $Cu + Ag^+$.

   Have students arrange the five positive ions in order of their decreasing strength as oxidizing agents. Compare to Table 24.1, Table of Standard Half-Cell Voltages.
2. Electrolysis of KI: **Teacher Demonstration Only**
   Electrolyze a 0.5 M KI solution in a U-tube. Use graphite electrodes and a 12 volt D.C. source. After a few minutes of operation, ask the students to identify the anode and cathode and the electrode products (I_2 and H_2). The brownish I_2 at the anode will be visible. A few drops of starch indicator may be used to prove its presence. Phenolphthalein may be added to the cathode to show that the surrounding solution is basic. Have students write the electrode reactions and the cell reaction.

V. Answers to Questions

1. In a *voltaic* cell, a spontaneous oxidation-reduction reaction produces electrical energy. In an *electrolytic* cell, electrical energy is used to make a non-spontaneous reaction occur.
2. The cathode is the electrode: (a) at which reduction occurs and (c) toward which positive ions move.
3. The anode is the electrode: (b) at which oxidation occurs and (d) toward which negative ions move.
4. $Zn + Cu^{2+} \rightarrow Zn^{2+} + Cu$
   a. Electrons enter the cell at the Cu electrode.
   b. Electrons leave the cell at the Zn electrode.
   c. The Zn^{2+} and Cu^{2+} ions move toward the Cu electrode (the cathode).
   d. The negative ions move toward the Zn electrode (the anode).
5. The electron transfer in a cell must take place through an external circuit. The Zn and Cu^{2+} ions are in separate compartments so that they do not make contact and short-circuit the cell. At the same time, the ions in solution must be able to move between electrodes in order to have a complete circuit. An open circuit would occur if the porous cup were replaced by a glass beaker. If it were removed, Zn and Cu^{2+} ions would come in contact giving a short circuit.
6. In some cells, one or both half reactions do not involve a metal. Cells of this type must use an inert electrode made of a conducting material which does not react with the molecules or ions around it. As metals are not involved in the $Cl_2 - Br^-$ cell, inert electrodes must be used in both the anode and cathode compartments. Graphite or platinum are acceptable choices for inert electrode materials.
7. The dry cell reactions are:
   a. Anode: $Zn(s) \rightarrow Zn^{2+}(aq) + 2e^-$
   b. Cathode:
   $$MnO_2(s) + 4\ NH_4^+(aq) + e^- \rightarrow Mn^{3+}(aq) + 2\ H_2O + 4\ NH_3(aq)$$
   c. Overall:
$$Zn(s) + 2\ MnO_2(s) + 8\ NH_4^+(aq) \rightarrow Zn^{2+}(aq) + 2\ Mn^{3+}(aq) + 8\ NH_3(aq) + 4\ H_2O$$
8. The anode is made of *lead*. The cathode contains an oxide of lead with the formula PbO_2. The water solution contains H_2SO_4 as an electrolyte.
9. The lead storage battery reactions are:
   a. Anode: $Pb(s) + SO_4^{2-}(aq) \rightarrow PbSO_4(s) + 2e^-$
   b. Cathode:
   $$PbO_2(s) + 4\ H^+(aq) + SO_4^{2-}(aq) + 2e^- \rightarrow PbSO_4(s) + 2\ H_2O$$
   c. Overall:
   $$Pb(s) + PbO_2(s) + 4\ H^+(aq) + 2\ SO_4^{2-}(aq) \rightarrow 2\ PbSO_4(s) + 2\ H_2O$$

10. By changing the reference standard to E°_{red} $2\ H^+ \rightarrow H_2 = +0.20$ V, all reduction potentials are increased by 0.20 V and all oxidation potentials are decreased by 0.20 V. The net effect is that the E° for the cell is unchanged. The cell voltage is independent of the reference standard.

    a. E°_{ox} $Zn \rightarrow Zn^{2+} = +0.56$ V

    b. E°_{ox} $H_2 \rightarrow 2\ H^+ = -0.20$ V

    c. $Zn \rightarrow Zn^{2+} + 2e^-$ $\qquad E^\circ_{ox} = +0.56$ V

    $2\ H^+ + 2e^- \rightarrow H_2$ $\qquad E^\circ_{red} = +0.20$ V

    $E^\circ_{cell} = +0.76$ V

11. For the reaction, $A \rightarrow A^{2+} + 2e^-$, where $E^\circ_{ox} = -2.00$ V:

    A is (d) a weak reducing agent.

12. $A^{2+} + 2e^- \rightarrow A$, $E^\circ_{red} = +2.00$ V

    A^{2+} is (a) a strong oxidizing agent.

13. a. E°_{ox} $Cu \rightarrow Cu^{2+} = -0.34$ V

    b. E°_{red} $Sn^{4+} \rightarrow Sn^{2+} = +0.15$ V

    c. E°_{ox} $Sn \rightarrow Sn^{2+} = +0.14$ V

14. In Table 24.1 (Standard Half-Cell Voltages), the strong oxidizing agents are located at (b), the lower left. Strong reducing agents are located at (c), the upper right.
15. To estimate the strength of a reducing agent, you look at the value of E°_{ox}. The strength of an oxidizing agent is related to its value of E°_{red}.
16. If a reaction has an E° of −1.61 V: (b) it is not spontaneous and (d) it can be carried out in an electrolytic cell.
17. When $CuCl_2$ solution is electrolyzed, copper is produced at the *cathode*. At the other electrode, called the *anode*, Cl_2 is produced.
18. When a NaCl solution is electrolyzed, the Na^+ ion is not reduced at the cathode because the water molecule is more easily reduced. The cathode products are H_2 and OH^- ions.

    $$2\ H_2O + 2e^- \rightarrow H_2(g) + 2\ OH^-(aq)$$

19. The situation is similar to that of Question 18. The Al^{3+} ion is not reduced; the cathode products are H_2 and OH^- ions.

VI. Solutions to Problems

1.

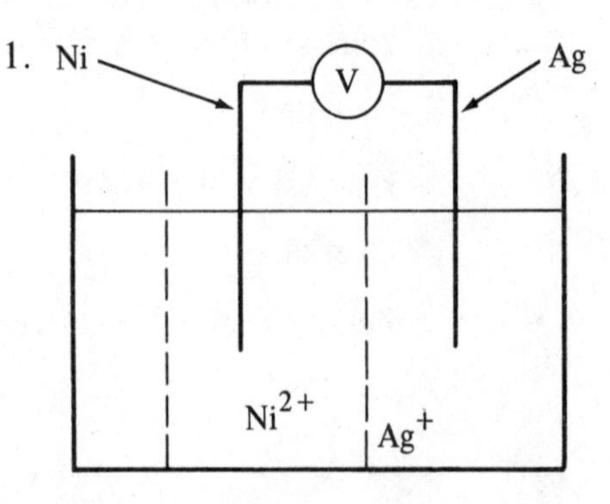

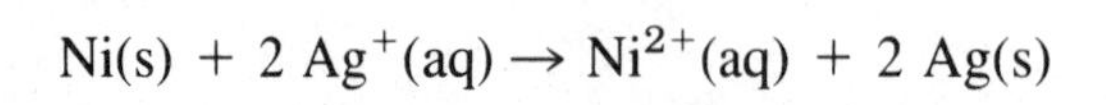

$Ni(s) + 2\ Ag^+(aq) \rightarrow Ni^{2+}(aq) + 2\ Ag(s)$

Ni anode, Ag cathode

Electrons move from Ni to Ag

2. 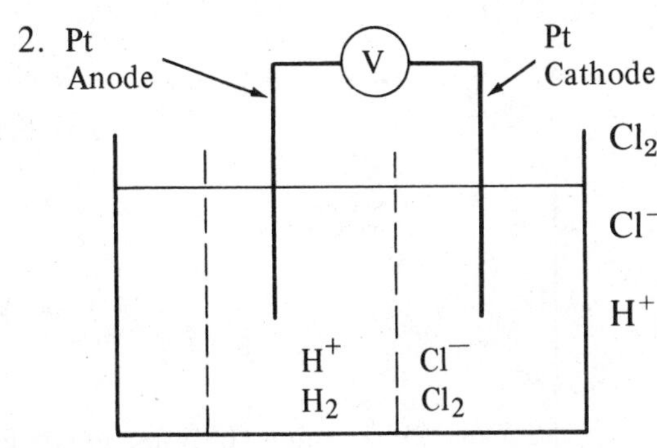

$Cl_2(g) + H_2(g) \rightarrow 2\ Cl^-(aq) + 2\ H^+(aq)$

Cl^- ions move to anode

H^+ ions move to cathode

3. $E°_{ox}$ Ni → Ni^{2+} = +0.25 V

$E°_{red}$ Ag^+ → Ag = +0.80 V

$E°_{cell}$ = +1.05 V

$E°_{ox}$ H_2 → H^+ = 0.00 V

$E°_{red}$ Cl_2 → Cl^- = +1.36 V

$E°_{cell}$ = +1.36 V

4. $MnO_2(s) + 4\ H^+(aq) + 2\ I^-(aq) \rightarrow Mn^{2+}(aq) + I_2(s) + 2\ H_2O$

$E°_{ox}$ I^- → I_2 = −0.53 V

$E°_{red}$ MnO_2 → Mn^{2+} = +1.23 V

$E°_{cell}$ = +0.70 V

5. Decreasing strength as reducing agents:
   Pb > H_2 > Cu > Cl^- (decreasing $E°_{ox}$)
6. Decreasing strength as oxidizing agents:
   Au^{3+} > O_2 > I_2 > H^+ (decreasing $E°_{red}$)
7. As all reactions in Problems 3 and 4 have a positive $E°_{cell}$, all of the reactions are spontaneous.
8. For the electrolysis of a $CaCl_2$ solution:

   Anode: $2\ Cl^-(aq) \rightarrow Cl_2(g) + 2e^-$

   Cathode: $2\ H_2O + 2e^- \rightarrow H_2(g) + 2\ OH^-(aq)$

   Overall: $2\ Cl^-(aq) + 2\ H_2O \rightarrow Cl_2(g) + H_2(g) + 2\ OH^-(aq)$

9. To react spontaneously with 1 M H^+ ions, the metal must have a positive $E°_{ox}$ value. Of the metals listed, (a) Al, (d) Mg, and (e) Ni^{2+} have a positive $E°_{ox}$ and will react with H^+.
10. For the electrolysis of a NiBr solution:

    Cathode: $Ni^{2+}(aq) + 2e^- \rightarrow Ni(s)$

    Anode: $2\ Br^-(aq) \rightarrow Br_2(l) + 2e^-$

    Overall: $Ni^{2+}(aq) + 2\ Br^-(aq) \rightarrow Ni(s) + Br_2(l)$

11.

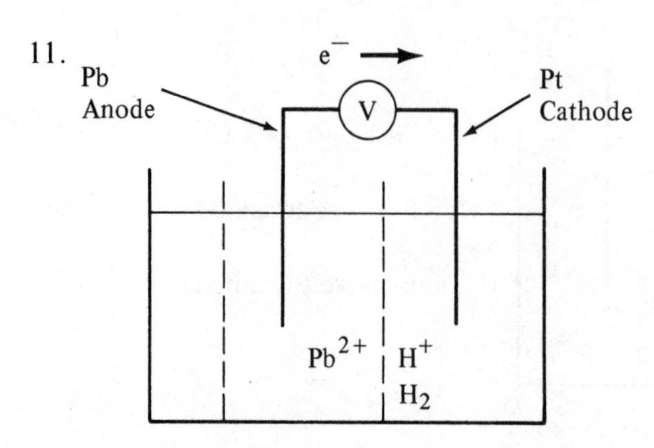

$Pb(s) + 2\ H^+(aq) \rightarrow Pb^{2+}(aq) + H_2(g)$

$E^\circ_{cell} = E^\circ_{ox} + E^\circ_{red}$

$E^\circ_{cell} = +0.13\ V + 0.00\ V = +0.13\ V$

Pb^{2+} and H^+ ions move to the cathode

12.

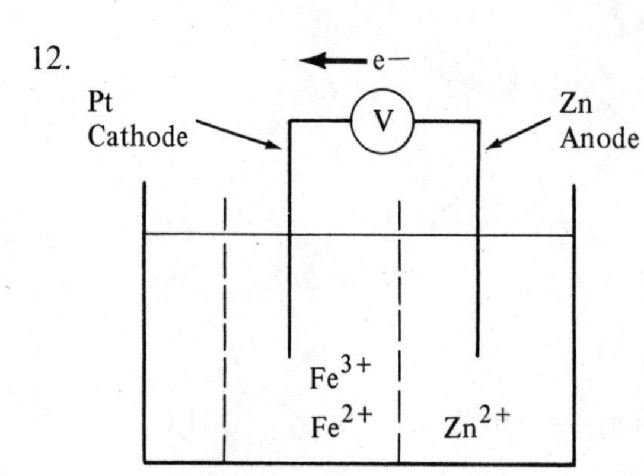

$Zn(s) + 2\ Fe^{3+}(aq) \rightarrow Zn^{2+}(aq) + 2\ Fe^{2+}(aq)$

$E^\circ_{cell} = E^\circ_{ox} + E^\circ_{red}$

$E^\circ_{cell} = +0.76\ V + 0.77\ V = +1.53\ V$

Zn^{2+}, Fe^{2+} and Fe^{3+} ions move to the cathode

13.

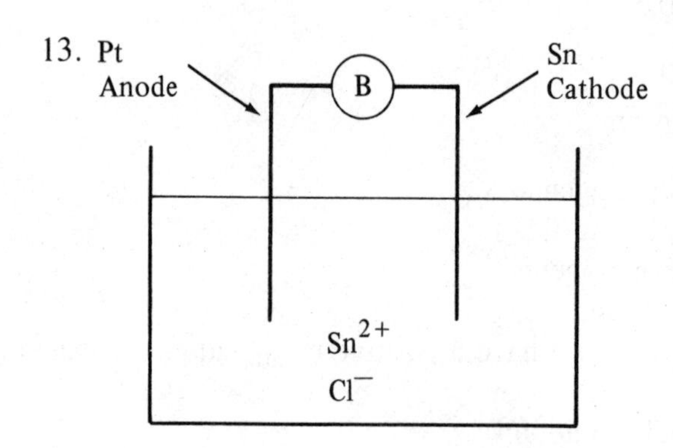

Anode: $2\ Cl^-(aq) \rightarrow Cl_2(g) + 2e^-$

Cathode: $Sn^{2+}(aq) + 2e^- \rightarrow Sn(s)$

Overall:

$Sn^{2+}(aq) + 2\ Cl^-(aq) \rightarrow Sn(s) + Cl_2(g)$

14.

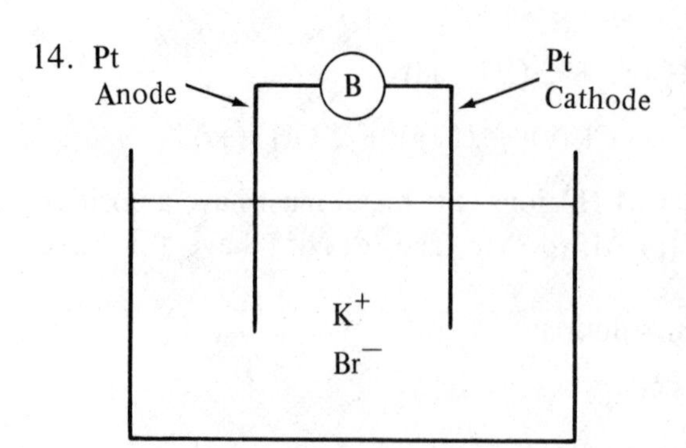

Anode: $2\ Br^-(aq) \rightarrow Br_2(l) + 2e^-$

Cathode: $2\ H_2O + 2e^- \rightarrow H_2(g) + 2\ OH^-(aq)$

Overall:

$2\ H_2O + 2\ Br^-(aq) \rightarrow H_2(g) + Br_2(l) + 2\ OH^-(aq)$

15. To react spontaneously with Cu^{2+} ($E°_{red} = +0.34$ V), the reactant must have an $E°_{ox}$ value greater than -0.34 V. Of those listed, (b) Ni ($E°_{ox} = +0.25$ V) and (c) H_2 ($E°_{ox} = 0.00$ V) will react with Cu^{2+}.

16. $2\ Na(s) + F_2(g) \rightarrow 2\ Na^+(aq) + 2\ F^-(aq)$

$$E°_{ox}\ Na \rightarrow Na^+ = +2.71\ V$$
$$E°_{red}\ F_2 \rightarrow F^- = \underline{+2.87\ V}$$
$$E°_{cell} = +5.58\ V$$

This is the theoretical voltage of a standard Na–F_2 cell. It has not been possible to build a cell of this type as both Na and F_2 react with the water in the electrolyte solutions.

17. a.

$$E°_{ox}\ I^- \rightarrow I_2 = -0.53\ V$$
$$E°_{red}\ Cl_2 \rightarrow Cl^- = \underline{+1.36\ V}$$
$$E°_{cell} = +\ .83\ V$$

b.

$$E°_{ox}\ I^- \rightarrow I_2 = -0.53\ V$$
$$E°_{red}\ Ni^{2+} \rightarrow Ni = \underline{-0.25\ V}$$
$$E°_{cell} = -0.78\ V$$

c.

$$E°_{ox}\ I^- \rightarrow I_2 = -0.53\ V$$
$$E°_{red}\ Zn^{2+} \rightarrow Zn = \underline{-0.76\ V}$$
$$E°_{cell} = -1.29\ V$$

d.

$$E°_{ox}\ I^- \rightarrow I_2 = -0.53\ V$$
$$E°_{red}\ Au^{3+} \rightarrow Au = \underline{+1.50\ V}$$
$$E°_{cell} = +0.97\ V$$

18. Reactions (a) $I^- - Cl_2$ and (d) $I^- - Au^{3+}$ have positive E° values and are therefore spontaneous reactions.
19. To react spontaneously with Pb^{2+} ($E°_{red} = -0.13$ V), the reactant must have an $E°_{ox}$ value greater than $+0.13$ V. Of those listed, (a) Ni ($E°_{ox} = +0.25$ V) and (d) Sn ($E°_{ox} = +0.14$ V) will react with Pb^{2+}.
20. a. All metals listed above H_2 (Pb to Li) are stronger reducing agents than H_2. There are 11 such metals.
    b. The species listed below Cl_2 (Au^{3+}, MnO_4^-, and F_2) are stronger oxidizing agents.

21.

$$E°_{ox}\ Cu \rightarrow Cu^{2+} = -0.34\ V$$
$$E°_{red}\ NO_3^-(H^+) \rightarrow NO = \underline{+0.96\ V}$$
$$E° = +0.62\ V$$

While Cu does not react with 1 M H^+ ($E°_{red} = 0.00$), it does react with 1 M HNO_3 due to the oxidizing strength of the NO_3^- ion.

22. The reaction to electroplate copper is:

$$Cu^{2+}(aq) + 2e^- \rightarrow Cu(s)$$

Using this equation: 1 mol Cu $\simeq$ 2 mol e's

$$\text{no. of e's} = 1.00 \text{ g Cu} \times \frac{1 \text{ mol Cu}}{63.5 \text{ g Cu}} \times \frac{2 \text{ mol e's}}{1 \text{ mol Cu}} \times \frac{6.02 \times 10^{23} \text{ e's}}{1 \text{ mol e's}}$$

$$= 1.90 \times 10^{22}$$

23. a. The water contains dissolved O_2 which slowly oxidizes the iron II ions to iron III ions.

$$E^\circ_{ox}\ O_2(H^+) \rightarrow H_2O = +1.23 \text{ V}$$

$$E^\circ_{red}\ Fe^{2+} \rightarrow Fe^{3+} = \underline{-0.77 \text{ V}}$$

$$E^\circ = +0.46 \text{ V}$$

b. A spontaneous reaction between Ag^+ and Fe^{2+} ions produces Ag and Fe^{3+} ions:

$$E^\circ_{ox}\ Fe^{2+} \rightarrow Fe^{3+} = -0.77 \text{ V}$$

$$E^\circ_{red}\ Ag^+ \rightarrow Ag = \underline{+0.80 \text{ V}}$$

$$E^\circ = +0.03 \text{ V}$$

c. Tin metal will reduce any Sn^{4+} (formed by oxidation) back to Sn^{2+}:

$$E^\circ_{ox}\ Sn \rightarrow Sn^{2+} = +0.14 \text{ V}$$

$$E^\circ_{red}\ Sn^{4+} \rightarrow Sn^{2+} = \underline{+0.15 \text{ V}}$$

$$E^\circ = +0.29 \text{ V}$$

24. $Cl_2(g) + 2\ Br^-(aq) \rightarrow 2\ Cl^-(aq) + Br_2(l)$

The voltage of the cell could be increased by:

(1) increasing the pressure of Cl_2

(2) increasing the concentration of Br^- ion

(3) decreasing the concentration of Cl^- ion

The cell voltage could be decreased by:

(1) decreasing the pressure of Cl_2

(2) decreasing the concentration of Br^- ion

(3) increasing the concentration of Cl^- ion

All of the above voltage changes can be predicted by applying LeChatelier's Principle. When a system is disturbed, the system will adjust (shift direction) so as to relieve the disturbance.

CHAPTER 25

Metals and Their Ores

I. Basic Skills

A. Qualitative Students should be able to:

1. Define an ore. List the major forms in which metallic ores occur.
2. a. Describe what is meant by the concentration, reduction, and refining of an ore.
   b. Describe the flotation and roasting processes.
   c. Distinguish between electrolytic and chemical reduction.
3. Describe how Na, Mg, and Al are produced by electrolysis.
4. a. Describe how iron is produced in a blast furnace.
   b. Describe how steel is produced from iron.
5. Describe how Zn, Hg, and Cu are produced from their sulfide ores.

B. Quantitative Students should be able to:

1. Determine the amount of metal which can be extracted from an ore (Examples 25.1 and 25.3).
2. Determine the expected lifetime of an ore (Table 25.1).
3. Write balanced chemical equations showing the:
   a. electrolysis of molten salts (Example 25.2).
   b. concentration of the magnesium in sea water.
   c. concentration of Al_2O_3 from bauxite.
   d. forming of slag in a blast furnace.
   e. the reduction of iron ore in a blast furnace.
   f. the roasting of a sulfide ore (Example 25.4).

II. Chapter Development

1. The principles of oxidation and reduction (Chapter 23) and electrolysis (Chapter 24) are extended quite naturally into the extraction of metals. Both electrolytic and chemical reduction methods are used. Representative metals have been selected, but due to space limitations some metals of interest had to be omitted.
2. The extraction processes given refer to actual industrial processes where energy costs and materials costs are paramount. The environmental problems caused by the waste products of these processes (SO_2 and trace metals) will be discussed in Chapter 26.

III. Problem Areas

1. A certain amount of difficulty will arise in writing the equations of this chapter. Students cannot be expected to know which compounds can be reduced chemically and which must be electrolyzed. In the same way, students cannot be expected to choose a suitable reducing agent or to know whether roasting reduces a sulfide ore to the metal or simply converts it to the oxide.

IV. Suggested Activities

1. Reactions of $Al(OH)_3$:
   Demonstrate the acidic and basic character of $Al(OH)_3$ and relate the reactions to the processing of bauxite ore (Section 25.3). Add 2–3 drops of 6 M NaOH to 5 ml of 1 M $Al(NO_3)_3$ to precipitate $Al(OH)_3$. Dissolve the precipitate with excess NaOH forming $Al(OH)_4^-$. Reform the precipitate by adding 6 M HNO_3 dropwise. Dissolve the precipitate with excess acid forming the Al^{3+} ion. Review the equations for the reactions.
2. Ore and Metals Exhibit:
   Obtain a variety of metal ore samples, locally or commercially. Exhibit the ores along with the concentrated compound and extracted metal. For example, show hematite ore with Fe_2O_3 and iron. Discuss the steps involved in extraction and some of the problems.

V. Answers to Questions

1. An ore is a mixture from which a metal can be extracted profitably. Aluminum can be extracted profitably from bauxite but not from granite.
2. a. Al: Al_2O_3 b. Fe: Fe_2O_3 c. Cu: Cu_2S
   d. Zn: ZnS e. Hg: HgS
3. Metal abundance in earth's crust: Al > Fe > Na
4. a. Cu global reserves: 310×10^9 kg
   b. Cu U.S. reserves: 28×10^9 kg
   c. Cu earth's crust: 4×10^{17} kg
5. 12 transition metals are listed in Table 25.1:
   Fe, Cu, Mn, Zn, Cr, Ni, Mo, W, Co, Ag, Hg, and Au
6. Water molecules are more readily reduced than Na^+ ions. For this reason, sodium metal cannot be obtained by the electrolysis of aqueous sodium ion solutions.
7. a. $NaCl(l) \rightarrow Na(l) + 1/2\ Cl_2(g)$

   b. $MgCl_2(l) \rightarrow Mg(l) + Cl_2(g)$

   c. $Al_2O_3(l) \rightarrow 2\ Al(l) + 3/2\ O_2(g)$

8. The electrolysis products, Na and Cl_2, must be kept from making contact to prevent them reacting spontaneously to form NaCl.
9. In making aluminum by electrolysis:
   a. the bauxite ore is treated with NaOH to separate the Al_2O_3 from the Fe_2O_3 and SiO_2 impurities. The Al_2O_3 dissolves by forming the complex ion, $Al(OH)_4^-$.
   b. cryolite, Na_3AlF_6, is mixed with the Al_2O_3 to form a mixture which has a reduced melting point.
   c. water solutions are not used because water is more readily reduced than Al^{3+} ions.
10. a. Aluminum is used in automobiles, in preference to steel, because its lower density decreases the auto's mass. This results in increased gasoline mileage.
    b. Aluminum is an excellent conductor of heat.
11. Coke is an impure form of carbon. It is used to produce the carbon monoxide required to reduce Fe_2O_3 to iron. Limestone is a form of calcium carbonate, $CaCO_3$. It is converted to calcium oxide, CaO, by heating and reacts with impurities to form "slag."
12. The reducing agent in the blast furnace is CO. It reduces the Fe_2O_3 to Fe.
13. Oxygen is needed in making steel to:
    a. convert Si, Mn, and P impurities to oxides which are removed in the slag.
    b. control the amount of carbon by burning part of it to CO_2.
14. The properties of steel may be varied by:
    a. varying the carbon content.
    b. heat treatment.
    c. adding small amounts of other metals such as Cr, Ni, and Mn.
15. Reactivity toward O_2:
    a. Na > Al   b. Zn > Hg   c. Cu > Au
16. "Roasting" is the process of heating a mineral in air. Sulfide minerals are commonly roasted to convert them to oxides or metals.
17. At roasting temperatures mercuric oxide is unstable and decomposes to free mercury, while the zinc oxide formed in roasting is stable.
18. Mercury is useful:
    a. in barometers because of its high density and low vapor pressure.
    b. in thermometers because it expands at a constant rate when heated.
    c. in electrical switches because of its electrical conductivity and ability to flow.
19. In the flotation process, the ore is crushed and mixed with oil, water, and detergent. Compressed air is then blown through it to form a froth. The sulfide particles are "wet" by the oil, rise to the surface, and become concentrated in the oil layer. The rocky material sinks to the bottom of the water.
20. The FeS and Cu_2S mixture from the flotation process is roasted in a limited amount of air:

$$2\ FeS(s) + 3\ O_2(g) \rightarrow 2\ FeO(s) + 2\ SO_2(g)$$

Sand, SiO_2, is added to form $FeSiO_3$ which forms a liquid slag:

$FeO(s) + SiO_2(s) \rightarrow FeSiO_3(l)$

The Cu_2S is later reduced to Cu by roasting at 1500°C:

$Cu_2S(s) + O_2(g) \rightarrow 2\ Cu(l) + SO_2(g)$

21. Blister copper is purified by electrolysis. The impure blister Cu is oxidized at the anode. Pure copper is formed at the cathode by reducing Cu^{2+} ions from the solution.

VI. Solutions to Problems

1. a. $\text{mass Fe} = 10^3\ \text{g Fe}_2\text{O}_3 \times \dfrac{111.7\ \text{g Fe}}{159.7\ \text{g Fe}_2\text{O}_3} = 699.4\ \text{g}$

   b. $\text{mass Zn} = 10^3\ \text{g ZnS} \times \dfrac{65.38\ \text{g Zn}}{97.44\ \text{g ZnS}} = 671.0\ \text{g}$

   c. $\text{mass Bi} = 10^3\ \text{g Bi}_2\text{S}_3 \times \dfrac{418.0\ \text{g Bi}}{514.2\ \text{g Bi}_2\text{S}_3} = 812.9\ \text{g}$

2. $ZnO(s) + CO(g) \rightarrow Zn(s) + CO_2(g)$ (Equation 25.18)

   a. $\text{mass CO} = 1.00\ \text{g ZnO} \times \dfrac{28.0\ \text{g CO}}{81.4\ \text{g ZnO}} = 0.344\ \text{g}$

   b. $\text{mass C} = 0.344\ \text{g CO} \times \dfrac{12.0\ \text{g C}}{28.0\ \text{g CO}} = 0.147\ \text{g}$

3. a. $2\ FeS(s) + 3\ O_2(g) \rightarrow 2\ FeO(s) + 2\ SO_2(g)$ (Equation 25.20)

   $\text{mass FeS} = 0.138\ (1000\ \text{g}) = 138\ \text{g}$

   $\text{mass O}_2 = 138\ \text{g FeS} \times \dfrac{3(32.00)\ \text{g O}_2}{2(87.91)\ \text{g FeS}} = 75.3\ \text{g}$

   b. $Cu_2S(s) + O_2(g) \rightarrow 2\ Cu(s) + SO_2(g)$ (Equation 25.22)

   $\text{mass Cu}_2\text{S} = 0.067\ (1000\ \text{g}) = 67\ \text{g}$

   $\text{mass O}_2 = 67\ \text{g Cu}_2\text{S} \times \dfrac{32.0\ \text{g O}_2}{159\ \text{g Cu}_2\text{S}} = 13\ \text{g}$

4. a. $2\ AlCl_3(l) \rightarrow 2\ Al(l) + 3\ Cl_2(g)$

   b. $BaCl_2(l) \rightarrow Ba(l) + Cl_2(g)$

c. $2\ KCl(l) \rightarrow 2\ K(l) + Cl_2(g)$

5. a. $2\ ZnS(s) + 3\ O_2(g) \rightarrow 2\ ZnO(s) + 2\ SO_2(g)$

b. $Cu_2S(s) + O_2(g) \rightarrow 2\ Cu(s) + SO_2(g)$

c. $2\ CoS(s) + 3\ O_2(g) \rightarrow 2\ CoO(s) + 2\ SO_2(g)$

d. $2\ As_2S_3(s) + 9\ O_2(g) \rightarrow 2\ As_2O_3(s) + 6\ SO_2(g)$

6. a. $Fe_2O_3(s) + 3\ CO(g) \rightarrow 2\ Fe(s) + 3\ CO_2(g)$

b. $ZnO(s) + CO(g) \rightarrow Zn(s) + CO_2(g)$

c. $SnO_2(s) + 2\ CO(g) \rightarrow Sn(s) + 2\ CO_2(g)$

d. $Cr_2O_3(s) + 3\ CO(g) \rightarrow 2\ Cr(s) + 3\ CO_2(g)$

7. Cu_2S and CuO contain almost exactly the same percentage of copper.

$$\text{For } Cu_2S\text{: \% Cu} = \frac{127.10 \text{ g Cu}}{159.16 \text{ g } Cu_2S} \times 100 = 79.86$$

$$\text{For CuO: \% Cu} = \frac{63.55 \text{ g Cu}}{79.55 \text{ g CuO}} \times 100 = 79.89$$

8. $Al_2O_3(l) \rightarrow 2\ Al(l) + 3/2\ O_2(g)$ (Equation 25.6)

$$\text{mass } O_2 = 1.00 \text{ g Al} \times \frac{3/2\ (32.0) \text{ g } O_2}{2(27.0) \text{ g Al}} = 0.889 \text{ g}$$

9. a. $MnO_2(s) + 2\ CO(g) \rightarrow Mn(s) + 2\ CO_2(g)$

b. $$\text{mass Mn} = 10^3 \text{ g } MnO_2 \times \frac{54.94 \text{ g Mn}}{86.94 \text{ g } MnO_2} = 631.9 \text{ g}$$

10. a. $2\ Sb_2S_3(s) + 9\ O_2(g) \rightarrow 2\ Sb_2O_3(s) + 6\ SO_2(g)$

b. $Sb_2O_3(s) + 3\ CO(g) \rightarrow 2\ Sb(s) + 3\ CO_2(g)$

c. $$\text{mass } Sb_2S_3 = 10^3 \text{ g Sb} \times \frac{339.7 \text{ g } Sb_2S_3}{243.5 \text{ g Sb}} = 1395 \text{ g}$$

11. a. $$\text{mass } TiO_2 = \frac{5.2\ (120 \text{ g})}{100} = 6.2 \text{ g}$$

b. $$\text{mass Ti} = 6.2 \text{ g } TiO_2 \times \frac{47.9 \text{ g Ti}}{79.9 \text{ g } TiO_2} = 3.7 \text{ g}$$

12. $Al_2O_3(s) + 3\ H_2O + 2\ OH^-(aq) \rightarrow 2\ Al(OH)_4^-(aq)$ (Equation 25.3)

mass $Al_2O_3 = 0.320\ (1000\ g) = 320\ g$

a. $\text{moles } OH^- = 320\ g\ Al_2O_3 \times \frac{1\ mol\ Al_2O_3}{102\ g\ Al_2O_3} \times \frac{2\ mol\ OH^-}{1\ mol\ Al_2O_3} = 6.27$

b. $\text{mass NaOH} = 6.27\ mols\ OH^- \times \frac{40.0\ g\ NaOH}{1\ mol\ OH^-} = 251\ g$

13. Life of Zn reserves: $\frac{120 \times 10^9\ kg}{5.3 \times 10^9\ kg/yr} = 23$ years

Life of Zn (earth's crust): $\frac{8 \times 10^{17}\ kg}{5.3 \times 10^9\ kg/yr} = 2 \times 10^8$ years

14. Energy required: $3 \times 10^9\ kg\ Al \times \frac{1\ mol\ Al}{0.027\ kg\ Al} \times \frac{1\ kwh}{1\ mol\ Al} = 1 \times 10^{11}$ kwh

15. 1 kg blister Cu contains: 990 g Cu, 2 g Ag, 0.4 g Au, and 0.1 g Pt.

value Cu $= 0.990\ kg\ Cu \times \frac{\$1.30}{kg\ Cu} = \$1.29$

value Ag $= 2 \times 10^{-3}\ kg\ Ag \times \frac{\$150}{kg\ Ag} = \$0.3$

value Au $= 4 \times 10^{-4}\ kg\ Au \times \frac{\$5 \times 10^3}{kg\ Au} = \$2$

value Pt $= 1 \times 10^{-4}\ kg\ Pt \times \frac{\$5 \times 10^3}{kg\ Pt} = \$0.5$

Total value = \$4/kg of blister copper.

16. a. $WO_3(s) + 3\ H_2(g) \rightarrow W(s) + 3\ H_2O$

b. volume $H_2 = 1.00\ g\ W \times \frac{3(22.4)\ \ell\ H_2}{184\ g\ W} = 0.365\ \ell$

CHAPTER 26

Chemistry of the Environment

I. Basic Skills

A. Qualitative Students should be able to:

1. a. Describe how the gaseous pollutants (SO_2, SO_3, CO, NO, and NO_2) enter the atmosphere.
   b. Describe the particular hazards of each gaseous pollutant.
2. a. Describe how the trace metals mercury and lead enter the environment.
   b. Describe the hazards of mercury and lead poisoning.
3. Explain the concentration unit of parts per million (ppm).
4. a. Explain what is meant by radioactive nuclei.
   b. Describe the three types of radioactivity and compare their properties.
   c. Give the nuclear symbols of alpha and beta particles.
5. a. Explain what is meant by the half-life of an isotope.
   b. Describe how half-life is related to the intensity of the radiation.
6. Describe the sources of radiation in the environment.

B. Quantitative Students should be able to:

1. Write balanced chemical equations showing how:
   a. the gaseous pollutants are produced.
   b. SO_2 can be removed from stack gases.
   c. O_3 is produced in the atmosphere.
2. Determine the amount of SO_2 produced by burning coal (Example 26.1).
3. Convert between ppm and mass or mass percent (Example 26.2).
4. Determine the symbol of the isotope formed in a nuclear decay reaction (Example 26.3).
5. Write nuclear equations for alpha decay (Example 26.3) and beta decay (Example 26.4).
6. Determine the fraction of a radioactive isotope which remains after a given number of half-lives (Example 26.5).

II. Chapter Development

1. The gaseous pollutants discussed in this chapter are by-products of combustion reactions used to produce energy. For this reason, the pollution problem and the energy problem are closely related. The energy requirements of our society and our energy sources are discussed in Chapter 27.

2. Pollution by solid particles and by unburned hydrocarbons is not discussed in detail. The ozone problem is limited to atmospheric ozone. Stratospheric ozone was previously discussed in Chapter 8.
3. Fission reactions and their radioactive products are discussed in Chapter 27.

III. Problem Areas

1. The concentration unit of ppm is being increasingly used as analytical methods become more sensitive. When the unit is applied to a solid or liquid mixture, it is based on the mass fraction. It is also frequently applied to gas mixtures, and in these cases it is based on mole fraction. The ppm unit is used in Experiment 39.
2. In teaching what happens in nuclear decay reactions, refer back to the graph of stable nuclei (Figure 2.3). Show how unstable nuclei can become more stable by nuclear decay. For example, when the n/p ratio is high, beta decay tends to take place. This can be explained as the conversion of a neutron to a proton and an electron.
3. In balancing nuclear equations, point out that the symbol represents only the nucleus of the isotope. Explain that both mass (mass number) and charge (atomic number) must be balanced. The difference between chemical and nuclear reactions is that electrons are not involved in nuclear reactions and that specific isotopes are involved.

IV. Suggested Activities

1. SO_2 and SO_3: **Teacher should do demonstration under ventilation hood.** Demonstrate the preparation and properties of the sulfur oxides. Burn sulfur in a deflagrating spoon in a wide mouth bottle containing about 20 ml of water. Swirl the bottle periodically to dissolve the SO_2 gas. Test the solution with litmus paper to show that it is acidic. Add 0.1 M $KMnO_4$ solution dropwise to show the reducing action of SO_2. The purple MnO_4^- ion is reduced to the colorless Mn^{2+} ion. The SO_2 is oxidized to SO_3 which in solution forms H_2SO_4. Add 1 M $BaCl_2$ dropwise forming a precipitate of $BaSO_4$ to show the presence of the SO_4^{2-} ion. Relate these reactions to environmental problems. Point out that $CaSO_4$ formed from the reaction of H_2SO_4 and the $CaCO_3$ of limestone is a more soluble compound.
2. Detection of Radiation: **Teacher Demonstration Only**
   Use a radiation detector (Geiger counter) to detect radiation. Begin by detecting the normal background radiation and discussing its sources. Use radioactive minerals or sources to detect radiation of higher intensity. Use paper and metal shields to show the different penetration powers of alpha, beta, and gamma radiation. Show how gamma radiation intensity decreases but is not stopped as additional shielding is added.

CAUTION: All nuclear radiation is inherently harmful. Use mineral specimens or sealed sources sufficiently weak (10 microcuries or less) that they can be handled with relative safety. Local or federal licenses are required for all unsealed sources or for sources of higher activity.

V. Answers to Questions

1. Gaseous air pollutants: SO_2, SO_3, NO, NO_2, and CO
2. a. Produced in auto engines: NO and CO
   b. Produced by burning coal: SO_2
   c. Produced by reaction of other pollutants with O_2: SO_3 and NO_2

3. $SO_3(g) + H_2O \rightarrow H_2SO_4(l)$ (Equation 26.3)

   $3\ NO_2(g) + H_2O \rightarrow 2\ HNO_3(l) + NO(g)$ (Equation 26.11)

4. a. $CaCO_3(s) \rightarrow CaO(s) + CO_2(g)$

   b. $CaO(s) + SO_2(g) + 1/2\ O_2(g) \rightarrow CaSO_4(s)$

5. $2\ CH_4(g) + 3\ O_2(g) \rightarrow 2\ CO(g) + 4\ H_2O$

6. The hemoglobin in the blood normally transports oxygen from the lungs to the tissues. Carbon monoxide molecules displace the O_2 molecules and rob the tissues of oxygen. A displacement of 20% of the hemoglobin molecules by CO can be fatal.

7. $2\ CO(g) + O_2(g) \rightarrow 2\ CO_2(g)$ (Equation 26.8)

8. $NO_2(g) \rightarrow NO(g) + O(g)$ (Equation 26.12)

   $O_2(g) + O(g) \rightarrow O_3(g)$ (Equation 26.13)

   NOTE: The first reaction is photochemical (caused by sunlight).

9. Poisonous trace metals: Be, Cd, Hg, and Pb
10. While mercury has a low vapor pressure (1.6×10^{-6} atm at 20°C), it increases with increasing temperature. For this reason, a mercury spill near a heat source is dangerous because more mercury will be vaporized and possibly inhaled.
11. a. $(CH_3)_2Hg$ b. $(C_2H_5)_4Pb$
12. Leaded gasoline, containing tetraethyl lead, is the primary source of lead compounds in the environment. The increased use of automobiles with catalytic converters (which require nonleaded gasoline) should decrease the lead pollution from this source. The use of lead pigments in interior paints has been stopped. Lead poisoning from eating of leaded paints should decrease as the older painted surfaces are replaced or repainted.

13. An alpha particle is made up of 2 *protons* and 2 *neutrons*. It is identical with the nucleus of a *helium* atom. A beta particle is identical with an *electron*.
14. A charged particle, moving through an electrical field, will be deflected toward the oppositely charged electrode. Positively charged alpha particles and negatively charged beta particles are bent in opposite directions. Gamma radiation, without mass or charge, is not deflected in an electrical field.
15. a. α-emission: The loss of an ^4_2He particle (2p and 2n) results in an atomic number decrease of 2.
    b. β-emission: The loss of an electron results in an atomic number increase of 1.
    c. γ-emission: There is no change in the atomic number.
16. a. α-emission: With the loss of an ^4_2He particle, the mass number decreases by 4.
    b. β-emission: With the loss of an electron, the mass number is unchanged.
    c. γ-emission: The mass number is unchanged.
17. Penetration power: $\gamma > \beta > \alpha$
18. The net effect of β emission is the loss of a neutron and the gain of a proton.

$$^1_0n \rightarrow {}^1_1p + {}^0_{-1}e$$

19. a. Isotope B, half-life of 10 min, gives off the more intense radiation. More nuclei are decaying in a unit of time.
    b. Isotope A, half-life of 10 yr, will last longer.
20. Yes. The decay rate of any radioactive isotope decreases as time passes. This can be seen in Table 26.1 for $^{131}_{53}I$. (A graph of this data is shown in Problem 17.)
21. Sources of radiation in the environment:
    a. cosmic radiation
    b. fall-out from the testing of nuclear weapons
    c. radioactive isotopes in the earth's crust
    d. nuclear reactors and their waste products
    e. x-ray machines

VI. Solutions to Problems

1. $4\ FeS_2(s) + 11\ O_2(g) \rightarrow 8\ SO_2(g) + 2\ Fe_2O_3(s)$ (Equation 26.1)

   a. 1 kg coal $\simeq$ 25 g FeS_2

$$\text{moles } FeS_2 = 25 \text{ g } FeS_2 \times \frac{1 \text{ mol } FeS_2}{120 \text{ g } FeS_2} = 0.21$$

   b. $$\text{moles } SO_2 = 0.21 \text{ mol } FeS_2 \times \frac{2 \text{ mol } SO_2}{1 \text{ mol } FeS_2} = 0.42$$

   c. $$\text{mass } SO_2 = 0.42 \text{ mol } SO_2 \times \frac{64 \text{ g } SO_2}{1 \text{ mol } SO_2} = 27 \text{ g}$$

2. a. $0.005 \text{ ppm Hg} = \dfrac{0.005 \text{ g Hg}}{10^6 \text{ g } H_2O}$ or 5×10^{-9} g Hg/g H_2O

b. $\text{mass Hg} = 1\ \ell\ H_2O \times \dfrac{10^3 \text{ g } H_2O}{1\ \ell\ H_2O} \times \dfrac{5 \times 10^{-9} \text{ g Hg}}{1 \text{ g } H_2O} = 5 \times 10^{-6} \text{ g}$

3. a. $1.00 \text{ mass \% Pb} = \dfrac{1.0 \text{ g Pb}}{100 \text{ g dust}} = \dfrac{0.010 \text{ g Pb}}{\text{g dust}}$

b. $\text{ppm} = \dfrac{0.010 \text{ g Pb}}{\text{g dust}} \times 10^6 \text{ g dust} = 1.0 \times 10^4$

4. a. ${}^{235}_{92}U \rightarrow {}^{4}_{2}He + {}^{231}_{90}Th$

b. ${}^{232}_{90}Th \rightarrow {}^{4}_{2}He + {}^{228}_{88}Ra$

5. a. ${}^{137}_{55}Cs \rightarrow {}^{0}_{-1}e + {}^{137}_{56}Ba$

b. ${}^{90}_{38}Sr \rightarrow {}^{0}_{-1}e + {}^{90}_{39}Y$

6. a. If half-life is 12 yr, after 12 yr one-half of the tritium remains:

mass ${}^{3}_{1}H = 0.500\ (2.40 \text{ g}) = 1.20 \text{ g}$

b. The number of elapsed half-lives is:

$n = \dfrac{36 \text{ yr}}{12 \text{ yr}} = 3$

The fraction of tritium which remains is:

fraction $= (1/2)^n = (1/2)^3$ or 0.125

mass ${}^{3}_{1}H = 0.125\ (2.40 \text{ g}) = 0.300 \text{ g}$

7. a. If half-life of ${}^{7}_{4}Be$ is 53 days, one-half of the isotope remains after this period.

b. $n = \dfrac{530 \text{ days}}{53 \text{ days}} = 10$

fraction $= (1/2)^{10} = 1/1024 = 9.8 \times 10^{-4}$

8. a. ${}^{40}_{19}K \rightarrow {}^{0}_{-1}e + {}^{40}_{20}Ca$

b. $n = \dfrac{5.2 \times 10^9 \text{ yr}}{1.3 \times 10^9 \text{ yr}} = 4$

fraction $= (1/2)^4 = 0.0625$

9. $^{28}_{13}\text{Al} \rightarrow\ ^{28}_{14}\text{Si} +\ ^{0}_{-1}\text{e}$ beta decay

10. $^{27}_{13}\text{Al} +\ ^{1}_{0}\text{n} \rightarrow\ ^{1}_{1}\text{H} +\ ^{27}_{12}\text{Mg}$

11. a. $n = \dfrac{40 \text{ min}}{20 \text{ min}} = 2$

$\text{fraction} = (1/2)^2 = 0.25$

$\text{fraction decayed} = 1.00 - 0.25 = 0.75$

b. $n = \dfrac{60 \text{ min}}{20 \text{ min}} = 3$

$\text{fraction} = (1/2)^3 = 0.125$

$\text{fraction decayed} = 1.000 = 0.125 = 0.875$

12. $\text{mass } ^{40}\text{K} = 50 \text{ kg} \times \dfrac{10^3 \text{ g}}{1 \text{ kg}} \times \dfrac{3 \text{ g } ^{40}\text{K}}{10^6 \text{ g}} = 0.150 \text{ g}$

13. $\text{mass S} = 10^6 \text{ g coal} \times \dfrac{0.5 \text{ g S}}{100 \text{ g coal}} = 5 \times 10^3 \text{ g or } 5 \times 10^3 \text{ ppm}$

14. $2\ C_8H_{18}(l) + 17\ O_2(g) \rightarrow 16\ CO(g) + 18\ H_2O$

$\text{mass CO} = 1.00 \text{ g } C_8H_{18} \times \dfrac{16(28.0) \text{ g CO}}{2(114) \text{ g } C_8H_{18}} = 1.96 \text{ g}$

15. $\dfrac{[O_2] \times [\text{Hem} \cdot \text{CO}]}{[\text{CO}] \times [\text{Hem} \cdot O_2]} = 200$

If 20% of the hemoglobin reacts with CO: $\dfrac{[\text{Hem} \cdot \text{CO}]}{[\text{Hem} \cdot O_2]} = \dfrac{0.20}{0.80} = 0.25$

$\dfrac{[\text{CO}]}{[O_2]} = \dfrac{[\text{Hem} \cdot \text{CO}]}{[\text{Hem} \cdot O_2]} \times \dfrac{1}{200} = \dfrac{0.25}{200} = 1.2 \times 10^{-3}$

16. a. $\text{mass Hg} = 2.4 \times 10^{-3} \text{ g } (CH_3)_2Hg \times \dfrac{201 \text{ g Hg}}{231 \text{ g } (CH_3)_2Hg}$
$= 2.1 \times 10^{-3} \text{ g}$

b. $\text{mass Hg} = 10^6 \text{ g fish} \times \dfrac{2.1 \times 10^{-3} \text{ g Hg}}{1.3 \times 10^3 \text{ g fish}} = 1.6 \text{ g or } 1.6 \text{ ppm}$

17.

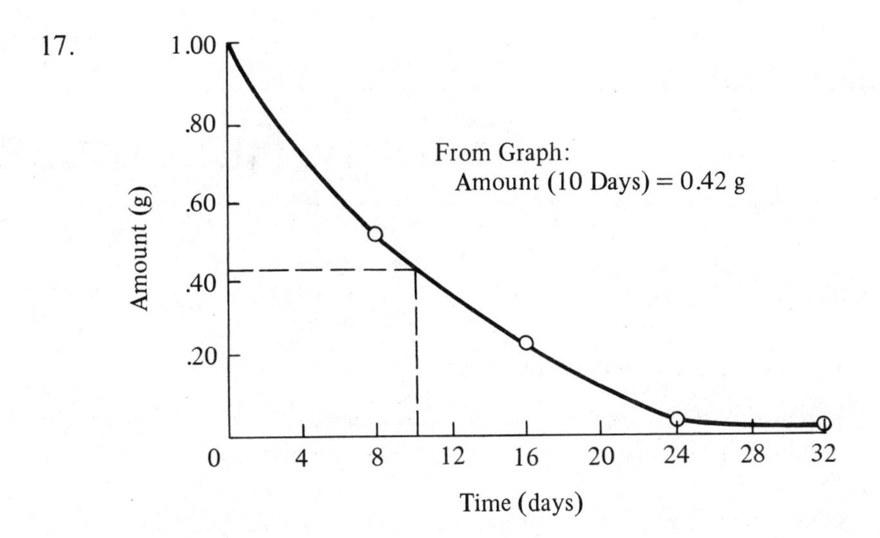

18. a. $\text{mass Pb} = 11 \text{ g paint} \times \dfrac{5.0 \text{ g } Pb_3O_4}{100 \text{ g paint}} \times \dfrac{622 \text{ g Pb}}{686 \text{ g } Pb_3O_4} = 0.50 \text{ g}$

b. $\text{mass Pb} = 10^6 \text{ g body} \times \dfrac{0.50 \text{ g Pb}}{10^4 \text{ g body}} = 50 \text{ g or } 50 \text{ ppm}$

CHAPTER 27

Energy Resources

I. Basic Skills

A. Qualitative Students should be able to:

1. Explain how U.S. energy consumption compares to world consumption and how the U.S. per capita consumption compares to the world average (Table 27.1).
2. a. Describe the major uses of energy in the U.S. (Table 27.2).
   b. Describe how energy consumption in the U.S. could be reduced.
3. Describe how the energy sources in the U.S. have changed since 1900 (Figure 27.3).
4. Describe the advantages and disadvantages of natural gas, petroleum, and coal as energy sources.
5. Describe how the energy change in a reaction is related to a change in mass.
6. a. Explain what is meant by nuclear fission.
   b. Describe the characteristics of fission reactions.
7. a. Describe how a fission reactor operates.
   b. Constrast fission power with conventional power.
   c. Describe how a breeder reactor operates and contrast it with conventional reactors.
8. a. Explain what is meant by nuclear fusion.
   b. Describe the characteristics of fusion reactions and contrast fusion power with fission power.
9. a. Describe the characteristics of solar energy.
   b. Describe the process of solar water heating.
   c. Explain what is meant by a solar cell and describe its operation.

B. Quantitative Students should be able to:

1. Determine the heating value of a fuel (Example 27.1).
2. Write balanced chemical equations showing how:
   a. coal or carbon is changed into gaseous fuels.
   b. CO and H_2 are converted into liquid hydrocarbons.
3. Determine the mass change associated with an energy change, or vice versa (Examples 27.2 and 27.3).
4. Write balanced nuclear equations for fission reactions (Example 27.4) and fusion reactions.

II. Chapter Development

1. Most of the chapter is devoted to the conventional fossil fuels and to nuclear energy. The alternatives of coal gasification and liquifaction, fusion, and solar energy are briefly examined. Other alternatives have been omitted.
2. While the material does not ''preach,'' an attempt is made to provide the student with a broad awareness of the energy problem. Ample material is printed daily to provide extensions and excursions in any direction on the subject of energy.
3. In comparing fuels, recall that heats of combustion are given in Table 4.3 and heating values are given in Table 4.4. The use of gasohol as a fuel was also discussed in Chapter 4.

III. Problem Areas

1. Students may have difficulty in understanding the interconversion of matter and energy. Until this time, they have subscribed to separate laws of conservation of mass and energy. The mass loss for nuclear reactions is significant and measurable. It can be calculated using a table of isotopic masses.

IV. Suggested Activities

1. Conventional Power vs Nuclear Power:
   Visit a conventional power plant and a nuclear power plant. (Arrange visits in advance.) Ask questions about power output, fuel consumption, and cost per kilowatt-hour. Discuss safety features. Discuss how the chemical or nuclear energy is converted to electrical energy. Discuss the methods used to dispose of the waste products.
2. Energy Survey:
   At the beginning of the chapter, have students complete an anonymous energy survey form. For example: find the cars per family, fuel used to heat the home, TV sets per home, distance to school, type of transportation used for school, number of passengers per car, and any other energy-related facts of interest. Tabulate the answers, compute averages, and discuss the results.

V. Answers to Questions

1. To reduce energy consumption by automobiles:
   a. produce automobiles which get increased gas mileage.
   b. use more efficient transportation methods (feet, bicycle, motorcycle, bus, or train).

c. drive fewer miles.

2. To reduce energy consumption by space heating:
   a. use more insulation.
   b. use thermal windows and have fewer windows.
   c. turn the thermostat down to 65°F (18°C) in the winter and to 80°F (27°C) in the summer, if air-conditioned.
3. The increased use of petroleum since 1900 is due to:
   a. the invention and use of the automobile.
   b. the conversion of space heating from coal and wood to oil and gas.
   c. the low cost of petroleum products.
4. United States petroleum reserves are estimated to last 20 years without imports. World petroleum reserves have an estimated lifetime of 80 years. Both of the above estimates are based on current consumption rates with no growth. New petroleum discoveries would not significantly change these estimates.
5. The major use of coal is to produce electrical energy.
6. The disadvantages of coal as a fuel are that it:
   a. is difficult to mine.
   b. is difficult to transport and handle.
   c. pollutes the atmosphere with SO_2 and solids.
7. Water gas, a mixture of CO and H_2, is made by heating solid carbon (coke) with steam:

   $$C(s) + H_2O(g) \rightarrow CO(g) + H_2(g)$$

   Synthesis gas is made in a similar manner by heating coal with steam. The product mixture contains about 40 mol % H_2, 15 mol % CO, 15 mol % CH_4, and 30 mol % CO_2.
8. The CO and H_2 gases from synthesis gas can be converted to liquid hydrocarbons when heated with the proper catalyst. The products can be refined to a low-grade gasoline for use in automobiles.
9. If a reaction absorbs energy, the mass of the products is greater than the mass of the reactants: Δm is positive.
10. The mass changes (Δm) and energy changes (ΔE) of nuclear reactions are both much greater than those of chemical reactions. (See the numerical comparison in Example 27.3.)
11. In nuclear fission, a heavy nucleus splits into two smaller nuclei. In nuclear fusion, two very small nuclei combine to form a heavier nucleus.
12. From Figure 27.6, it can be seen that the isotopes of small mass number have a larger mass (per particle) than those which immediately follow. In fusion, two nuclei combine to form a single nucleus with a smaller average mass. As the mass change (Δm) is negative, energy is released. A similar situation exists for isotopes of heavier mass such as $^{235}_{92}U$. In fission, the heavy isotope is split into two nuclei of smaller average mass. As before, Δm is negative and energy is released.
13. Isotopes which undergo fission in a reactor are:

    (a) $^{235}_{92}U$ and (d) $^{239}_{94}Pu$

14. The neutron production rate in a reactor controls the rate at which energy is produced. A neutron rate which is too high may melt the fuel elements. Control rods, which capture neutrons, are used to adjust the neutron production rate.
15. In most reactors, the heat from fission is used to heat water. This heat is then transferred through a closed loop to another system in which water is converted to steam. The steam is then used to operate a conventional turbine-generator whose output is electrical energy.
16. Advantages of using nuclear energy to produce electrical energy are:
    a. the small amount of fuel required:

    $$1\ \text{g}\ ^{235}\text{U} \simeq 3\ \text{tons coal}$$

    b. the reduced air pollution; no gaseous combustion products.
    c. that it helps to conserve fossil fuels.
17. The major problem associated with nuclear energy is the radioactivity of the fission products. Any accident, man-made or natural, could release radioactive material into the environment. In addition, the radioactive waste products must be safely stored for hundreds of years.
18. A breeder reactor is designed to convert non-fissionable $^{238}_{92}U$ into fissionable $^{239}_{94}Pu$:

    $$^{238}_{92}\text{U} + ^{1}_{0}\text{n} \rightarrow ^{239}_{94}\text{Pu} + 2\ ^{0}_{-1}\text{e}$$

    Ideally, a breeder reactor produces as much or more fuel than it consumes.
19. Breeder reactor vs conventional nuclear reactor:

    a. Fuel used: $^{238}_{92}U$ vs $^{235}_{92}U$

    b. Safety: Breeder reactors, by their nature, are more difficult to control. In addition, the $^{239}_{94}Pu$ produced is extremely toxic if released into the atmosphere.
20. The advantages of fusion are that:
    a. the fuel supply, deuterium, is almost unlimited.
    b. the fusion products, helium nuclei, are not radioactive.
    The major problem to be solved in the development of fusion energy is related to its very large activation energy. Temperatures in the range of 10^6 to 10^7 K are required to fuse hydrogen nuclei. Energies of this order are difficult to attain, difficult to sustain, and difficult to contain.
21. Solar energy is produced by: (c) fusion. Solar cells and windmills convert solar energy into other forms.
22. Solar energy is difficult to use directly because it is a diffuse energy source ($3000\ \text{kcal/m}^2 \cdot \text{day}$) and its supply is not constant.
23. Solar energy is most readily used to: (a) heat water.
24. In solar cells, the radiant energy from the sun causes certain atoms in the cell to lose electrons. The electrons can be made to flow, producing a voltage of about 0.5 V. Solar cells are not yet economical because their efficiency

is low and the materials cost is high.

25. The fraction of total energy supplied by the year 2000 is expected to change as follows:
    a. petroleum; decrease b. coal; increase
    c. solar energy, increase d. water power; no change
    e. fission; increase f. fusion; no change
    NOTE: As these answers are predictions only, some of them are debatable.

VI. Solutions to Problems

1. a. $\text{heat value } (H_2) = \dfrac{68.3 \text{ kcal}}{2.016 \text{ g } H_2} = 33.9 \text{ kcal/g}$

   b. $\text{heat value (CO)} = \dfrac{67.7 \text{ kcal}}{28.01 \text{ g CO}} = 2.42 \text{ kcal/g}$

2. $\text{heat evolved} = 0.5 \text{ mol } H_2 \times \dfrac{68.3 \text{ kcal}}{1 \text{ mol } H_2} + 0.5 \text{ mol CO} \times \dfrac{67.7 \text{ kcal}}{1 \text{ mol CO}}$

   $\text{heat evolved} = 34.2 \text{ kcal} + 33.8 \text{ kcal} = 68.0 \text{ kcal}$

3. $\text{heat value} = \dfrac{1376 \text{ kcal}}{2 \text{ mol } C_4H_{10}} \times \dfrac{1 \text{ mol } C_4H_{10}}{58.12 \text{ g } C_4H_{10}} = 11.84 \text{ kcal/g } C_4H_{10}$

4. a. $\Delta m = \dfrac{\Delta E}{2.15 \times 10^{10} \text{ kcal/g}}$

   $\Delta m = \dfrac{-6.0 \times 10^{6} \text{ kcal}}{2.15 \times 10^{10} \text{ kcal/g}} = -2.8 \times 10^{-4} \text{ g}$

   b. $\text{mass of products} = 1.00000 \text{ g} - 0.00028 \text{ g} = 0.99972 \text{ g}$

5. a. $\Delta E = 2.15 \times 10^{10} \text{ kcal/g} \times \Delta m$

   $\Delta E = 2.15 \times 10^{10} \text{ kcal/g } (1.2 \times 10^{-10} \text{ g}) = 2.6 \text{ kcal}$

   b. $\Delta E = 2.15 \times 10^{10} \text{ kcal/g } (-1.6 \times 10^{-3} \text{ g}) = -3.4 \times 10^{7} \text{ kcal}$

6. ${}^{239}_{94}\text{Pu} + {}^{1}_{0}\text{n} \rightarrow {}^{90}_{37}\text{Rb} + {}^{147}_{57}\text{La} + 3\ {}^{1}_{0}\text{n}$

7. ${}^{3}_{1}\text{H} + {}^{6}_{3}\text{Li} \rightarrow {}^{9}_{4}\text{Be}$

8. $\text{heat evolved} = 0.3 \text{ mol } H_2 \times \dfrac{68.3 \text{ kcal}}{1 \text{ mol } H_2} + 0.3 \text{ mol CO} \times \dfrac{67.7 \text{ kcal}}{1 \text{ mol CO}}$

   $\text{heat evolved} = 20.5 \text{ kcal} + 20.3 \text{ kcal} = 40.8 \text{ kcal}$

9. $$\text{heat value } (CH_3OH) = \frac{174 \text{ kcal}}{32.0 \text{ g}} = 5.44 \text{ kcal/g}$$

10. $$\text{heat value (CO)} = \frac{68 \text{ kcal}}{1 \text{ mol CO}} \times \frac{1 \text{ mol CO}}{28.0 \text{ g CO}} = 2.4 \text{ kcal/g}$$

$$\text{heat value (C)} = \frac{94 \text{ kcal}}{1 \text{ mol C}} \times \frac{1 \text{ mol C}}{12.0 \text{ g C}} = 7.8 \text{ kcal/g}$$

11. ${}^{235}_{92}U + {}^{1}_{0}n \rightarrow {}^{140}_{53}I + {}^{92}_{39}Y + 4\ {}^{1}_{0}n$

12. ${}^{2}_{1}H + {}^{3}_{1}H \rightarrow {}^{4}_{2}He + {}^{1}_{0}n$

13. If 0.1% of the mass is "lost," then $\Delta m = -10^{-3}$ g/g of reactants

$\Delta E = 2.15 \times 10^{10}$ kcal/g $\times -10^{-3}$ g $= -2 \times 10^{7}$ kcal

14. a. mass of products = 4.0015 g + 221.9703 g = 225.9718 g

Δm = mass of products − mass of reactants

Δm = 225.9718 g − 225.9771 g = −0.0053 g

b. $\Delta E = 2.15 \times 10^{10}$ kcal/g (−0.0053 g) $= -1.1 \times 10^{8}$ kcal

15. a. $$\text{world energy/yr} = 3.61 \times 10^{9} \text{ persons} \times 84 \times \frac{10^{6} \text{ kcal}}{\text{yr} \cdot \text{person}}$$

$$= 3.0 \times 10^{17} \frac{\text{kcal}}{\text{yr}}$$

b. $$\text{world energy/yr} = 3.61 \times 10^{9} \text{ persons} \times \frac{60 \times 10^{6} \text{ kcal}}{\text{yr} \cdot \text{person}}$$

$$= 2.2 \times 10^{17} \frac{\text{kcal}}{\text{yr}}$$

16. a. $$\text{Fraction of transportation energy used by autos} = \frac{2.75 \times 10^{15} \text{ kcal/yr}}{4.22 \times 10^{15} \text{ kcal/yr}} = 0.652$$

b. $$\text{Fraction of space heating energy used in homes} = \frac{1.88 \times 10^{15} \text{ kcal/yr}}{3.03 \times 10^{15} \text{ kcal/yr}} = 0.620$$

17. Consumption of oil and gas = 0.75 (1.7×10^{16} kcal/yr)
= 1.3×10^{16} kcal/yr

$$\text{Lifetime of oil and gas} = \frac{2.5 \times 10^{17} \text{ kcal}}{1.3 \times 10^{16} \text{ kcal/yr}} = 19 \text{ yr}$$

18. $$\text{Lifetime of one day of solar energy} = \frac{1.5 \times 10^{18} \text{ kcal}}{4.9 \times 10^{16} \text{ kcal/yr}} = 31 \text{ yr}$$

19. $$\text{Oil cost}/10^6 \text{ kcal} = 44 \text{ gal} \times \frac{\$0.50}{\text{gal}} = \$22$$

$$\text{Wood cost}/10^6 \text{ kcal} = 0.40 \text{ cord} \times \frac{\$75}{\text{cord}} = \$30$$

20. $8\ CO(g) + 17\ H_2(g) \rightarrow C_8H_{18}(l) + 8\ H_2O$

21. The rate for "n" years can be found by multiplying the initial rate by 1.03 (for 3%) and 1.01 (for 1%) "n" times.

Time (yrs)	Rate (3% increase)	Rate (1% increase)
0	1.7×10^{16} kcal	1.7×10^{16} kcal
5	2.0×10^{16} kcal	1.8×10^{16} kcal

22. Δm = mass of products − mass of reactants

$\Delta m = 4.00150 \text{ g} - 2(2.01355 \text{ g}) = -0.02560 \text{ g}$

$\Delta E = 2.15 \times 10^{10} \text{ kcal/g } (-0.02560 \text{ g}) = -5.50 \times 10^8 \text{ kcal}$

$$\frac{\Delta E}{\text{g deuterium}} = \frac{-5.50 \times 10^8 \text{ kcal}}{4.03 \text{ g deuterium}} = -1.36 \times 10^8 \text{ kcal/g}$$

23. $$\text{Collector area } (m^2) = \frac{2 \times 10^5 \text{ kcal/day}}{(0.80)(3 \times 10^3 \text{ kcal/m}^2 \cdot \text{day})} = 80 \text{ m}^2$$

GUIDE TO THE LABORATORY

Introduction to the Laboratory Guide

Since the time of Lavoisier, chemistry has been a laboratory science. It is especially important that beginning students spend a fair portion of their time in the laboratory. Only in this way can they learn what chemistry is and what chemists do. It is quite common to be told by former students that not only did they learn from their time in the laboratory, but that they remember it as an experience which they enjoyed. The authors hope that your laboratory, using this manual, will provide many enjoyable learning experiences.

Selecting experiments for a laboratory manual is a difficult task and one in which compromises must be made. We know that it is not possible to satisfy all teachers, either in the experiments selected or in the procedures by which they are carried out. To that extent, more experiments are provided than will normally be done in a school year. Extensions are provided for most experiments, and in some, alternative procedures are suggested.

Our primary objective in the Laboratory Manual is to demonstrate and reinforce the principles presented in the text, **Chemistry** by Masterton, Slowinski, and Walford. While we have tried to be somewhat innovative in doing this, we have also been quite traditional. In agreement with the text we have placed an increased emphasis on descriptive chemistry, both inorganic and organic. Quite a few quantitative experiments are also included as this aspect of chemistry must also be stressed. Many of the quantitative experiments use unknowns to stimulate student interest.

We have made an effort to provide a set of experiments which are relatively inexpensive to perform. For example, the use of silver nitrate has been avoided except as required in qualitative analysis. In many instances household chemicals have been specified which can be purchased locally. (It is our experience that student interest picks up when working with recognizable commercial products.) No special equipment or instruments are required. This is a large advantage when converting from one laboratory manual to another.

We have tried to provide the student with a thorough discussion of the principles involved in each experiment. We hope the preliminary discussions, the procedures, and the follow-up questions are sufficiently clear and complete that the student can accomplish the stated objectives of the experiment with a minimum amount of direct assistance. A sample data table is provided for most experiments to help students in organizing and completing their collection of data. The specific format of the laboratory report is left to the teacher's discretion.

The Laboratory Guide, which follows, includes the information which we believe would be of benefit to the teacher. It is organized in such a way that you should be able to rapidly locate the information you need. Each experiment in the guide is organized by the headings which are discussed below. The authors welcome comments from teachers as to suggestions for changes or additions.

TIMING

An estimated time for each experiment is provided. Most experiments can be completed by most students in a single laboratory period (45–50 minutes). Some, as noted, can be conveniently broken into parts. Successful completion within time limits usually requires prior preparation by both the students and the teacher.

A correlation of the text chapters with the laboratory experiments is included on pages 215 to 217. This can be used as an overall planning guide. There is some flexibility, and possible variations in placement are suggested under this heading for certain experiments.

PRELIMINARY STUDY

Students should read the Discussion and the Procedures before coming to the laboratory and starting the experiment. Each experiment also contains a Preliminary Study section which suggests further preparation. In particular, sections of the text which apply to the experiment are referenced. In many cases questions or practice problems are given which the student should attempt before lab time. The solutions to the practice problems are given in this section of the Laboratory Guide.

EXPERIMENT NOTES

An effort is made in this section to provide hints for preparing for the experiment and for performing it. These are expected to be most useful for the new teacher or for one doing it for the first time. If any special instructions are required for preparing reagents they will appear in this section. Most quantitative experiments include an estimate of the average experimental error which can be expected.

MATERIALS NEEDED

The apparatus (hardware and glassware), chemicals, and supplies needed are listed in this section. The number of items and amounts of chemicals needed are listed on a per experiment basis (individuals or pairs). The chemicals marked with an asterisk can be purchased locally, usually at a savings. Balances and ring stands are not listed as they are considered to be always available. In some cases the requirements can be reduced by sharing. (Everyone doesn't need their personal voltmeter, for example.) A complete materials list for all experiments begins on page 343.

All materials needed for an experiment should be assembled or prepared in advance. Many schools use student lab assistants to help in this work. If possible, separate sets of reagents should be available at each desk. Most teachers prefer to provide special set-ups, such as glass tubing in stoppers, rather than expend student time to make them and risk an accident. Schools vary in how the necessary hardware and glassware are distributed. Some issue them as needed for the particular experiment, others have stocked individual student drawers, and still others use a combination of the two systems. Use whatever system works best for you. Students should be made aware of the cost of the equipment they use and should be responsible for taking care of it.

SAMPLE DATA TABLE

A set of typical student data is provided for each experiment. It is not intended to convey the idea that this is the only acceptable set of data. In some cases the data varies considerably with the unknown which is being analyzed. Student data should be recorded directly on a data table during the progress of the experiment.

CALCULATIONS AND QUESTIONS

A complete set of answers is provided for this section of each experiment. The calculations are based on the data provided in the accompanying data table. The questions are usually answered from the viewpoint of the teacher though other expected answers may be given. Student answers may vary considerably. Try to get them to answer questions in a logical, readable, and grammatically correct form. Particularly check that student answers are consistent with their recorded observations and data.

EXTENSIONS

Extensions are provided for most experiments. They are primarily provided for the faster or above average student. They also provide a source of investigations for those who simply want to do "extra" work. In some cases you may want to assign an extension in addition to the experiment. The materials needed for an extension (if any) are not listed in the Materials section and must be separately provided. A brief summary of answers or expected results is provided in most cases.

SPECIAL MESSAGE ON SAFETY IN THE LABORATORY

Whenever your students are working in the laboratory, it is important that you keep in mind the procedures students should use to ensure that the experiment being performed is done safely. Students tend to be unaware of possible hazards, and need to be reminded if they are doing something that is potentially dangerous.

As a general rule, it is advisable to have students wear safety glasses or goggles whenever they are working in the laboratory. We have indicated those experiments where safety eyewear is essential, but you may wish to have it used at all times and thereby establish a standard procedure.

Since you may not be familiar with some of these experiments, it would be a good idea if you personally carry out such experiments before they are redone by the class. That way you will be aware of what to expect and will be able to note procedures that could be dangerous if done improperly. We have attempted to indicate in the procedures where caution must be used, but there is nothing like doing the experiment to learn where one needs to be careful.

In some of the experiments, particularly those involving organic compounds, the liquids used may give off appreciable amounts of vapor. While small amounts of such vapors are not likely to be bothersome, it is important to have good ventilation in the laboratory when such experiments are performed. At the end of the laboratory period, students should routinely wash their hands and faces.

The experiments we have included in this manual are safe if performed as directed. This means that students should follow directions and never perform unauthorized experiments. Violation of this rule can be serious, and intentional offenders should be banished from the laboratory. In a similar vein, roughhousing in the laboratory cannot be tolerated, since it could easily result in spilled chemicals, broken glass, or fires, all of which are dangerous.

If you can instill in your students a respect for chemicals and laboratory equipment, you will do them a favor as far as safety goes, and will probably make them more effective chemistry students in the bargain.

CORRELATION BETWEEN TEXT AND LABORATORY MANUAL

Chapter Number	Title	Experiment Number	Title
1	An Introduction to Chemistry	1	Laboratory Techniques: Making Measurements
		2	Laboratory Techniques: Making Chemical Products
		3	Relationships Between Variables: Mass and Volume of a Liquid
2	Atoms, Molecules, and Ions	4	Counting Particles and Finding Their Relative Masses
		5	Finding the Size of a Molecule and a Value for Avogadro's Number
3	Chemical Formulas and Equations	6	Determining the Simplest Formula of a Compound
		7	Observing Chemical Reactions and Writing Equations
4	Energy Changes	8	The Specific Heat of Liquids and Solids
		9	Heat Effects of Chemical Reactions
5	The Physical Behavior of Gases	10	Determining the Molecular Mass of a Gas
		11	Volume-Mass Relations in Chemical Reactions
6	The Chemical Behavior of Gases	12	The Preparation and Properties of Pure Oxygen
7	The Periodic Table	13	Periodicity and Predictions of Properties

Chapter Number	Title	Experiment Number	Title
8	Chemical Behavior of the Main Group Elements	14	Some Chemical Properties of the Main Group Elements
9	Electronic Structure of Atoms	15	Understanding Electron Charge Density Diagrams
10	Ionic Bonding	16	Analysis of a Hydrate
11	Covalent Bonding	17	Covalent Bonding and Molecular Structure
12	Liquids and Solids	18	Solid-Liquid Phase Changes
		19	Relationships Between Physical Properties and Chemical Bonding in Solids
13	Water Solutions	20	The Effect of Temperature on Solubility
		21	Preparation and Properties of Solutions
14	Organic Chemistry: Hydrocarbons	22	Isomerism in Organic Chemistry
		23	Properties of Hydrocarbons
15	Organic Chemistry: Oxygen Compounds	24	Organic Functional Groups
		25	Organic Syntheses
16	Organic Chemistry: Polymers	26	Preparation and Properties of Polymers
17	Rate of Reaction	27	Rates of Chemical Reactions
18	Chemical Equilibrium	28	Systems in Chemical Equilibrium

Chapter Number	Title	Experiment Number	Title
19	Acids and Bases	29	Measurement of pH with Acid-Base Indicators
		30	Properties of Acids and Bases
20	Precipitation Reactions	31	Ionic Precipitation Reactions
21	Quantitative Analysis	32	Acid-Base Titrations
		33	Testing Consumer Products
22	Qualitative Analysis	34	Developing a Qualitative Analysis Scheme
		35	Qualitative Analysis of the Group I ions: Ag^+, Pb^{2+}, Hg_2^{2+}
23	Oxidation-Reduction Reactions	36	Analysis by an Oxidation-Reduction Titration
24	Electrochemistry	37	Voltaic Cells
25	Metals and Their Ores	38	Winning a Metal from its Ore
26	Chemistry of the Environment	39	Trace Analysis by Colorimetry

Laboratory Techniques: Making Measurements

TIMING

The experiment requires about 45 minutes. It should be done early during Chapter 1. It is recommended that the following day be used to review the calculations.

PRELIMINARY STUDY

3. $\% \text{ Error} = \dfrac{12.5 - 13.6}{12.5} \times 100 = \dfrac{1.1}{12.5} \times 100 = 8.8\%$

NOTE: Percent errors are given in absolute values, that is, without regard to sign.

EXPERIMENT NOTES

1. While many students will already know how to use a laboratory balance it is best to review the procedures in a Prelab. The most common error is to make an incorrect reading. Be certain that students can read the measurement correctly and that they record it to the correct number of decimal places. Begin the first day to teach respect for the proper use and care of the balance.
2. Aluminum foil (supermarket variety) is 1.0×10^{-3} inch (2.5×10^{-3} cm) thick. Tin foil is 1.5×10^{-3} inch (3.8×10^{-3} cm) thick. Expect an experimental error in the thickness determination of about 10%.
3. One piece metal slugs (5–20 cm^3) are most convenient for finding the density of a metal. If other forms are used the student should use enough sample to get an appreciable water displacement. Appropriate sample sizes can be prepared in advance and placed in labeled test tubes. (The same metal samples can be saved and used as specific heat specimens in Experiment 8.) When the water displacement is less than 10 cm^3, the volume measurement and the resulting density calculations are limited to two significant figures.
4. In Procedure 4, use any convenient object whose mass is 10–100 g. (A toy makes a nice conversation piece.) The actual mass should be predetermined on an analytical balance, if possible. Have students place their data on the board and initial it. When a result is obviously incorrect the object should be reweighed. Using a centigram balance, classes of beginning students obtain a precision error of $\pm$ 0.03 g.

MATERIALS NEEDED

Apparatus	Chemicals
metric rule	*Al foil (15 cm × 10 cm)
graduated cylinder, 50 ml	tin foil (10 cm × 5 cm)
	metal unknowns (5–20 cm^3)
	(See Table 1.1 for suggestions.)

SAMPLE DATA TABLE

Thickness of a Metal Foil:

metal __Al__ density __2.70 g/cm^3__ mass __0.95 g__
length __15.15 cm__ width __9.91 cm__

Density of a Metal:

unknown number __Pb__ volume of water (initial) __30.0 cm^3__
mass __64.90 g__ volume of water (final) __35.5 cm^3__

Precision of Balance:

mass of class object __28.56 g__

CALCULATIONS AND QUESTIONS

1. a. area = l × w = 15.15 cm × 9.91 cm = 150 cm^2 (3 s.f.)

   b. $\text{volume} = \frac{\text{mass}}{\text{density}} = \frac{0.95\text{ g}}{2.70\text{ g/cm}^3} = 0.35\text{ cm}^3$

   c. $\text{thickness} = \frac{\text{volume}}{\text{area}} = \frac{0.35\text{ cm}^3}{150\text{ cm}^2} = 2.3 \times 10^{-3}\text{ cm}$

   d. $\text{thickness (in.)} = 2.3 \times 10^{-3}\text{ cm} \times \frac{1\text{ inch}}{2.54\text{ cm}} = 9.1 \times 10^{-4}\text{ in}$

2. a. volume metal = 35.5 cm^3 − 30.0 cm^3 = 5.5 cm^3

   b. $\text{density} = \frac{\text{mass}}{\text{volume}} = \frac{64.90\text{ g}}{5.5\text{ cm}^3} = 12\text{ g/cm}^3$

   c. As the metal doesn't look like silver, it is probably lead.

   $\%\text{ Error} = \frac{11.35 - 12}{11.35} \times 100 = \frac{1}{11.35} \times 100 = 10\%$

d. density (lb/ft^3) $= \frac{12\ g}{cm^3} \times \frac{1\ lb}{454\ g} \times \frac{(2.54\ cm)^3}{1\ in^3} \times \frac{(12\ in)^3}{1\ ft^3} = 750\ \frac{lb}{ft^3}$

3. a. Class data on weighing class object:

Mass (g)	Error
28.59	0.01
28.58	0.00
28.56	0.02
28.59	0.01
28.61	0.03
28.56	0.02
28.57	0.01
28.59	0.01
228.65 Total	0.11 Total

sample error calculation:
error = 28.59 − 28.58 = 0.01

$$\text{average mass} = \frac{228.65}{8} = 28.58\ g$$

b. $\text{average error} = \frac{0.11}{8} = 0.01$ (uncertainty is ± 0.01 g)

EXTENSIONS

1. The thickness error should be less using a micrometer.
2. If the metal piece is large enough, the volume obtained from linear measurements (and therefore, the density) is more precise than the volume obtained by displacement.
3. The precision obtained on a 50 ml graduated cylinder is about ± 0.2 cm^3.

Laboratory Techniques: Making Chemical Products

TIMING

Most students complete three of the four preparations during a 50 minute period. Allow students to select in advance the preps which they prefer to do. While the experiment is recommended for Chapter 1, it will fit in anytime before the recommended Chapter 3 experiments.

EXPERIMENT NOTES

1. The experiment is intended to teach laboratory techniques in an enjoyable manner by the preparation of familiar products. Circulate among the students helping them to develop good laboratory procedures. While quantitative data is not required, point out that the same techniques will be used in quantitative experiments.

2. The experiment offers many opportunities to follow the correct safety procedures as discussed in the introduction of this manual. Begin now to teach and enforce safe methods.

3. Special care should be taken in the dispensing of concentrated sulfuric acid. If the acid is poured from a glass-stoppered bottle, sooner or later a drop of acid will run down the outside of the bottle. If not wiped off (a hazardous task itself) the next student using the bottle may be burned. One solution which works is to use a separate medicine dropper stored in a flask next to the acid bottle. (The dropper should be fitted with a glass tube which reaches to the bottom of the acid bottle.) In this procedure the student removes the stopper (holding it as shown in Figure 2(a) of the Introduction), uses the dropper to dispense the acid, and returns it to the flask when finished. The acid set-up should be on the instructor's table where its use can be closely watched.

4. A simple data table using column headings of "Reactants" and "Observations" is sufficient for this experiment.

MATERIALS NEEDED

Apparatus	Chemicals	
test tubes, small (3)	salicylic acid	(0.5 g)
test tubes, regular (1)	methyl alcohol	(3 ml)
beaker, 250 ml	H_2SO_4, conc.	(0.5 ml)
beaker, 100 ml	stearic acid	(3 g)
burner	glycerol	(1 ml)
stirring rod	K_2CO_3	(0.1 g)
thermometer	K_2CrO_4	(0.3 g)
funnel	$Pb(NO_3)_2$	(0.5 g)
wire gauze	$CaSO_4 \cdot 2H_2O$	(10 g)
iron ring	acetone	(2 ml)
wash bottle	*linseed oil	(0.5 ml)
spatula	*turpentine	(0.5 ml)
evaporating dish	(or mineral spirits)	
glass plate	*cottonseed oil	(0.5 ml)
crucible tongs	(or equivalent)	
heat resistant pad		

Supplies

filter paper
wood splints

QUESTIONS

1. The answers to this question will vary considerably, but most students will restrict themselves to simple properties: ''Methyl salicylate flavor smells like Lifesavers'', ''Moisturizing cream is smooth'', ''Lead chromate pigment is yellow'', and ''Plaster of Paris becomes hard when water is added.'' This offers an opportunity to again discuss physical and chemical properties.
2. No attempt will be made here to describe the range of possible advertisements. The point should be made, though, that chemical products satisfy a need. Does the advertisement show how the properties of the product relate to the need?
3. See the ''Introduction to the Laboratory'' for procedures.
4. (a) filtration (b) evaporation
5. See the ''Introduction to the Laboratory,'' pages ix–x.

EXTENSIONS

1. a. Stirring with a thermometer might break the thermometer and spill mercury.

b. Evaporating a salt solution to dryness might cause the salt to char or decompose to a different product. It also causes spattering and a loss of product.
c. An unattended burner can cause boiling over or heating to dryness. If the flame goes out, unburned gas will be given off by the burner.
d. Many solutions are corrosive and cause damage to skin or clothing when not wiped up.
e. Discarding solids in sinks can cause unexpected reactions when other chemicals are present or can clog the pipes and cause expensive repairs.
f. Returning unused solutions to stock bottles frequently causes contamination and spoils the stock solution.
g. Drinking water from a beaker is extremely hazardous as the beaker may contain a residue of a toxic chemical.
h. Protective goggles should be worn to protect the eyes from chemical splashes or sparks produced by reactions.

Relationships Between Variables: Mass and Volume of a Liquid

TIMING

The experiment requires about 40 minutes. Early finishers can begin calculations and graphing. The experiment is recommended for Chapter 1.

EXPERIMENT NOTES

1. The experimental techniques required are simple; weighing and measuring volume. If working in pairs, have one student measure the mass while the other adjusts the volume of the liquid. Have students trade jobs when doing the unknown.

2. While the data is quite accurate, the density determination is limited to two significant figures by the volume measurement. Class data on the density of water can be collected and analyzed as in Experiment 1.

3. As this is the first experiment in which data is graphed, begin here to require correct graphing procedures. Help on graphing is given in Appendix 1 of the text. Because of their experiences in algebra some students will obtain an incorrect slope by counting squares. Others will need help in jumping from the graph to a description of the relationship to a mathematical equation. Graphing procedures and relationships should be covered in Prelab. The relationship between a student's mass and height (mass $=$ k $\times$ height) is a helpful analogy.

4. A student may recognize the slope as being the density of the liquid, develop the correct equation, and yet be unable to apply it. Stress the value of knowing relationships so that they may be used to make valid predictions. Using a liquid's density to determine the volume needed will be a common laboratory operation.

5. Discuss the meaning of an experimental variable and the need to control all variables except the two which are being measured. Discuss the possible existence of other variables (temperature, humidity, purity) and their effect on the data.

MATERIALS NEEDED

Equipment	Unknowns	10 ml/student
10 ml graduated cylinder	hexane	0.660 g/cm^3
medicine dropper	ethyl alcohol	0.789 g/cm^3
	methyl alcohol	0.791 g/cm^3
Supplies	toluene	0.867 g/cm^3
graph paper	cottonseed oil	0.917 g/cm^3
	glycerol	1.26 g/cm^3
	CCl_3CH_3	1.34 g/cm^3
	butyl alcohol	1.40 g/cm^3

SAMPLE DATA TABLE

mass of graduated cylinder: 6.50 g unknown liquid: ethyl alcohol

Total Volume of Water	Mass of Grad. Cyl. and Water	Total Mass of Water
2.0 cm^3	8.52 g	2.02 g
4.0	10.51	4.01
6.0	12.51	6.01
8.0	14.49	7.99
10.0	16.50	10.00

Total Volume of Unknown	Mass of Grad. Cyl. and Unknown	Total Mass of Unknown
2.0 cm^3	8.10 g	1.60 g
4.0	9.69	3.19
6.0	11.24	4.74
8.0	12.82	6.32
10.0	14.42	7.92

CALCULATIONS AND QUESTIONS

1. See data table.

2.

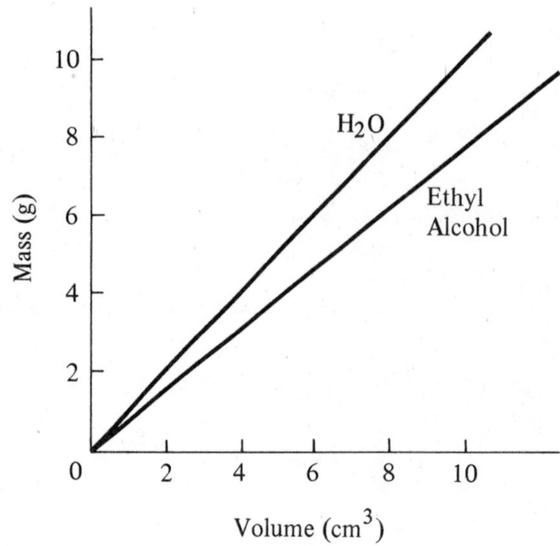

3.

Vol (cm^3)	1.0	3.0	5.0
H_2O (g)	1.0	3.0	5.0
unknown (g)	0.80	2.4	4.0

4. The mass of any volume of water in grams is numerically equal to its volume. The mass of any volume of unknown in grams is numerically equal to 0.80 times its volume.
5. mass (H_2O) = 1.0 × volume (H_2O)
   mass (unk) = 0.80 × volume (unk)
6. The constants of 1.0 and 0.80 must have units of g/cm^3 for the units of the equation to be consistent. The constant term describes the mass of one cm^3 of the liquid or its density. That is, the density of water is 1.0 g/cm^3 and that of the unknown liquid is 0.80 g/cm^3.
7. mass = density × volume

$$\text{mass }(H_2O) = 1.0\ \frac{g}{cm^3}\ (14\ cm^3) = 14\ g \qquad \text{mass (unk)} = 0.80\ \frac{g}{cm^3}\ (24\ cm^3) = 19\ g$$

8. volume = mass/density

$$\text{volume }(H_2O) = \frac{68\ g}{1.0\ g/cm^3} = 68\ cm^3 \qquad \text{volume (unk)} = \frac{29\ g}{0.80\ g/cm^3} = 36\ cm^3$$

Volume measurements are simpler and more rapid than mass measurements. Mass measurements may be preferable when greater precision is required.

EXTENSIONS

1. Temperature-density data for water appears in the *Handbook of Chemistry and Physics*. Greater precision is required to detect the temperature-density relationship for a

liquid than is available in this experiment. Point out the usefulness of the effect in mercury thermometers.

2. a. The temperature conversion equation, °F = 1.8°C + 32, is of the same form as $y = ax + b$. This indicates a linear relationship where a is the slope and b is the y-intercept.
   b. It is unlikely that any relationship exists here.
   c. For a rectangle, area = length × width. This is an inverse relationship of the same form as xy = a constant.
3. Most population data show an exponential relationship of the form $y = ax^n$.

Counting Particles and Finding Their Relative Masses

TIMING

The experiment requires about 45 minutes. It should be done early in Chapter 2 so that it will reinforce the text material.

EXPERIMENT NOTES

1. Comparing atoms to beans helps in teaching the concepts of relative mass and the mole. The student discovers that the relative mass in grams (a "bunch") of each bean type contains the same number of beans. Therefore, beans and atoms can be "counted" by mass measurements instead of actual counting.
2. Use 4–5 types of beans of varying sizes. Use one very small bean, such as a lentil, to make the number of beans in a bunch equal to 20 or more. If the smallest bean is too heavy a bunch will be only 4 or 5 beans. The beans are inexpensive and can be stored almost indefinitely. Bean shooters should be banned on the day of this experiment.

MATERIALS NEEDED

Chemicals:
*dry beans (lima, kidney, pinto, white, lentil, etc.): ¼ to 3 lbs
Supplies:
*paper cups (large enough to hold 100 of the largest bean)

SAMPLE DATA TABLE

	(white) Bean 1	(lima) Bean 2	(lentil) Bean 3	(kidney) Bean 4	(pinto) Bean 5
Mass of Beans and Cup	20.45 g	116.68 g	8.75 g	57.74	37.66 g
Mass of Cup	3.88 g	3.88 g	3.88 g	3.88 g	3.88 g
Mass of Beans	16.57 g	112.80 g	4.87 g	53.86 g	33.78 g
Average Mass	0.1657 g	1.1280 g	0.0487 g	0.5386 g	0.3378 g
Relative Mass	3.40	23.2	1.00	11.1	6.94
No. of Beans in Relative Mass	21	20–21	21	20–21	21

CALCULATIONS AND QUESTIONS

Part I

1. a. 21 beans/bunch (The answer depends upon the lightest bean used.)
   b. A relative mass in grams contains about the same number of beans (20–21) for each type of bean.
   c. A bunch of any type of bean will have a mass in grams numerically equal to its relative mass. That is, a bunch of beans with a relative mass of 3.5 will weigh 3.5 g, and so on.

2. a. $1000 \text{ beans} \times \dfrac{1 \text{ bunch}}{21 \text{ beans}} = 48 \text{ bunches}$

   b. You would weigh out a sample of the heaviest bean (lima) with a mass equal to that of 48 bunches:

   $$48 \text{ bunches} \times \frac{23.2 \text{ g}}{1 \text{ bunch}} = 1100 \text{ g}$$

3. a. $\text{No. of beans} = 15 \text{ bunches} \times \dfrac{21 \text{ beans}}{1 \text{ bunch}} = 310 \text{ (2 s.f.)}$

   (The answer is independent of the type of bean.)

   b. $\text{No. of beans} = 3.0 \text{ g of beans} \times \dfrac{21 \text{ beans}}{1.2 \text{ g of beans}} = 52$

4. $\text{mass} = 35 \text{ bunches limas} \times \dfrac{23.2 \text{ g}}{1 \text{ bunch limas}} = 810 \text{ g}$

Part II

1. a. 6.0×10^{23} H atoms/mole
   b. Each type of unit, atom, ion or molecule contains 6.0×10^{23} units in each mole.
   c. 12.0 g C/mole C; 18.0 g H_2O/mole H_2O; 55.8 g Fe/mole Fe

2. a. $\dfrac{12 \times 10^{23} \text{ atoms He}}{6.0 \times 10^{23} \text{ atoms/mole}} = 2.0 \text{ moles He}$

   $$\frac{6 \times 10^{20} \text{ atoms O}}{6 \times 10^{23} \text{ atoms/mole}} = 1 \times 10^{-3} \text{ mole O}$$

   $$\frac{1000 \text{ molecules } H_2O}{6.0 \times 10^{23} \text{ molecules/mole}} = 1.7 \times 10^{-21} \text{ mole } H_2O$$

   b. Weigh out 18 g (1 mole) of H_2O.

3. a. $15 \text{ moles } CH_4 \times \dfrac{6.0 \times 10^{23} \text{ molecules}}{1 \text{ mole}} = 9.0 \times 10^{24} \text{ molecules}$

b. $3.0\ g\ H_2O \times \dfrac{6.0 \times 10^{23}\ \text{molecules}}{18\ g\ H_2O} = 1.0 \times 10^{23}$ molecules

c. $2.0\ \text{moles Fe} \times \dfrac{6.0 \times 10^{23}\ \text{atoms}}{1\ \text{mole}} = 1.2 \times 10^{24}$ atoms

d. $4.0\ g\ Fe \times \dfrac{6.0 \times 10^{23}\ \text{atoms}}{56\ g\ Fe} = 4.3 \times 10^{22}$ atoms

4. $20\ \text{moles C} \times \dfrac{12\ g\ C}{1\ \text{mole}} = 240\ g$

5. $\dfrac{0.0487\ g}{1\ \text{lentil bean}} \times \dfrac{6.02 \times 10^{23}\ \text{beans}}{1\ \text{mole}} = 2.93 \times 10^{22}\ g/\text{mole}$

$$\frac{6 \times 10^{27}\ g\ \text{(mass of earth)}}{3 \times 10^{22}\ g\ \text{(mass of 1 mole lentil beans)}} = 2 \times 10^{5}$$

That is, 2×10^{5} moles of lentil beans would have the same mass as the earth.

EXTENSIONS

1. $\dfrac{100\ g\ \text{limas}}{1\ \text{bunch}} \times \dfrac{1\ \text{bean}}{1.1280\ g\ \text{lima}} = \dfrac{89\ \text{beans}}{\text{bunch}}$

$$\frac{89\ \text{beans}}{1\ \text{bunch}} \times \frac{0.0487\ g\ \text{lentils}}{\text{bean}} = \frac{4.3\ g\ \text{lentils}}{\text{bunch}}$$

Alternative solution: $\dfrac{100\ g\ \text{lima beans}}{1\ \text{bunch}} \times \dfrac{1.00\ g\ \text{lentil}}{23.2\ g\ \text{lima}} = 4.31\ g\ \dfrac{\text{lentils}}{\text{bunch}}$

2. $\dfrac{10\ g\ H}{1\ \text{mole}} \times \dfrac{6.0 \times 10^{23}\ \text{atoms}}{1.0\ g\ H} = \dfrac{6.0 \times 10^{24}\ \text{atoms}}{\text{mole}}$ (New Avogadro No.)

As the ‘‘new’’ mole is 10 times larger than the old mole:
1 mole C = 120 g C and 1 mole CH_4 = 160 g CH_4

Finding the Size of a Molecule and a Value for Avogadro's Number

TIMING

The experiment requires about 30 minutes. The remaining time is spent on calculations. The experiment is a good wind-up for Chapter 2.

PRELIMINARY STUDY

2. $\text{Area (film)} = \dfrac{\pi d^2}{4} = \dfrac{\pi\,(10\text{ cm})^2}{4} = 79\text{ cm}^2$

   $\text{Area (molecule)} = (\text{length})^2 = (25 \times 10^{-8}\text{ cm})^2 = 6.2 \times 10^{-14}\text{ cm}^2$

   $\text{No. of molecules} = \dfrac{79\text{ cm}^2}{6.2 \times 10^{-14}\text{ cm}^2/\text{molecule}} = 1.3 \times 10^{15}$

3. $\text{mass} = 0.040\text{ cm}^3 \times \dfrac{0.100\text{ g}}{10^3\text{ cm}^3} = 4.0 \times 10^{-6}\text{ g}$

EXPERIMENT NOTES

1. The experiment is a valuable exercise in working with both very large and very small numbers. It also provides considerable experience in working with equations and converting units. While all equations are given in the Discussion, some students will need help in carrying out the calculations.

2. The original assumption made in the experiment is that the oleic acid molecule is cubic. With this assumption, expect to get answers of around 10^{22} for Avogadro's Number. If Extension 1 is used (width = ⅕ length), the number will be 25 times larger, or around 10^{23}.

3. Results can be improved by using a large tray. Trays are usually available from biology. In an alternative procedure for measuring area, the bottom of the tray is marked off in centimeter squares and the area obtained by counting squares.

MATERIALS NEEDED

Apparatus	Chemicals
shallow tray (18 in × 24 in minimum)	sulfur, powdered (3 g)
capillary pipet (about 1 mm dia)	stearic acid in cyclohexane, 0.15 g/ℓ
metric rule	cyclohexane (2 ml)
graduated cylinder, 10 ml	

SAMPLE DATA TABLE

	Film with 1 drop	Film with 2 drops	Film with 3 drops
Diameter 1	4.5 cm	6.0 cm	8.5 cm
Diameter 2	5.0 cm	7.0 cm	9.0 cm
Diameter 3	5.0 cm	8.0 cm	8.0 cm
Avg. Diameter	4.8 cm	7.0 cm	8.5 cm

Concentration of stearic acid solution 0.15 g/ℓ
Calibration of pipet 46 drops/cm^3

CALCULATIONS AND QUESTIONS

1. a. Area = $\pi d^2/4$

$$\text{Area (1 drop film)} = \frac{\pi(4.8\text{ cm})^2}{4} = 18\text{ cm}^2$$

Area (2 drop film) = 38 cm^2 Area (3 drop film) = 57 cm^2

b. Area/drop (1 drop film) = 18 cm^2

Area/drop (2 drop film) = 38 cm^2/2 = 19 cm^2

Area/drop (3 drop film) = 57 cm^2/3 = 19 cm^2

As the area per drop is approximately equal for all films, the area increases in direct proportion to the number of drops of solution added. As each drop contains about the same number of molecules of stearic acid, then the number of molecules on the surface also increases directly as the area increases. These results support the assumption that the stearic acid is in a monomolecular layer. (NOTE: If the film becomes too large for the size of the tray the data will not be consistent.)

c. average area/drop = 19 cm^2

2. $$\text{volume soln/drop} = \frac{1\text{ cm}^3}{46\text{ drops}} = 0.022\text{ cm}^3$$

3. mass stearic acid/drop $= 0.022 \text{ cm}^3 \times \dfrac{0.15 \text{ g}}{10^3 \text{ cm}^3} = 3.3 \times 10^{-6} \text{ g}$

4. volume stearic acid/drop $= \dfrac{3.3 \times 10^{-6} \text{ g}}{0.85 \text{ g/cm}^3} = 3.9 \times 10^{-6} \text{ cm}^3$

5. thickness $= \dfrac{3.9 \times 10^{-6} \text{ cm}^3}{19 \text{ cm}^2} = 2.1 \times 10^{-7} \text{ cm}$

   (NOTE: The film thickness value is also equal to the molecule's length.)

6. volume of molecules $= (2.1 \times 10^{-7} \text{ cm})^3 = 9.3 \times 10^{-21} \text{ cm}^3$

7. no. of molecules $= \dfrac{3.9 \times 10^{-6} \text{ cm}^3}{9.3 \times 10^{-21} \text{ cm}^3} = 4.2 \times 10^{14}$

8. moles stearic acid/drop $= \dfrac{3.3 \times 10^{-6} \text{ g}}{284 \text{ g/mol}} = 1.2 \times 10^{-8} \text{ mol}$

9. Avogadro's Number $= \dfrac{4.2 \times 10^{14} \text{ molecules}}{1.2 \times 10^{-8} \text{ mol}} = \dfrac{3.5 \times 10^{22} \text{ molecules}}{\text{mole}}$

   The experimental value obtained is less than the theoretical value of 6.0×10^{23}. The major assumptions made which might affect our value are (1) that the stearic acid film is truly monomolecular, and (2) that the shape of the molecule is cubic.

EXTENSIONS

1. If $W = \frac{1}{5}(L)$: volume $= L \times W^2 = L \times (\frac{1}{5} L)^2 = \dfrac{L^3}{25}$

   As the length of the molecule is equal to the film thickness:

   volume of molecules $= \dfrac{(2.1 \times 10^{-7} \text{ cm})^3}{25} = 3.7 \times 10^{-22} \text{ cm}^3$

2. No. of molecules $= \dfrac{3.9 \times 10^{-6} \text{ cm}^3}{3.7 \times 10^{-22} \text{ cm}^3} = 1.1 \times 10^{16}$

   Avogadro's No. $= \dfrac{1.1 \times 10^{16} \text{ molecules}}{1.2 \times 10^{-8} \text{ mol}} = \dfrac{9.2 \times 10^{23} \text{ molecules}}{\text{mol}}$

   In this case, the value obtained is closer to the theoretical value. This would support the assumption that the stearic acid molecule is not cubic, but longer than it is wide. The accuracy of this method depends upon knowing the geometry of the molecule.

3. volume/mol $= \dfrac{58.45\ \text{g}}{2.165\ \text{g/cm}^3} = 27.00\ \text{cm}^3$

edge length (molar cube) $= (27.00\ \text{cm}^3)^{1/3} = 3.000\ \text{cm}$

ions/edge length $= \dfrac{3.000\ \text{cm}}{2.814 \times 10^{-8}\ \text{cm/ion}} = 1.066 \times 10^8$

ions/molar cube $= (1.066 \times 10^8)^3 = 1.211 \times 10^{24}$

As there are two ions (Na^+ and Cl^-) per formula unit (NaCl), the number of NaCl units is half of the number of ions. This value is equivalent to Avogadro's Number.

$$\frac{1.211 \times 10^{24}\ \text{ions/mole}}{2\ \text{ions/NaCl unit}} = 6.055 \times 10^{23}\ \frac{\text{NaCl units}}{\text{mol NaCl}}$$

Determining the Simplest Formula of a Compound

TIMING

The experiment requires about 45 minutes. The experiment should be done during Chapter 3.

PRELIMINARY STUDY

3. mass S = 1.00 g − 0.87 g = 0.13 g

$$\text{moles S} = \frac{0.13 \text{ g}}{32.1 \text{ g/mol}} = 0.0040$$

$$\text{moles Ag} = \frac{0.87 \text{ g}}{108 \text{ g/mol}} = 0.0081$$

$$\frac{\text{moles Ag}}{\text{moles S}} = \frac{0.0081}{0.0040} = 2.0 \quad \text{Simplest formula} = Ag_2S$$

4. $4\ Al(s) + 3\ O_2(s) \rightarrow 2\ Al_2O_3(s)$

EXPERIMENT NOTES

1. This is the student's first quantitative experiment. Stress the importance of good technique, especially in the use of the balance. When determining a mass by difference, the same balance should be used for both measurements.
2. The drying of the empty crucible in Procedure 1 can be omitted in low humidity areas unless it has been recently washed with water. If time permits, a second heating to constant mass should be done.
3. The experimental error, based on percent Mg, is about 5%. As this will give a Mg:O mole ratio of 0.9 to 1.1, most students will obtain the correct formula.
4. A common technique error at this stage is insufficient heating of the crucible. Remind students that the hottest part of the flame is the tip of the inner cone. If the flame is too high a reducing atmosphere will be present in the crucible, instead of an oxidizing atmosphere, and the magnesium will not react.

MATERIALS NEEDED

Apparatus	Chemicals
burner	Mg, ribbon (50 cm)
crucible and cover	(Mg turnings may also be used.)
crucible tongs	
iron ring	
clay triangle	
heat resistant pad	
stirring rod	
wash bottle	

SAMPLE DATA TABLE

mass of crucible + lid	16.21 g
mass of crucible + lid + Mg	16.56 g
mass of crucible + lid + product	16.80 g

CALCULATIONS AND QUESTIONS

1. a. mass Mg = 16.56 g − 16.21 g = 0.35 g

   mass product = 16.80 g − 16.21 g = 0.59 g

   b. mass O = 0.59 g − 0.35 g = 0.24 g

   c. $$\text{moles Mg} = \frac{0.35 \text{ g Mg}}{24.3 \text{ g Mg/mol}} = 0.014$$

   $$\text{moles O} = \frac{0.24 \text{ g O}}{16.0 \text{ g O/mol}} = 0.015$$

   d. $$\frac{\text{mol Mg}}{\text{mol O}} = \frac{0.014}{0.015} = 0.93$$

   The Mg:O atom ratio is the same as the Mg:O mol ratio.

   e. Mg:O = 1:1 (rounded to nearest ratio of integers)

   The simplest formula of magnesium oxide is MgO.

2. a. $$\underset{\text{(experimental)}}{\% \text{ Mg}} = \frac{\text{mass Mg}}{\text{mass product}} \times 100 = \frac{0.35 \text{ g}}{0.59 \text{ g}} \times 100 = 59$$

   b. $$\underset{\text{(theoretical)}}{\% \text{ Mg}} = \frac{\text{mass Mg}}{\text{mass MgO}} \times 100 = \frac{24.30 \text{ g}}{40.30 \text{ g}} \times 100 = 60.30$$

   c. $$\% \text{ error} = \frac{60.30 - 59}{60.30} \times 100 = \frac{1}{60} \times 100 = 2\%$$

3. $2\ Mg\ (s) + O_2(g) \rightarrow 2\ MgO\ (s)$

4. $3\ Mg\ (s) + N_2(g) \rightarrow Mg_3N_2(s)$

5. $Mg_3N_2(s) + 3\ H_2O \rightarrow 3\ MgO\ (s) + 2\ NH_3\ (g)$

EXTENSIONS

2. a. Unreacted Mg will cause the percent Mg to be high.

   b. Spilled product will cause the percent Mg to be high.

   c. Unevaporated water will cause the percent Mg to be low.

   d. Unreacted Mg_3N_2 will cause the percent Mg to be high.

   This is best seen by comparing the FM of Mg_3N_2 (101) to that of an equivalent amount of oxide, 3 MgO (121). As the Mg_3N_2 has a lower mass, the mass of the product will be low, and the percent Mg will be high.

Observing Chemical Reactions and Writing Equations

TIMING

While the experiment can be completed in one period, disorganized or unprepared students may not complete all procedures. To save time, Procedures 2 and/or 3 may be omitted. Alternatively, the experiment may be stopped after Procedure 3 or 4a and continued the next period. The experiment is best performed during Chapter 3.

PRELIMINARY STUDY

2. a. $4\ Na(s) + O_2(g) \rightarrow 2\ Na_2O(s)$

   b. $Na_2O(s) + H_2O \rightarrow 2\ Na^+(aq) + 2\ OH^-(aq)$

EXPERIMENT NOTES

1. The experiment relates the observing of chemical reactions to the writing of equations to describe the reactions. A variety of techniques are required in carrying out the reactions. Carefully observe and correct improper procedures, especially if Experiment 2 was not performed.
2. Keep reaction discussions at a basic level. Detailed information on precipitation, acid-base, and oxidation-reduction reactions should come later. In most cases, students are provided the formulas of products or can easily deduce them from a previous procedure.
3. In order to save time and avoid accidents, it is recommended that the stoppered delivery tube (Figure 7–1) be prepared in advance for the students.
4. A simple data table using headings of ''Reactants,'' ''Products,'' and ''Observations'' is sufficient for this experiment.

MATERIALS NEEDED

Apparatus	Chemicals	
crucible	copper (wool or turnings)	(small wad)
clay triangle	H_2SO_4, 3 M; 167 ml conc./ℓ	(15 ml)

Apparatus (continued)	Chemicals (continued)	
iron ring	NaOH, 6 M; 240 g/ℓ	(5 ml)
burner	$CuSO_4 \cdot 5\ H_2O$	(1 g)
crucible tongs	$CuSO_4$, 0.1 M; 25 g hydrate/ℓ	(5 ml)
beaker, 50 ml (2)	*iron nails	(1)
funnel	$CuCO_3$	(2 g)
stirring rod	$Ca(OH)_2$, 1 g/ℓ (Satd)	(20 ml)
test tubes, small (4)	charcoal, powdered	(0.5 g)
test tube, regular		
test tube holder	Supplies	
test tube rack	filter paper	
wash bottle		
right angle glass tube (See Note 3)		
1-hole 00 stopper		
test tube clamp		

SAMPLE DATA TABLE

Procedure	Reactants	Products	Observations
1.(a)	Cu, O_2; (heat)	CuO	product is black solid
1.(b)	CuO, H^+	Cu^{2+}, H_2O	light blue soln is formed
1.(c)	Cu^{2+}, OH^-	$Cu(OH)_2$	light blue ppt is formed
1.(d)	$Cu(OH)_2$, H^+	Cu^{2+}, H_2O	light blue soln is formed
1.(e)	$Cu(OH)_2$; (heat)	CuO, H_2O	black solid is formed
2.(a)	$CuSO_4 \cdot 5\ H_2O$; (heat)	$CuSO_4$, H_2O	white solid formed; H_2O condensed on tube
2.(b)	$CuSO_4$, H_2O	$CuSO_4 \cdot 5\ H_2O$	blue solid formed; heat evolved
3.	Fe, Cu^{2+}	Cu, Fe^{2+}	nail has Cu coating; soln color lightens
4.(a)	$CuCO_3$, H^+	Cu^{2+}, CO_2, H_2O	gas is evolved which forms a ppt in limewater; soln is light blue
(b)	$CuCO_3$; (heat)	CuO, CO_2	gas is evolved which forms a ppt in limewater; residue is black solid
(c)	CuO, C; (heat)	Cu, CO_2	gas is evolved which forms a ppt in limewater; Cu flakes in residue

CALCULATIONS AND QUESTIONS

1. (1a) $2\ Cu(s) + O_2(g) \rightarrow 2\ CuO(s)$

   (1b) $CuO(s)\ 2\ H^+(aq) \rightarrow Cu^{2+}(aq) + H_2O$

   (1c) $Cu^{2+}(aq) + 2\ OH^-(aq) \rightarrow Cu\ (OH)_2(s)$

   (1d) $Cu(OH)_2(s) + 2\ H^+(aq) \rightarrow Cu^{2+}(aq) + 2\ H_2O$

   (1e) $Cu(OH)_2(s) \rightarrow CuO(s) + H_2O$

   (2a) $CuSO_4{\cdot}5\ H_2O(s) \rightarrow CuSO_4(s) + 5\ H_2O$

   (2b) $CuSO_4(s) + 5\ H_2O \rightarrow CuSO_4{\cdot}5\ H_2O(s)$

   (3) $Fe(s) + Cu^{2+}(aq) \rightarrow Fe^{2+}(aq) + Cu(s)$

   (4a) $CuCO_3(s) + 2\ H^+(aq) \rightarrow Cu^{2+}(aq) + CO_2(g) + H_2O$

   (4b) $CuCO_3(s) \rightarrow CuO(s) + CO_2(g)$

   (4c) $2\ CuO(s) + C(s) \rightarrow 2\ Cu(s) + CO_2(g)$

2. a. CuO, $Cu(OH)_2$, and $CuCO_3$ all react with acidic solution, dissolving and forming the Cu^{2+} ion.
   b. A $CuSO_4$ solution contains Cu^{2+} and SO_4^{2-} ions and is blue. An H_2SO_4 solution contains H^+ and SO_4^{2-} ions and is colorless. Hence, the blue color is due to the Cu^{2+} ion.
   c. $CuCO_3$ evolves CO_2 gas when an acid is added while $CuSO_4$ hydrate does not. When heated, $CuCO_3$ turns black (CuO) and evolves CO_2 gas while $CuSO_4$ hydrate turns white and evolves H_2O vapor.

EXTENSIONS

1. Copper carbonate, $CuCO_3$, has low solubility in water. When solutions of $CuSO_4$ and Na_2CO_3 are mixed, a precipitate of copper carbonate would be expected:

   $Cu^{2+}(aq) + CO_3^{2-}(aq) \rightarrow CuCO_3(s)$

   As Na_2SO_4 has high solubility, the Na^+ and SO_4^{2-} ions remain in solution and are not written in the equation.

The Specific Heat of Liquids and Solids

TIMING

The experiment requires about 45 minutes. Early finishers can make a second run in Part II. The experiment should be done early in Chapter 4 and prior to Experiment 9.

PRELIMINARY STUDY

2. $Q(H_2O) = C \times m \times \Delta t = \dfrac{1.00 \text{ cal}}{\text{g}\cdot{}^\circ\text{C}} \times 200 \text{ g} \times 12.5^\circ\text{C} = 2500 \text{ cal}$

EXPERIMENT NOTES

1. The subject of calorimetry, including the concept of specific heat, is discussed in an optional section at the end of Chapter 4 of the text. Preceding the text with these ideas should not present problems. If the student understands the calorie and the Law of Conservation of Energy the experiment will be understandable.

2. Care should be taken that the organic liquids used in Part I are not contaminated so that they can be reused by succeeding classes. Do not use liquids with high vapor pressure or whose vapors are harmful. The liquid should not be brought to a boil.

3. The specific heats of organic liquids are not easily found in the literature. In addition, most do not remain constant but tend to increase with temperature. The liquids used in this experiment have specific heat values of 0.6 to 0.7 cal/g·°C over the temperature range of 20°C–50°C. The error caused by using beakers of up to 2 grams difference in mass is not significant.

4. Experimental errors of about 10% can be expected in the specific heats of solids in Part II. The error is reduced by using a larger mass of metal (10 g–50 g). Commercial specific heat specimens of equal mass are available. (If the metal's identity is on the specimen it can be removed on a lathe.)

MATERIALS NEEDED

Apparatus	Chemicals	
electric hot plate	Organic liquid (Pt. I)	(100 g)
beakers, 150 ml (2)		
beaker, 250 ml		Density (g/cm^3)
beaker, 400 ml	1—propanol	0.780
graduated cylinder, 100 ml or 50 ml	*2—propanol (rubbing alcohol)	0.785
stirring rods (2)	ethylene glycol	1.11
watch glasses (2)	glycerol	1.26
thermometers (2)	Metal (Pt. II)	(10–50 g)
*styrofoam cups (2)		Sp. Ht. (cal/g·°C)
	aluminum	0.215
	bismuth	0.0296
	chromium	0.112
	copper	0.092
	lead	0.0310
	magnesium	0.243
	tin	0.052
	zinc	0.093

SAMPLE DATA TABLE

Part I: unknown liquid (ethylene glycol)

density of unknown	1.11 g/cm^3	vol. of 100 g of unknown	90.1 cm^3
temp. of H_2O (initial)	32.6°C	temp. of unk (initial)	26.2°C
temp. of H_2O (final)	53.4°C	temp. of unk (final)	59.5°C
Δt of H_2O	20.8°C	Δt of unknown	33.3°C

Part II: unknown metal (Mg)

mass of metal	10.90 g	temp. of H_2O (initial)	26.2°C
mass of H_2O	60.0 g	temp. of H_2O (final)	29.5°C
temp. of H_2O (bath)	99.5°C	Δt of H_2O	3.3°C
Δt of metal	70.0°C		

CALCULATIONS AND QUESTIONS

1. $Q(H_2O) = 1.00\ \frac{\text{cal}}{\text{g}\cdot{}^\circ\text{C}} \times 100\ \text{g} \times 20.8^\circ\text{C} = 2080\ \text{cal}$

2. $\text{Sp.Ht.} = \frac{2080\ \text{cal}}{100\ \text{g} \times 33.3^\circ\text{C}} = 0.625\ \text{cal/g}\cdot{}^\circ\text{C}$

3. $$Q(H_2O) = \frac{1.00 \text{ cal}}{\text{g}\cdot°\text{C}} \times 60.0 \text{ g} \times 3.3°\text{C} = 200 \text{ cal (2 s.f.)}$$

4. $$\text{Sp.Ht.} = \frac{200 \text{ cal}}{10.90 \text{ g} \times 70.0°\text{C}} = 0.26 \text{ cal/g}\cdot°\text{C}$$

EXTENSIONS

1. Heat lost to surroundings and heat absorbed by the calorimeter both cause the temperature change of the water (Δt_{H_2O}) to be low. This causes the Q of water to be low. As the final temperature of the water is low the temperature change of the metal (Δt metal) will be high. Both of these effects result in a specific heat value that is smaller than the true value.

2. If: $C_m \times AM = 6.0$ cal/°C

$$AM = \frac{6.0 \text{ cal/°C}}{0.26 \text{ cal/g}\cdot°\text{C}} = 23 \text{ g (Mg)}$$

$$C_m = \frac{6.0 \text{ cal/°C}}{24.3 \text{ g}} = 0.25 \text{ cal/g}\cdot°\text{C}$$

$$\% \text{ Error} = \frac{0.25 - 0.26}{0.25} \times 100 = 4\%$$

Heat Effects of Chemical Reactions

TIMING

The experiment runs about 45 minutes. A second run of Part II can be made if time permits. The experiment should be performed during Chapter 4.

PRELIMINARY STUDY

2. $\Delta H/g = \dfrac{-17.8 \text{ kcal}}{2.50 \text{ g}} = -7.12 \text{ kcal/g}$

$$\text{moles alcohol} = \frac{2.50 \text{ g}}{46.0 \text{ g/mol}} = 0.0543$$

$$\Delta H/\text{mol} = \frac{-17.8 \text{ kcal}}{0.0543 \text{ mol}} = -328 \text{ kcal/mol}$$

EXPERIMENT NOTES

1. If students have done Experiment 8 they will be familiar with calorimetry procedures and calculations. If not, some prelab discussion of the techniques is advised.
2. You may want the students to provide the cans used in Part I. We use beer cans for the small can and large fruit juice cans (No. 3 tall) for the large can. The holes in the small can may be punched with an ice pick or nail and widened with a file. The draft holes in the large can are made with a beer can opener.
3. The hydrocarbons in candle wax have a heating value of about 10 kcal/g. The design of the experiment is such that efficiencies of 50%–70% can be expected. It is important that the water can be kept at the tip of the candle flame throughout the experiment.
4. The heat effect of dissolving a solid in a liquid, ΔH_{soln}, is not specifically discussed in Chapter 4. This should present no problem, however.

   This type of reaction has been chosen because of the ease of performing and reasonably good results obtained. It is also interesting in that the ΔH can be endothermic or exothermic.
5. Pure, dry salts must be used in Part II for best results. Experimental errors will average between 5%–10% depending upon the salt used.

MATERIALS NEEDED

Apparatus	Chemicals	
can, 12 oz (See Note 2.)	*candle	
can, No. 3 tall (See Note 2.)	Pt. II unknown	(5 g)
glass plate		
graduated cylinder, 100 ml or 50 ml	Compound	ΔH_{soln} (cal/mol)
	NH_4NO_3	6200
stirring rods (2)	$CaCl_2$	−18,000
thermometer	KNO_3	8500
iron ring	$NaNO_3$	5000
beaker, 250 ml	Na_2CO_3	−5600
*styrofoam cups (2)	$NaC_2H_3O_2$	−4100

SAMPLE DATA TABLE

Part I

mass of candle (initial)	50.04 g	water temp. (initial)	23.2°C
mass of candle (final)	48.82 g	water temp. (final)	56.2°C
mass of candle (burned)	1.22 g	Δt of water	33.0°C
		mass of water	200.0 g

Part II

mass of paper	2.16 g	water temp. (initial)	24.0°C
mass of paper & compound	6.97 g	water temp. (final)	38.6°C
mass of compound	4.81 g	Δt of water	14.6°C
formula of compound	$CaCl_2$	mass of water	50.0 g

CALCULATIONS AND QUESTIONS

1. a. $Q(H_2O) = \dfrac{1.00 \text{ cal}}{\text{g}\cdot°\text{C}} \times 200.0 \text{ g} \times 33.0°\text{C} = 6600 \text{ cal (3 s.f.)}$

   b. The combustion reaction is exothermic. The temperature of the water increased, indicating that it absorbed energy given off by the burning of the candle.

   c. $\text{heat(cal/g)} = \dfrac{6600 \text{ cal}}{1.22 \text{ g candle}} = 5410 \text{ cal/g}$

   d. $\text{heat required} = \dfrac{1.00 \text{ cal}}{\text{g}\cdot°\text{C}} \times 10^3 \text{ g} \times 75°\text{C} = 7.5 \times 10^4 \text{ cal}$

   $\text{mass of candle} = 7.5 \times 10^4 \text{ cal} \times \dfrac{1 \text{ g}}{5410 \text{ cal}} = 14 \text{ g}$

e. $\% \text{ efficiency} = \dfrac{-5.4 \text{ kcal/g}}{-10 \text{ kcal/g}} \times 100 = 54\%$

f. $C_{20}H_{42}(s) + 30.5\ O_2(g) \rightarrow 20\ CO_2(g) + 21\ H_2O(l)$

2. a. $Q(H_2O) = \dfrac{1.00 \text{ cal}}{\text{g}\cdot{}^\circ\text{C}} \times 50.0 \text{ g} \times 14.6^\circ\text{C} = 730 \text{ cal (3 s.f.)}$

b. The solution reaction is exothermic because the temperature of the water increased. (In endothermic reactions the water temperature decreases.)

c. The sign of ΔH for exothermic reactions is negative ($-$).

d. $\text{moles } CaCl_2 = \dfrac{4.81 \text{ g } CaCl_2}{111 \text{ g } CaCl_2/\text{mol}} = 0.0433$

e. $\Delta H/\text{mol} = \dfrac{-730 \text{ cal}}{0.0433 \text{ mol}} = -16{,}900 \text{ cal/mol}$

EXTENSIONS

1. Heat is lost mostly in heating the water can and heating the surrounding air. A sealed, insulated system would eliminate the heat lost to the surroundings. In this case pure oxygen gas would have to be supplied internally for the reaction. It is also necessary to know how much heat is absorbed by the calorimeter assembly. A reaction with a known ΔH can be carried out in the calorimeter to determine this value. Specifically: heat absorbed (calorimeter) = heat of reaction − heat absorbed (water).
2. The procedures and calculations are carried out in the same manner as the unknowns of Part II. Qualitative analysis procedures can be used to identify the salts in the packs if not noted on the label.

Determining the Molecular Mass of a Gas

TIMING

The experiment can be completed in 50 minutes if preparations are made as explained in Note 3 below. It should be done after the gas laws have been covered in Chapter 5.

PRELIMINARY STUDY

2. First, find the volume under standard conditions (Equation 10.1):

$$V_2 = 0.500\ \ell \times \frac{730 \text{ mm Hg}}{760 \text{ mm Hg}} \times \frac{273 \text{ K}}{293 \text{ K}} = 0.447\ \ell$$

Second, using the molar volume find the moles of gas (Equation 10.3):

$$\text{moles of gas} = 0.447\ \ell \times \frac{1 \text{ mol}}{22.4\ \ell} = 2.00 \times 10^{-2}$$

Last, find the molar mass (Equation 10.4):

$$\text{MM} = \frac{0.60 \text{ g}}{2.00 \times 10^{-2} \text{ mol}} = 30 \text{ g/mol (2 s.f.)}$$

EXPERIMENT NOTES

1. There is no real ''unknown'' in this experiment as students know the identity of the gas and its molecular mass. However, it is useful from the viewpoint that new techniques of working with a gas are involved and it requires the application of the gas laws. It is also convenient, in that high pressure gas cylinders are not required.
2. The experiment can be easily reversed. That is, the student can begin with the molar mass of the gas and experimentally determine the molar volume. A third alternative is to substitute the data into the Ideal Gas Law and solve for R, the gas constant.
3. To save time and accidents, the three stopper assemblies should be prepared in advance. If the budget permits, assemblies of this type should be saved and stored for future use.
4. The experimental set-ups should be checked and approved before the acid is added to the generator. In particular, make certain that the stopper to the collection flask is loose.

5. Students may work on calculations 1 and 4 during the time that the gas is being generated.
6. The density of air, in grams per liter of air, is tabulated here for your use. Alternatively, students may find this information in the *Handbook of Chemistry and Physics:* Density of Air (g/ℓ) at Different Temperatures and Pressures.

Density of Air in Grams per Liter at Various Temperatures and Pressures

Pressure (mm)	*Temperature* 15°C	20°C	25°C	30°C
600	0.97	0.95	0.94	0.92
610	0.98	0.97	0.95	0.93
620	1.00	0.98	0.97	0.95
630	1.02	1.00	0.98	0.97
640	1.03	1.01	1.00	0.98
650	1.05	1.03	1.01	1.00
660	1.06	1.05	1.03	1.01
670	1.08	1.06	1.04	1.03
680	1.10	1.08	1.06	1.04

Pressure (mm)	*Temperature* 15°C	20°C	25°C	30°C
690	1.11	1.09	1.07	1.06
700	1.13	1.11	1.09	1.07
710	1.14	1.12	1.10	1.09
720	1.16	1.14	1.12	1.10
730	1.18	1.16	1.14	1.12
740	1.19	1.17	1.15	1.13
750	1.21	1.19	1.17	1.15
760	1.23	1.21	1.19	1.16
770	1.24	1.22	1.20	1.18

MATERIALS NEEDED

Apparatus

Florence flask, 250 ml
stopper, 1-hole, with glass tube (to fit Florence flask)
medicine dropper bulb
wide mouth bottle
stopper, 2-hole, with right angle glass tube & thistle tube (to fit wide mouth bottle)
test tube clamp
drying tube
stopper, 1-hole, with glass tube (to fit drying tube)
rubber tubing (2 pieces)
graduated cylinder, 50 ml
graduated cylinder, 250 ml
thermometer
barometer

Chemicals

$CaCO_3$ (marble chips) (20 g)
(marble chips can be reused)
HCl, 6M; 500 ml conc./ℓ (50 ml)
$CaCl_2$, anhydrous (10 g)

SAMPLE DATA TABLE

mass of flask stopper and air	122.30 g	air density	1.14 g/ℓ
mass of flask stopper and CO_2	122.43 g	temp. of CO_2, T_1	26.0°C
		lab pressure, P_1	733 mm Hg
		vol of flask, V_1	257 cm^3

CALCULATIONS AND QUESTIONS

1. a. $\text{mass air} = 257\ cm^3 \times \frac{1\ \ell}{10^3\ cm^3} \times \frac{1.14\ g}{1\ \ell} = 0.293\ g$

   b. $\text{mass empty flask} = 122.30\ g - 0.29\ g = 122.01\ g$

   c. $\text{mass } CO_2 = 122.43\ g - 122.01\ g = 0.42\ g$

2. a. $P_1 = 733\ \text{mm Hg} \times \frac{1\ atm}{760\ \text{mm Hg}} = 0.964\ atm$

   $V_1 = 257\ cm^3 \times \frac{1\ \ell}{10^3\ cm^3} = 0.257\ \ell$

   $T_1 = 26.0°C + 273 = 299\ K$

   b. $V_2 = V_1 \times \frac{P_1}{P_2} \times \frac{T_2}{T_1}$

   $V_2 = 0.257\ \ell \times \frac{0.964\ atm}{1\ atm} \times \frac{273\ K}{299\ K} = 0.226\ \ell$

   c. $\text{moles } CO_2 = 0.226\ \ell \times \frac{1\ mol}{22.4\ \ell} = 0.0101$

3. a. $MM = \frac{0.42\ g}{0.0101\ mol} = 42\ g/mol$

   b. True MM CO_2 = 44

   $\%\ Error = \frac{44-42}{44} \times 100 = 5\%$

4. a. $\text{moles } CO_2 = 20.0\ g\ CaCO_3 \times \frac{1\ mol\ CaCO_3}{100.1\ g\ CaCO_3} \times \frac{1\ mol\ CO_2}{1\ mol\ CaCO_3} = 0.200$

   b. $V_2\ (\text{molar volume at } 25°C) = 22.4\ \ell \times \frac{298\ K}{273\ K} = 24.5\ \ell$

   $\text{volume } CO_2 = 0.200\ mol \times \frac{24.5\ \ell}{mol} = 4.90\ \ell$

EXTENSIONS

1. The equation for the decomposition of H_2O_2 is:

$$2\ H_2O_2(aq) \rightarrow 2\ H_2O(1) + O_2(g)$$

   The amount of H_2O_2 specified will produce about 1 ℓ of O_2 gas at room conditions. More of the peroxide solution may be needed if it is weak.
2. The mass of 250 ml of H_2 collected would barely be detectable on a centigram balance.

$$\text{mass } H_2 = 0.25\ \ell \times \frac{1\ \text{mole}}{24.5\ \ell} \times \frac{2.02\ \text{g}}{1\ \text{mole } H_2} = 0.021\ \text{g}$$

   The experimental error would be decreased by collecting larger samples of the gas (there is a practical limit here) or by using a more sensitive balance, say one whose precision is $\pm$ 0.0001 g.

Volume-Mass Relations in Chemical Reactions

TIMING

The experiment requires about 40 minutes. It should be done at the completion of Chapter 5.

PRELIMINARY STUDY

2. $P_{dry\ gas} = P_{total} - P_{H_2O}$ (Equation 11.3)

$$P_{dry\ gas} = 730 \text{ mm Hg} - 21 \text{ mm Hg} = 709 \text{ mm Hg}$$

$$P_{atm} = 709 \text{ mm Hg} \times \frac{1 \text{ atm}}{760 \text{ mm Hg}} = 0.933 \text{ atm}$$

EXPERIMENT NOTES

1. The students have used the combined gas law previously in Experiment 10. As the H_2 gas in this experiment is collected over water they will have the opportunity to use Dalton's Law of partial pressures. The data allows the student to determine the H_2:Mg mole ratio and the charge on the magnesium ion. The average experimental error in the volume obtained is about 3%. This allows most students to obtain the correct reaction ratio.
2. The length of magnesium ribbon required varies with the linear density of the ribbon and the laboratory's altitude. One gram of Mg produces about 1 ℓ of H_2 at room conditions (1 atm and 25°C). For example, if 40 ml (0.04 ℓ) of H_2 is desired, then 0.04 g Mg should be used. If the linear density of the ribbon is 0.800 g/m, then about 5 cm lengths of ribbon should be used. The amount must be reduced proportionately at higher altitudes (lower pressures). Pieces of approximately correct length should be cut in advance. The students should measure their length to ± 0.5 mm.
3. If possible, the linear density of ribbon should be determined to 3 significant figures on an analytical balance by weighing precisely measured pieces (15 cm–20 cm) and calculating the average linear density. If only a centigram balance is available, then lengths of one meter or more must be weighed.
4. Burets may be substituted for eudiometers (gas tubes) if they are not available. If used, the volume between the 50 ml mark and the stopcock must first be determined.

This is done by filling the buret with water to the 50 ml mark and draining it into a 10 ml graduated cylinder. The volume of gas collected is then found by reading the buret, subtracting that reading from 50.0 ml, and adding that to the uncalibrated volume found above.

5. You may need to discuss further why the water levels must be equalized in Procedure 4(b). The alert students will notice that the volume varies as the tube is raised or lowered.

MATERIALS NEEDED

Apparatus	Chemicals	
metric rule	Mg ribbon	(5 cm)
eudiometer, 50 ml	(see Note 2)	
(see Note 4)	Cu wire, fine	(20 cm)
test tube clamp	(may be reused)	
graduated cylinder, 10 ml	HCl, 6M; 500 ml conc./ℓ	(10 ml)
beaker, 50 ml		
beaker, 600 ml		
stopper, 1-hole		
(to fit gas tube)		
graduated cylinder, 1 ℓ or 500 ml		
thermometer		
barometer		

SAMPLE DATA TABLE

length of Mg ribbon	45.0 mm	mass of Mg/meter	0.816 g/m
water temperature	26.4°C	barometric pressure	743 mm Hg
volume of gas	40.45 ml		

CALCULATIONS AND QUESTIONS

1. From Table 11.1: P_{H_2O} at 26.4°C = 25.8 mm Hg

   P_{H_2} = 743 mm Hg − 26 mm Hg = 717 mm Hg

2. $V_2 = V_1 \times \frac{P_1}{P_2} \times \frac{T_2}{T_1} = 0.04045\ \ell \times \frac{717 \text{ mm Hg}}{760 \text{ mm Hg}} \times \frac{273 \text{ K}}{299 \text{ K}} = 0.0348\ \ell$

3. a. moles $H_2 = 0.0348\ \ell \times \frac{1 \text{ mol}}{22.4\ \ell} = 1.55 \times 10^{-3}$

   b. mass Mg $= 45.0 \text{ mm} \times \frac{1 \text{ m}}{10^3 \text{ mm}} \times \frac{0.816 \text{ g}}{1 \text{ m}} = 0.0367 \text{ g}$

c. $\text{moles Mg} = 0.0367 \text{ g Mg} \times \dfrac{1 \text{ mole Mg}}{24.31 \text{ g Mg}} = 1.51 \times 10^{-3}$

d. $\dfrac{\text{moles } H_2}{\text{mole Mg}} = \dfrac{1.55 \times 10^{-3}}{1.51 \times 10^{-3}} = 1.03 \text{ or } 1{:}1$

4. $Mg(s) + 2\ H^+(aq) \rightarrow H_2(g) + Mg^{2+}(aq)$

   NOTE: The H^+ ion coefficient must be 2 so as to maintain the H_2:Mg ratio of 1:1 as found in the experiment. The charge on the magnesium ion must be +2 in order to conserve charge.

EXTENSIONS

1. From PV = n RT:

$$n = \frac{PV}{RT} = \frac{717 \text{ mm Hg} \times \dfrac{1 \text{ atm}}{760 \text{ mm Hg}} \times 0.04045\ \ell}{0.0821\ \dfrac{\ell\cdot\text{atm}}{\text{mol}\cdot\text{K}} \times 299 \text{ K}} = 1.55 \times 10^{-3}$$

   The values obtained by the two methods are the same.

2. The equation for the reaction of aluminum with an acid is:

   $2\ Al(s) + 6\ H^+(aq) \rightarrow 3\ H_2(g) + 2\ Al^{3+}(aq)$

   The correct H_2:Al mole ratio is 3:2 or 1.5:1. More acid is used because the oxide coating on Al tends to slow down the reaction. About 0.03 g of Al will produce about 40 ml of gas at room conditions.

The Preparation and Properties of Pure Oxygen

TIMING

This experiment takes about 40 minutes. If it is modified by doing the extension, more time is required. It should still be completed in a single period by most students. The experiment is designed for Chapter 6 but can also be performed during Chapter 5.

PRELIMINARY STUDY

2. a. $\text{moles } O_2 = 12.3 \text{ g } KClO_3 \times \dfrac{1 \text{ mol } KClO_3}{123 \text{ g } KClO_3} \times \dfrac{3 \text{ mol } O_2}{2 \text{ mol } KClO_3} = 0.150$

   b. $\text{volume } O_2 = 0.150 \text{ mol } O_2 \times \dfrac{22.4\ \ell}{1 \text{ mol}} = 3.36\ \ell$

EXPERIMENT NOTES

1. The setup for collecting a gas by water displacement is new to the student and should be approved by you before heating begins. Check the mixing of the $KClO_3$ and MnO_2, the clamp placement, and all connections. The reaction mixture should not make contact with the stopper.
2. If desired, the experiment can be converted to a quantitative experiment in which the percent of oxygen in $KClO_3$ is determined. The procedure is outlined in the Extension. The decision to do this must be made in advance so that the necessary weighings are made.
3. Potassium chlorate is a strong oxidizing agent and forms explosive mixtures with combustible materials. The $KClO_3$ used should be tested in advance by the teacher before being used in a class experiment or demonstration.
4. If a small amount of water is not left in the bottles of oxygen in Part I, the exothermic reactions of Part II may crack the bottle. If the reaction mixture is heated too strongly, some chlorine dioxide gas will be produced. If this happens, Procedure 4 of Part II will give misleading results.
5. A simple data table using headings of ''Reactants'' and ''Observations'' is sufficient for this experiment. If the extension is done, use a data table similar to that shown in the solution to the extension.
6. If your school district policy or some other higher authority precludes the student use of $KClO_3$, then the experiment can still be done advantageously as a teacher demonstration.

MATERIALS NEEDED

Apparatus	Chemicals	
test tube, large	$KClO_3$	(10 g)
stopper, 1-hole, with right angle glass tube (to fit test tube)	MnO_2	(3 g)
rubber tubing (2 pieces)	*steel wool	(small wad)
wide mouth bottles (4)	charcoal, lump	(1 piece)
glass plates (4)	**Supplies**	
burner	glass wool (may be reused)	(small wad)
test tube clamp	*cotton	(small wad)
crucible tongs		
deflagrating spoon		
pneumatic trough		

CALCULATIONS AND QUESTIONS

1. a. $$\text{moles } O_2 = 10 \text{ g } KClO_3 \times \frac{1 \text{ mol } KClO_3}{123 \text{ g } KClO_3} \times \frac{3 \text{ mol } O_2}{2 \text{ mol } KClO_3} = 0.12$$

   b. $$\text{volume } O_2 \text{ (0°C)} = 0.12 \text{ mol} \times \frac{22.4\ \ell}{1 \text{ mol}} = 2.7\ \ell$$

   $$\text{volume } O_2 \text{ (25°C)} = 2.7\ \ell \times \frac{298 \text{ K}}{275 \text{ K}} = 2.9\ \ell$$

2. Pure oxygen gas is odorless and colorless. (As explained in Note 4 some odor may be detected.) Oxygen gas has low solubility in water and displaces it in the experiment.
3. a. The combustion reactions take place much more rapidly in pure oxygen than in air. This is evidenced by the more rapid release of energy.
   b. Inflammable means combustible or burnable. Oxygen is not "burned" in combustion reactions but it is the necessary reactant to burn the other substances, cotton, wood, etc. The correct statement is that oxygen supports combustion.
4. a. $C_6H_{10}O_5(s) + 6\ O_2(g) \rightarrow 6\ CO_2(g) + 5\ H_2O(l)$

   b. $4\ Fe(s) + 3\ O_2(g) \rightarrow 2\ Fe_2O_3(s)$

   $C(s) + O_2(g) \rightarrow CO_2(g)$
5. Iron (III) oxide is a solid with low solubility in water. Carbon dioxide is a gas with high solubility in water. In general, metals produce ionic oxides with high melting

points (solids) and non-metals produce molecular oxides with low melting points (gases).

EXTENSIONS

1. In determining the percent oxygen, the decomposition must be carried to completion. Additional time is required for this and for the four weighings. The theoretical percent oxygen in $KClO_3$ is 39.2.

SAMPLE DATA TABLE

mass test tube	30.30 g	mass test tube + $KClO_3$ + MnO_2	43.50 g
mass test tube + $KClO_3$	40.40 g	mass test tube + residue	39.60 g
mass $KClO_3$	10.10 g	mass oxygen released	3.90 g

$$\%\ \text{oxygen} = \frac{3.90\ \text{g O}}{10.10\ \text{g KClO}_3} \times 100 = 38.6$$

$$\%\ \text{error} = \frac{39.2 - 38.6}{39.2} \times 100 = 1.5$$

Periodicity and Predictions of Properties

TIMING

Students may not complete Procedure 3 during a single lab period. If not, it can be completed as a home assignment. The experiment should be done toward the completion of Chapter 7 but also fits in with Chapter 8.

EXPERIMENT NOTES

1. Even though this is a ''dry'' lab, it develops considerable interest among the students. The concept of a periodic property becomes much clearer after one is graphed.

2. In making predictions from a graph, point out the distinction between an interpolation (Procedure 1) and an extrapolation (Procedure 2). Extending a curve into the wild blue yonder is usually less reliable than estimating between points.

3. Locating the mystery elements helps in learning their properties and their periodic table locations. Students may have to do some library work in identifying an element. A blank table, which can be reproduced, is included. (See back pages of the guide.)

MATERIALS NEEDED

Supplies

graph paper (2)
blank periodic table

MYSTERY ELEMENT PERIODIC TABLE

M (H)																	B (He)
H (Li)														E (N)		P (F)	A (Ne)
	K (Mg)											T (Al)	S (Si)		R (S)	Q (Cl)	D (Ar)
G (K)	J (Ca)						W (Fe)			U (Cu)				F (As)		N (Br)	
																O (I)	C (Xe)
										X (Au)	V (Hg)						
I (Fr)	L (Ra)																

		Y (U)									Z (Md)		

CALCULATIONS AND QUESTIONS

1. a. The graph shows three cycles in which the atomic radii generally decrease to a minimum and then increase sharply to a maximum. The peaks are occupied by Group 1 metals, Li, Na, and K, and each peak is successively higher. Beginning with the first peak (Li), each of the first two cycles contains eight elements. The third cycle, beginning with K, contains 18 elements.

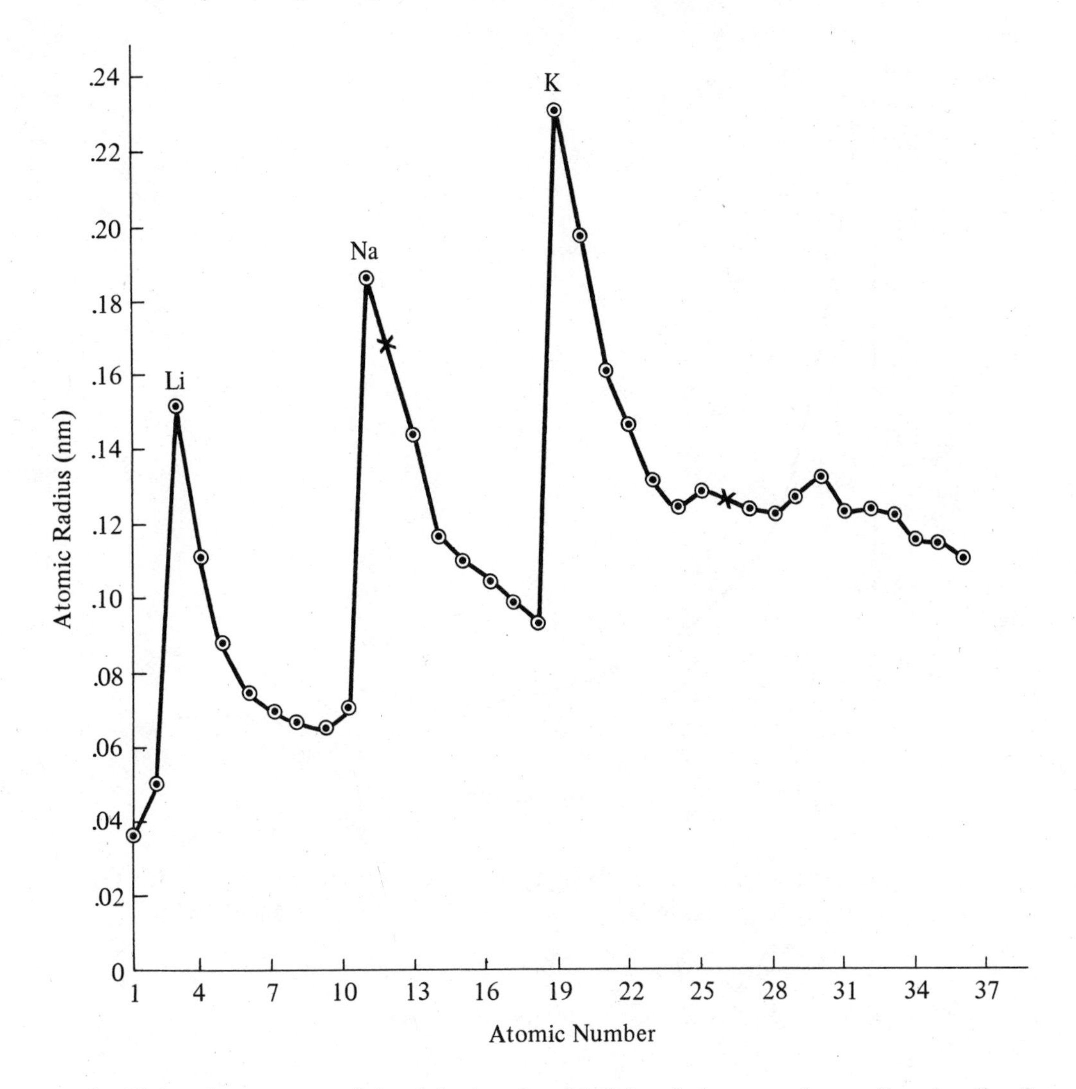

b. Using the average of the right-hand and left-hand elements the predicted radius for Mg is:

$$\text{Mg} = \frac{0.186 \text{ nm} + 0.143 \text{ nm}}{2} = 0.164 \text{ nm}$$

c. $Fe = \dfrac{0.129 \text{ nm} + 0.125 \text{ nm}}{2} = 0.127 \text{ nm}$

(The actual values are 0.160 nm for Mg and 0.126 nm for Fe.)

2. a. The curve shows that the melting points of the Group 1 elements decrease as the atomic number increases. The relationship is nonlinear as the slope of the curve decreases and seems to approach zero.

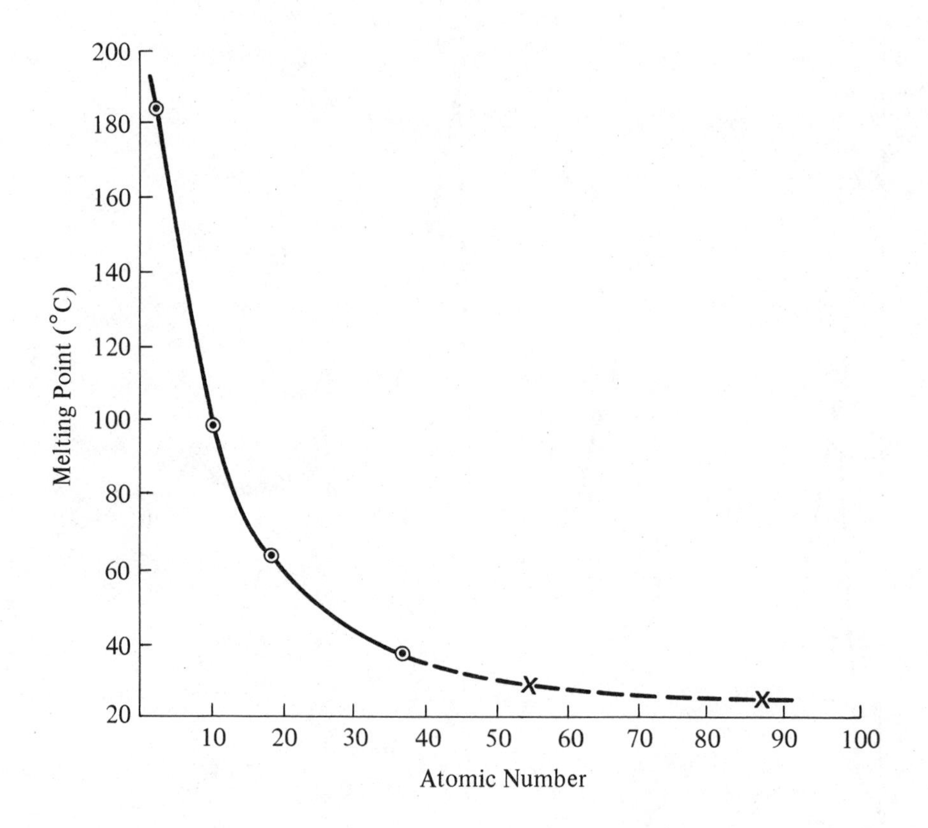

b. Using the extrapolated curve, the predicted melting points are 29°C for Cs and 26°C for Fr. (The actual melting points are 28.5°C for Cs and 27°C for Fr.)

3. a. MP (HF) c. EM_3 (NH_3) e. None

b. TQ_3 ($AlCl_3$) d. I_2R (Na_2S) f. JN_2 ($CaBr_2$)

EXTENSIONS

1. The graphs show that both melting points and boiling points are periodic properties. There is a surprise however, as the peaks are the Group 4 elements, C and Si. The minimums are the noble gases.

Some Chemical Properties of the Main-Group Elements

TIMING

The experiment requires about 70 minutes and some students may not finish. If necessary, it can be stopped at Part IV and completed the next period. It should be done during Chapter 8.

EXPERIMENT NOTES

1. The experiment exposes the student to a series of reactions typical of the main-group elements. The student should look for group properties and group trends. The experiment further develops the skills of making observations on reactions and describing them with balanced chemical equations.
2. The instructor should demonstrate Part I using small pieces of the alkali metals. Students should observe from a safe distance with goggles on. Once in a while an alkali metal will explode in water and splatter nearby. Part I can be done on the day before as a Prelab.
3. Trichloroethane, CCl_3CH_3, is recommended for dissolving the halogens in place of carbon tetrachloride in Part IV. This solvent is believed to be much less toxic than CCl_4.
4. A simple data table using headings of ''Reactants'' and ''Observations'' is sufficient for this experiment.

MATERIALS NEEDED

Apparatus	Chemicals	
beakers, 400 ml (3 for Part I)	Li, Na, and K	(for Part I)
test tubes, small (5)	phenolphthalein; 0.05 g/100 ml 50% C_2H_5OH	
test tube, large	Ca	(small piece)
wide mouth bottle	$Pb(NO_3)_2$, 0.1 M; 33 g/ℓ	(1 ml)
stopper (to fit bottle)	NaI	(2 g)
deflagrating spoon	MnO_2	(1 g)
burner	H_2SO_4, 3 M; 167 ml conc./ℓ	(10 ml)
test tube clamp	CCl_3CH_3	(5 ml)
test tube holder	NaCl, 0.1 M; 5.8 g/ℓ	(2 ml)
stirring rod	NaBr, 0.1 M; 10 g/ℓ	(2 ml)

Apparatus	Chemicals	
beaker, 100 ml	NaI, 0.1 M; 15 g/ℓ	(2 ml)
beaker, 250 ml	Cl_2–H_2O (saturated)	(1 ml)
evaporating dish	sulfur	(5 g)
wire gauze	**Supplies**	
iron ring		
test tube rack	cellophane tape	
spatula	filter paper	
	corks (3 to fit small test tube)	

CALCULATIONS AND QUESTIONS

1. a. chemical reactivity: K > Na > Ca

   As reactivity increases within a group and decreases within a period, Mg would be least reactive of the four elements.

   b. $2\ K(s) + 2\ H_2O \rightarrow 2\ K^+(aq) + 2\ OH^-(aq) + H_2(g)$

   $Ca(s) + 2\ H_2O \rightarrow Ca^{2+}(aq) + 2\ OH^-(aq) + H_2(g)$

   c. The phenolphthalein indicator turned pink indicating the presence of OH^- ions.
   d. The reaction of Ca with water produced a cloudy mixture. This indicates that some $Ca(OH)_2$ was precipitating due to its low solubility.
2. The Group I metals reacted with water in a chemical change in which H_2 gas was produced. The iodine dissolves slightly in water giving a brownish-colored solution. Sulfur neither reacts with water nor dissolves in it.
3. a. $S(s) + O_2(g) \rightarrow SO_2(g)$

   b. Sulfur dioxide is a gas with a pungent, disagreeable odor and is soluble in water. The precipitate obtained on addition of the $Pb(NO_3)_2$ solution had to be produced by a solute in the water layer.

   For the instructor: $SO_2(g) + H_2O \rightleftharpoons H_2SO_3(aq) \rightleftharpoons 2\ H^+(aq) + SO_3{}^{2-}\ (aq)$

   $$Pb^{2+}(aq) + SO_3^{2-}(aq) \rightarrow PbSO_3(s)$$

   c. When heated, the S_8 molecules in crystalline sulfur are broken open and long chain polymers are formed. The polymer solidifies when quenched in water. The product is known as plastic sulfur.
4. a. $2\ I^-(aq) + 4\ H^+(aq) + MnO_2(s) \rightarrow I_2(s) + Mn^{2+}(aq) + 2\ H_2O$

   b. The NaI furnishes the I^- ions that are converted to I_2. The sulfuric acid provides the H^+ ions for making water.
   c. When I_2 is warmed slightly in a test tube purple vapor can be seen. This indicates that I_2 has a relatively high vapor pressure compared to most solids. (It also indicates that I_2 sublimes.)
   d. Iodine is more soluble in CCl_3CH_3 than in water. This implies that the molecular structure of I_2 is more similar to that of CCl_3CH_3 than to H_2O.

e. $Cl_2(aq) + 2\ I^-(aq) \rightarrow I_2(aq) + 2\ Cl^-(aq)$
f. Add chlorine water and CCl_3CH_3 to the unknown and shake well. The color of the CCl_3CH_3 layer on the bottom of the tube indicates the negative ion originally present in the unknown: colorless, Cl^-; amber, Br^-; purple-pink, I^-.

EXTENSIONS

Chlorine gas is produced by the reaction:

$$OCl^-(aq) + 2\ H^+(aq) + Cl^-(aq) \rightarrow Cl_2(g) + H_2O$$

The gas will bleach (whiten) many dyes in colored cloth or paper.

Understanding Electron Charge Density Diagrams

TIMING

The experiment requires about 20 minutes. The remainder of the period can be spent in graphing the data. The experiment should be done during Chapter 9.

EXPERIMENT NOTES

1. The experiment is a two-dimensional analogy of the problem of locating the electron in a three-dimensional hydrogen atom. A "time-lapse" picture of an electron cloud (or orbital) is obtained by aiming darts at the bulls-eye of a target. The hit distribution will be similar to that of the 1 electron in the hydrogen atom.

2. An experiment in which darts are thrown requires a certain amount of self-discipline on the part of the students. Reduce temptation by collecting the darts immediately upon completion of the experiment.

3. It is possible that once in a while a set of data will produce nontypical graphs. This is the way that probability laws work. Have the students drop 200 darts, or 500 darts. Eventually, probability wins!

4. A sample target, which may be reproduced, is provided. The first ring has a 1 cm radius, the second ring has a 2 cm radius, and so on. (See back pages of the guide.)

MATERIALS NEEDED

Apparatus	Supplies
target board (2 ft × 2 ft)	target paper
(plywood or wallboard)	masking tape
dart	

SAMPLE DATA TABLE

Number of Concentric Rings	Average Distance of Ring from Bullseye (cm)	Area of Concentric Ring (cm^2)	Number of Hits in Ring	Number of Hits per Unit Area (hits/cm^2)
1	0.5	3.1	5	1.6
2	1.5	9.4	12	1.3
3	2.5	16	16	1.0
4	3.5	22	23	1.0
5	4.5	28	18	0.64
6	5.5	35	10	0.29
7	6.5	41	7	0.17
8	7.5	47	4	0.085
9	8.5	53	4	0.075
10	9.5	60	1	0.017

CALCULATIONS AND QUESTIONS

1. a.

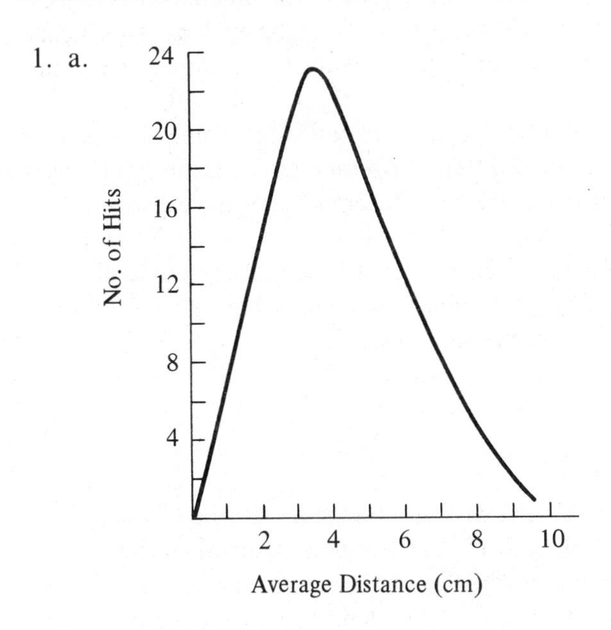

b. There is a 23% probability of a dart hit in Ring 4. As Ring 4 has the highest probability, the most likely distance for a dart to fall is 3.5 cm from the bulls-eye.

c. The dart curve (above) and Figure 15.2(a) for the hydrogen electron have similar shapes. The electron curve shows the probability of locating the electron in a thin spherical shell at varying distances from the nucleus. The probability is zero at the nucleus, it increases to a maximum at 0.05 nm, and decreases from then on.

2. a.

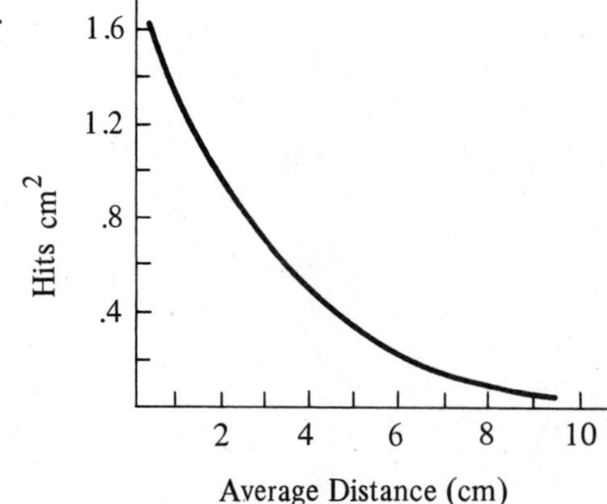

b. The probability of a hit in a given area, say hits/cm^2, is large at small distances and decreases as the distance from the bullseye increases. The most probable location for a 1 cm square to be hit is at the bullseye.

c. The hit density curve (above) and Figure 15.2(b) for the hydrogen electron have similar shapes. The electron curve shows the probability of finding the electron in a small volume at a particular distance from the nucleus. The probability is highest at the nucleus and decreases with increasing distance from the nucleus. This means that within a small volume the electron is most likely to be found at the nucleus.

3. a. The radius of maximum probability for a dart hit will vary somewhat among students. Different students will have different average errors in dropping darts due to eye differences, arm differences, etc. Different darts may also produce different errors.

b. No. The 2s electron in Li is in the n=2 energy level. The most probable distance from the nucleus increases as the value of n increases, and decreases with increasing nuclear charge for electrons with the same value of n.

EXTENSIONS

1. a. Beginning at 12 o'clock and going clockwise around the target, the hits per quadrant are: Q-1, 21; Q-2, 34; Q-3, 30; Q-4, 15. The two bottom quadrants have more hits than expected while the two top quadrants have fewer hits than expected. This indicates a directional effect in that the dart hits tend to concentrate below the bullseye.

b. In Figure 15.2(a) there is no directional effect for electron probability. The electron probability has spherical symmetry.

Analysis of a Hydrate

TIMING

The experiment requires about 45 minutes. Classes with longer lab periods may want to do a second heating and weighing. The experiment is recommended for Chapter 10 but is also suitable for Chapter 21, Quantitative Analysis.

PRELIMINARY STUDY

2. mass H_2O = 3.00 g − 2.56 g = 0.44 g

$$\text{moles } H_2O = \frac{0.44 \text{ g } H_2O}{18.0 \text{ g } H_2O/\text{mol}} = 0.024$$

$$\text{moles } BaCl_2 = \frac{2.56 \text{ g } BaCl_2}{208.2 \text{ g } BaCl_2/\text{mol}} = 0.0123$$

$$\frac{\text{moles } H_2O}{\text{mole } BaCl_2} = \frac{0.024}{0.0123} = 2.0$$

Hydrate formula: $BaCl_2 \cdot 2\ H_2O$

EXPERIMENT NOTES

1. Some hydrates do not give good results in this experiment. If you intend to use a hydrate other than those recommended it should be checked in advance for suitability.

MATERIALS NEEDED

Apparatus

crucible and cover
iron ring
clay triangle
burner
crucible tongs
heat resistant pad

Chemicals

hydrate unknown (3 g)
$BaCl_2 \cdot 2\ H_2O$
$Na_2MoO_4 \cdot 2\ H_2O$
$K_2CO_3 \cdot 3/2\ H_2O$
$CaSO_4 \cdot 2\ H_2O$
$Na_2CO_3 \cdot H_2O$
$Na_2CrO_4 \cdot 4\ H_2O$
$CuSO_4 \cdot 5\ H_2O$

SAMPLE DATA TABLE

Anhydrous salt formula	$CuSO_4$
Mass of empty dry crucible and cover	18.71 g
Mass of crucible, cover and hydrate	21.74 g
Mass of crucible, cover and residue	20.64 g

CALCULATIONS AND QUESTIONS

1. a. mass H_2O = 21.74 g – 20.64 g = 1.10 g

   b. moles $H_2O = \dfrac{1.10 \text{ g } H_2O}{18.0 \text{ g } H_2O/\text{mol}} = 0.0611$

   c. mass $CuSO_4$ = 20.64 g – 18.71 g = 1.93 g

   d. moles $CuSO_4 = \dfrac{1.93 \text{ g } CuSO_4}{160 \text{ g } CuSO_4/\text{mol}} = 0.0121$

2. a. $\dfrac{\text{moles } H_2O}{\text{mole } CuSO_4} = \dfrac{0.0611}{0.0121} = 5.05 \cong 5{:}1$

   b. Hydrate formula: $CuSO_4 \cdot 5\ H_2O$

3. a. The moles of H_2O would be low and the moles of anhydrous salt would be high. The ratio would be low.

   b. The moles of water would be high because all mass lost would be assumed to be water from the hydrate. The moles of anhydrous salt would be low because the mass of the wet crucible would be used to determine the mass of anhydrous salt. The ratio would be high.

   c. The anhydrous salt will begin to absorb water from the air as soon as it is cool. This means that all of the water is not removed at the time of weighing. The results are the same as in question 3(a).

EXTENSION

The behavior of $CaCl_2$ in air is described in Chapter 13 in the optional section labeled ''Deliquescence.''

Covalent Bonding and Molecular Structure

TIMING

Procedures 1 and 2 require about 50 minutes for most students. Procedure 3, using unknowns, requires an additional 10–15 minutes on the second day. The experiment is done during Chapter 11.

EXPERIMENT NOTES

1. This is the first of three experiments using molecular models. The ideas of molecules, bonding and structure become much clearer through the use of models. If you have space-filling models compare some of these to the ball and stick models.
2. Use balls with four holes at tetrahedral angles to represent all atoms which obey the octet rule. Using sticks to represent the nonbonding electron pairs as well as the bonds, gives a model which is consistent with the Lewis structure.
3. The atomic radii and the bond lengths are not to scale with ball and stick models. The bond angles are all 109° when using the four hole balls. The repulsion of nonbonding electron pairs decreases this angle in pyramidal and bent arrangements.

MATERIALS NEEDED

Apparatus

ball and stick molecular model kit (at least one per two students)

SAMPLE DATA TABLE

<table>
<tr><th>Molecule</th><th>Lewis Structure</th><th>Geometry</th><th>Polarity</th></tr>
<tr><td>CH_4</td><td>H
|
H−C−H
|
H</td><td>tetrahedral</td><td>nonpolar</td></tr>
<tr><td>CH_2Cl_2</td><td>H
|
H−C−:C̤̈l:
|
:C̤l:</td><td>tetrahedral</td><td>polar</td></tr>
</table>

Molecule	*Lewis Structure*	*Geometry*	*Polarity*
CH_4O	H–C(H)(H)–Ö–H	tetrahedral at C bent at C–O–H	polar
H_2O	H–Ö–H	bent	polar
H_3O^+	H–Ö(–H)–H$^+$	pyramidal	polar
HF	H–:F̤:	linear	polar
NH_3	H–N̈(–H)–H	pyramidal	polar
H_2O_2	H–Ö–Ö–H	bent at each O (nonplanar)	polar
N_2	:N≡N:	linear	nonpolar
Cl_2	:C̤̈l–C̤̈l:	linear	nonpolar
C_2H_4	(H)₂C=C(H)₂	planar	nonpolar
CH_2O	(H)₂C=Ö	planar	polar
C_2H_2	H–C≡C–H	linear	nonpolar
SO_2	:Ö–S̈=Ö	bent	polar
SO_4^{2-}	[:Ö–S(–:Ö:)(–:Ö:)–Ö:]$^{2-}$	tetrahedral	nonpolar
CO_2	Ö=C=Ö	linear	nonpolar

Each student should be given two unknowns (one molecule and one ion) for Procedure 3.

Molecules: S_8, N_2H_4, P_4, Cl_2O, PCl_3, CH_3NH_2, SO_3, NH_2OH, C_3H_6, C_2H_6

Ions: NO_3^-, CO_3^{2-}, ClO_4^-, PO_4^{3-}, ClO_2^-, SiO_3^{2-}, NH_2^-, SO_3^{2-}, NH_4^+, NO_2^-

LEWIS STRUCTURES OF UNKNOWNS

Molecules:

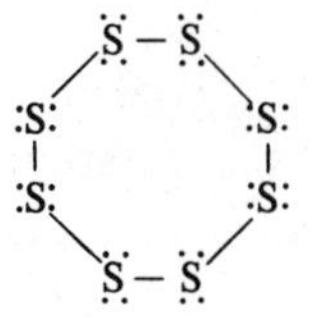

S_8 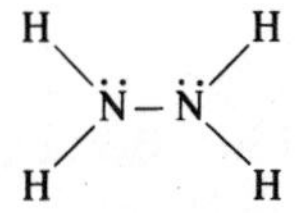 N_2H_4 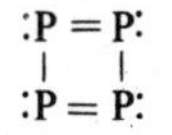 P_4 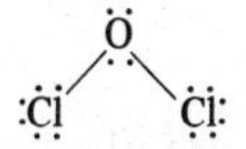 Cl_2O 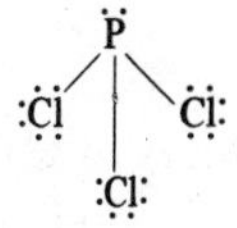 PCl_3

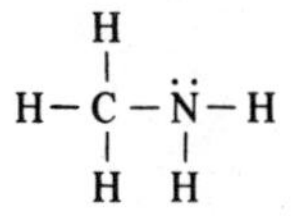

CH_3NH_2 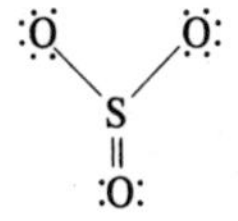 SO_3 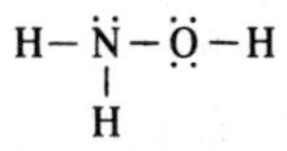 NH_2OH 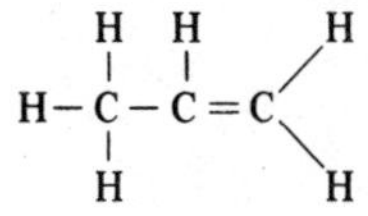 C_3H_6 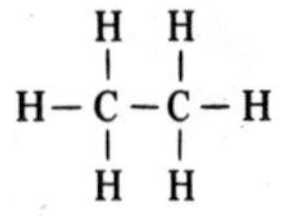 C_2H_6

Ions:

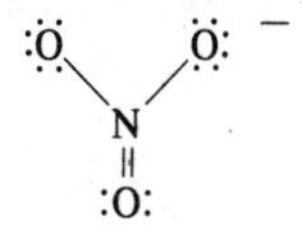

NO_3^- 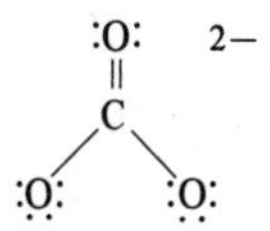 CO_3^{2-} 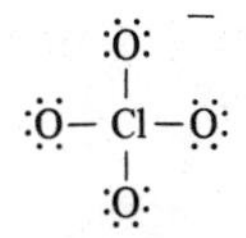 ClO_4^- 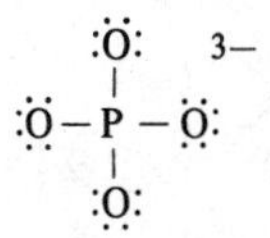 PO_4^{3-} 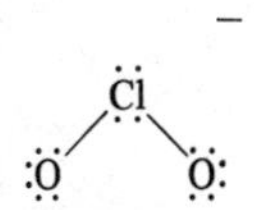 ClO_2^-

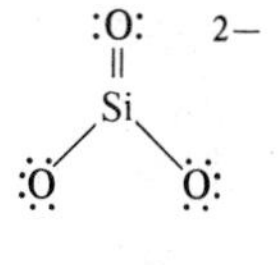

SiO_3^{2-} 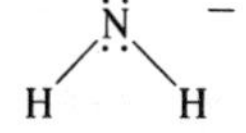 NH_2^- 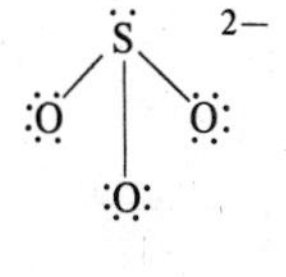 SO_3^{2-} 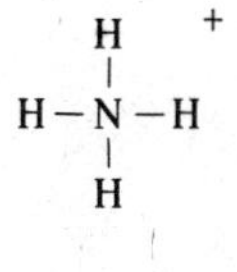 NH_4^+ 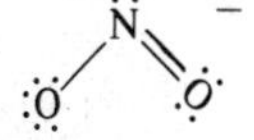 NO_2^-

EXTENSIONS

Cl Cl	Cl H	H Cl
C=C	C=C	C=C
H H	H Cl	H Cl
(1)	(2)	(3)

All isomers of $C_2H_2Cl_2$ are planar with bond angles of 120°. Because there is no rotation about C–C double bonds, isomers (1) and (2) are different structures. Isomers (1) and (3) are polar while isomer (2) is nonpolar.

Solid-Liquid Phase Changes

TIMING

The experiment, when done in pairs, requires about 50 minutes. The experiment should be done early in Chapter 13.

PRELIMINARY STUDY

2. $\Delta t = 25.8°C - 24.3°C = 1.5°C$

$$\text{heat (cal)} = \frac{1.00 \text{ cal}}{\text{g°C}} \times 100 \text{ g} \times 1.5°C = 150 \text{ cal (2 s.f.)}$$

$$\Delta H_{fus} = \frac{150 \text{ cal}}{8.0 \text{ g}} = 19 \text{ cal/g}$$

EXPERIMENT NOTES

1. Organic solids of low melting point are used to study solid-liquid phase changes. Two properties, melting point and heat of fusion, are used to identify the solid. If possible, different unknowns should be used throughout the class.

2. If the rate of cooling in Part I is too rapid the plateau on the cooling curve may be missed. Melting points within ± 1°C of the true values should be obtained.

3. The heat of fusion determination in Part II often produces results that are too high. If the tube is placed in the water too soon the extra heat of the tube and sample will be transferred to the water. The average experimental error is about ± 10%.

4. If labeled, the test tubes of unknown used in Part I can be corked and saved for reuse the following year. Clean the used tubes with alcohol as the unknowns are not water soluble.

5. Use adequate ventilation to minimize exposure to vapors of the molten samples. Students should avoid getting chemicals on skin. Because of its toxicity, phenol should not be used as an unknown.

MATERIALS NEEDED

Apparatus	Chemicals	
beakers, 400 ml (2)	unknown in test tube	(15 g)
iron ring	urethane	M.P. 49°C
wire gauze	thymol	M.P. 51°C
thermometers (2)	p-dichlorobenzene	M.P. 53°C
burner	naphthalene	M.P. 80°C
test tube clamp		
styrofoam cup		
test tube, large		
beaker tongs		
graduated cylinder, 100 ml		

SAMPLE DATA TABLE

Part I:

mass of test tube 30.29 g   mass of test tube and unknown 46.30 g
mass of unknown 16.01 g   unknown number naphthalene

Time (min)	Temp (°C)	Time (min)	Temp (°C)
0	80.0	3	79.0
0.5	79.9	3.5	78.0
1	79.8	4	76.5
1.5	79.7	4.5	74.0
2	79.5	5	69.4
2.5	79.3		

Time when the temperature of the solid started to drop: 3 min

Part II:

volume of water 150 ml   temp of water (initial) 25.0°C
temperature change (Δt) 4.0°C   temp of water (final) 29.0°C

CALCULATIONS AND QUESTIONS

1.

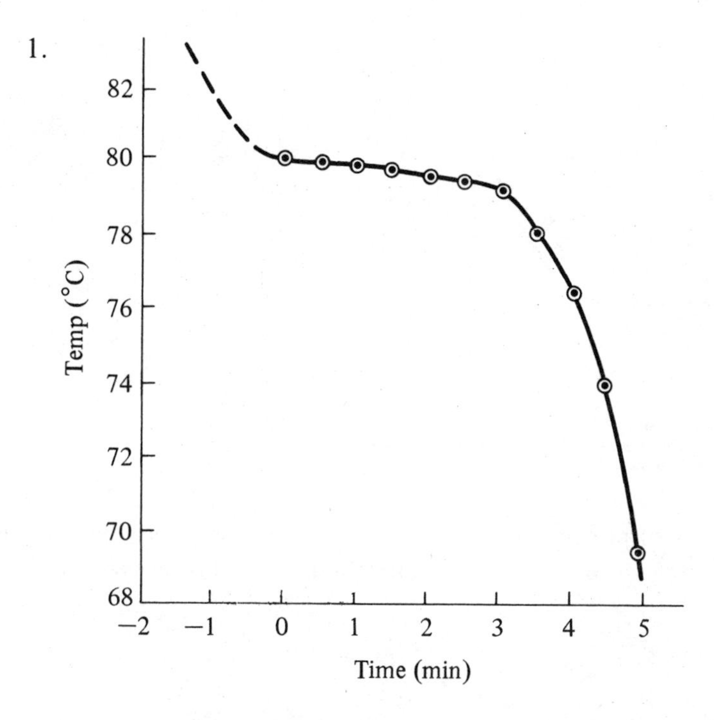

2. The temperature drops as the liquid cools until the time when the first crystals form. As solidification takes place the temperature tends to remain constant and the curve forms a plateau. (This indicates that heat is being evolved by the phase change.) After solidification is complete the temperature drops rapidly as the solid is cooled.
3. The melting point of the solid is 79–80°C. (This is the estimated temperature of the plateau of the curve.) Using Table 18-1, the solid is identified as naphthalene.
4. Increasing the amount of unknown would not affect its melting point. (It would lengthen the plateau portion of the cooling curve.)

5. $\text{Heat} = 150\text{ g} \times \dfrac{1.00\text{ cal}}{\text{g°C}} \times 4.0\text{°C} = 600\text{ cal (2 s.f.)}$

6. a. $\Delta H_{fus}\text{ (cal/g)} = \dfrac{600\text{ cal}}{16.01\text{ g}} = 38\text{ cal/g}$

   b. The heat of fusion value is between that of naphthalene (35 cal/g) and that of urethane (41 cal/g). As the melting point obtained was 79–80°C the unknown is most likely to be naphthalene.

7. a. Increasing the amount of unknown would increase the heat effect in calories.

   b. The heat effect in calories/gram would remain unchanged.

EXTENSIONS

1.

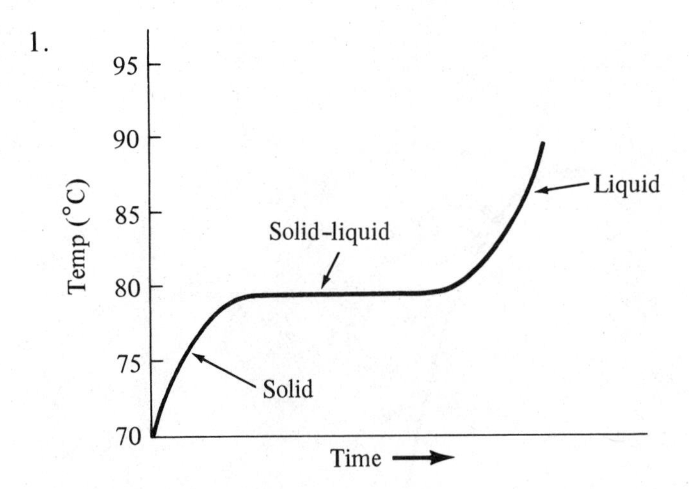

2. This experiment produces surprising results as the hot water freezes faster than cold water. See *Scientific American,* Sept. 1977, p. 246, for a detailed discussion of this paradox.

Relationships Between Physical Properties and Chemical Bonding in Solids

TIMING

The experiment requires about 50 minutes. It fits in best at the completion of Chapter 12.

EXPERIMENT NOTES

1. The general differences between the different classes of solids are investigated using a typical member of each class. Students should be reminded when doing an unknown that exceptions to some of the rules do exist:

   a. Some ionic compounds have low solubility and, therefore, low conductivity. Their melting point range is such that some can be melted by a burner flame while others cannot.

   b. Some molecular compounds are soluble in water due to hydrogen bonding. Some of these also ionize slightly.

   c. Some metals dissolve in water (Groups 1 and 2) by reacting with it and form a conducting solution.

   For these reasons students must look at the whole picture when analyzing an unknown and not base their conclusion on a single test result.

2. In Procedure 2, students should not heat any substance after it has melted. If students have difficulty in melting NaCl in an evaporating dish, have them heat a small amount in a small test tube over the hottest part of the flame.

3. Each student does not need individual test samples of the solids and solutions needed for the conductivity tests of Procedure 4. Prepare labeled beakers of the solids and the trichloroethane solutions and allow students to share these in order to save on the amounts of chemicals used.

MATERIALS NEEDED

Apparatus	Chemicals	
evaporating dish	NaCl	(5 g)
test tubes, small (5)	*p-dichlorobenzene	(5 g)
wire gauze	iron turnings	(5 g)
iron ring	SiO_2	(5 g)
burner	CCl_3CH_3	(20 ml)
conductivity tester		
beakers, 100 ml (2)		

Suggested Unknowns (5 g each)

Formula	Type	MP (°C)
$NaNO_3$	ionic	307
$NaC_2H_3O_2$	ionic	324
NH_4Cl	ionic	340
$CaCl_2$	ionic	772
CaO	ionic	2580
benzoic acid	molecular	122
urea	molecular	132
salicylic acid	molecular	160
sucrose	molecular	185
SiC	covalent network	2700
Si	covalent network	1410

SAMPLE DATA TABLE

Property Tested	*Observations* NaCl	$C_6H_4Cl_2$	SiO_2	Fe
hardness	brittle	soft	hard, scratches	hard
volatility		odor of mothballs		
melting	just melts	melts at low temp.	doesn't melt	doesn't melt
solubility				
a. in H_2O	soluble	insoluble	insoluble	insoluble
b. in CCl_3CH_3	insoluble	soluble	insoluble	insoluble
conductivity				
a. solids	no	no	no	yes
b. H_2O soln	yes	no	no	no
c. CCl_3CH_3 soln	no	no	no	no

QUESTIONS

1. a. The softest substance, $C_6H_4Cl_2$, is molecular. Molecular solids are held together by the relatively weak Van der Waals forces. Molecular solids are also relatively volatile for the same reason. The hardest substance, SiO_2, is a covalent network solid. This type of solid is a network of atoms bonded to each other by strong covalent bonds.
   b. No. A substance may be volatile and yet not have an odor. The nose can only detect certain molecules. Carbon monoxide gas, for example, is not detectable.
2. Based on melting behavior, the order of melting points is:

   $SiO_2 \cong Fe > NaCl > C_6H_4Cl_2$

   NOTE: The actual melting points are SiO_2, 1600°C–1700°C; Fe, 1535°C; NaCl, 801°C; $C_6H_4Cl_2$, 53°C.

3. a. Only NaCl dissolves appreciably in water and only $C_6H_4Cl_2$ dissolves appreciably in trichloroethane.
   b. Ionic substances tend to be soluble in water. (NOTE: Molecular solids of low molecular mass have some solubility in water if they form hydrogen bonds with water. See Chapter 13.)
4. The only solids which conduct electricity are metallic. Water solutions of ionic solids also conduct due to presence of ions which are free to move.
5. NaCl, ionic $C_6H_4Cl_2$, molecular
   Fe, metallic SiO_2, covalent network

EXTENSIONS

1. a. I_2, molecular Na_2SO_4, ionic
   $CaCl_2$, ionic   SiC, covalent network
   SO_3, molecular

   b. MP: SiC $>$ Na_2SO_4 $\cong$ $CaCl_2$ $>$ I_2 $>$ SO_3
   (2700°C) (884°C) (772°C) (114°C) (32°C)

   The melting point of I_2 is expected to be higher than that of SO_3 due to the larger mass of the I_2 molecule. In molecular solids the attractive forces (Van der Waals) increase with increasing mass of the molecule.

The Effect of Temperature on Solubility

TIMING

The experiment requires about 30 minutes. The remainder of the period can be spent in drawing the solubility curve. The experiment should be done early in Chapter 13.

PRELIMINARY STUDY

1. $$\text{solubility} = 100 \text{ g } H_2O \times \frac{0.94 \text{ g } Na_2SO_4}{20 \text{ g } H_2O} = 4.7 \text{ g } Na_2SO_4$$

EXPERIMENT NOTES

1. The experiment brings out the dependence of solubility upon temperature. It also exposes the student to the solubility unit (g solute/100 g H_2O) used in the *Handbook of Chemistry and Physics*. Some teachers may want to use the Extension as a starting point to discuss how a solid can be separated by fractional crystallization.

2. The experiment should be done as a class project with results posted on the board. Assign masses of $K_2Cr_2O_7$ within a range of 4 to 12 grams in integral values. The theoretical saturation temperatures are:

$\frac{\text{g } K_2Cr_2O_7}{100 \text{ g } H_2O}$	20	25	30	35	40	45	50	55	60	65	70	75
temp (°C)	36	42	47	52	57	61	65	69	73	77	81	84

3. Students tend to report the temperature at which the first crystal is sighted and this value is usually low. With good mixing a more accurate result is obtained by noting the temperature at which a substantial amount (~ 5%) of the solid has crystallized.

4. Solubility curves may be produced in a similar manner for other salts such as KNO_3 and NH_4Cl. Consult the solubility curves in the text for the solubility ranges to be assigned.

MATERIALS NEEDED

Apparatus

beaker, 400 ml
test tube, large
iron ring
wire gauze
graduated cylinder, 10 ml
stirring rod
burner
test tube clamp
thermometer

Chemicals

$K_2Cr_2O_7$ (4–12 g each)

Supplies

graph paper

SAMPLE DATA TABLE

	mass of $K_2Cr_2O_7$	satn temp (°C)	g $K_2Cr_2O_7$/100 g H_2O
1.	4.00 g	30	20.0 g
2.	5.00 g	37	25.0 g
3.	6.00 g	44	30.0 g
4.	7.00 g	50	35.0 g
5.	8.00 g	55	40.0 g
6.	9.00 g	60	45.0 g
7.	10.00 g	64	50.0 g
8.	11.00 g	68	55.0 g
9.	12.00 g	72	60.0 g

CALCULATIONS AND QUESTIONS

1. See Data Table.
2. a.

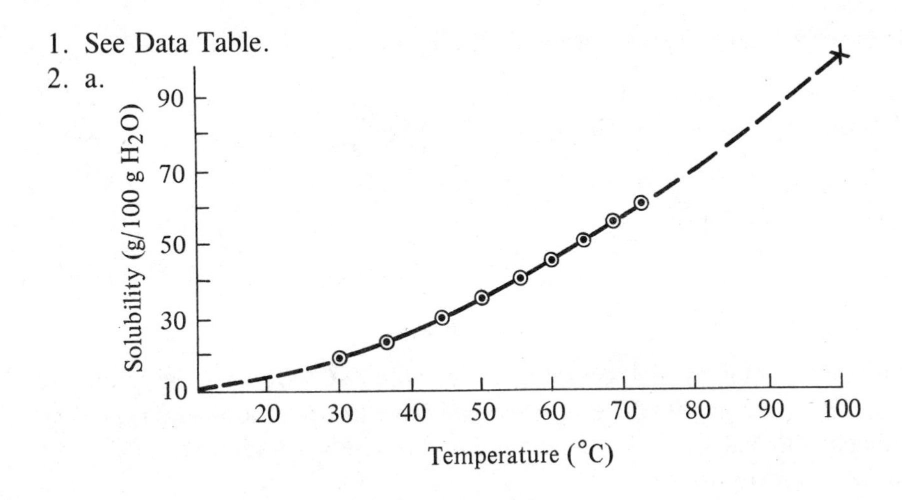

b. The solubility of $K_2Cr_2O_7$ increases as the temperature increases. (The relationship is non-linear.)

3. Using the experimental solubility curve:
   a. solubility at 100°C is about 101 g/100 g H_2O
   b. solubility at 10°C is about 11 g/100 g H_2O
   c. solubility at 25°C is about 17 g/100 g H_2O

$$\text{moles } K_2Cr_2O_7 = \frac{17 \text{ g } K_2Cr_2O_7}{294 \text{ g } K_2Cr_2O_7/\text{mol}} = 0.058 \text{ mol}$$

$$\text{mass of soln} = 100 \text{ g } H_2O + 17 \text{ g } K_2Cr_2O_7 = 117 \text{ g}$$

$$\text{volume of soln} = \frac{117 \text{ g soln}}{1.1 \text{ g/ml}} = 110 \text{ ml or } 0.11 \ \ell$$

$$\text{molarity} = \frac{0.058 \text{ mol}}{0.11 \ \ell} = 0.53 \text{ mol}/\ell = 0.53 \text{ M}$$

   d. solubility at 50°C is about 35 g/100 g H_2O

$$\text{mass \% } (K_2Cr_2O_7) = \frac{35 \text{ g } K_2Cr_2O_7}{135 \text{ g soln}} \times 100 = 26\%$$

   e. mass of $K_2Cr_2O_7$ = solubility (100°C) − solubility (10°C)

$$= 101 \text{ g} - 11 \text{ g} = 90 \text{ g}$$

$$\% \ K_2Cr_2O_7 \text{ crystallized} = \frac{90 \text{ g}}{101 \text{ g}} \times 100 = 89\%$$

EXTENSIONS

1. Using the experimental solubility curve for $K_2Cr_2O_7$:

$$\text{At 100°C: mass } H_2O = 90 \text{ g } K_2Cr_2O_7 \times \frac{100 \text{ g } H_2O}{101 \text{ g } K_2Cr_2O_7} = 89 \text{ g}$$

Using Figure 20.1 for NaCl:

$$\text{At 100°C: mass } H_2O = 10 \text{ g NaCl} \times \frac{100 \text{ g } H_2O}{40 \text{ g NaCl}} = 25 \text{ g}$$

This means that the solution will become saturated with $K_2Cr_2O_7$ when the amount of water is reduced to 89 grams. (The water would have to be reduced to 25 g before being saturated with NaCl.) If the temperature of this solution is reduced to 10°C the solubilities of the two salts are:

$$\text{solubility } (K_2Cr_2O_7) = 89 \text{ g } H_2O \times \frac{11 \text{ g } K_2Cr_2O_7}{100 \text{ g } H_2O} = 9.8 \text{ g}$$

$$\text{solubility (NaCl)} = 89 \text{ g } H_2O \times \frac{37 \text{ g NaCl}}{100 \text{ g } H_2O} = 33 \text{ g}$$

This means that the $K_2Cr_2O_7$ will crystallize out at 10°C while the NaCl remains in solution.

$$\text{mass } K_2Cr_2O_7 \text{ recovered} = 90 \text{ g} - 9.8 \text{ g} = 80 \text{ g}$$

$$\% \text{ recovered} = \frac{80 \text{ g } K_2Cr_2O_7}{90 \text{ g } K_2Cr_2O_7} \times 100 = 89\%$$

Preparation and Properties of Solutions

TIMING

The experiment requires about 50 minutes. Slow students may not finish both parts of Part III. The experiment should be done during Chapter 13.

PRELIMINARY STUDY

2. a. $\text{mass sucrose} = 25.0 \text{ ml soln} \times \dfrac{2.00 \text{ mol sucrose}}{10^3 \text{ ml soln}} \times \dfrac{342 \text{ g sucrose}}{1 \text{ mol sucrose}} = 17.1 \text{ g}$

   b. $\text{moles } CaCl_2 = 25.0 \text{ ml soln} \times \dfrac{2.00 \text{ mol } CaCl_2}{10^3 \text{ ml soln}} = 0.0500 \text{ mol}$

   $\text{volume soln} = 0.0500 \text{ mol } CaCl_2 \times \dfrac{10^3 \text{ ml soln}}{3.00 \text{ mol } CaCl_2} = 16.7 \text{ ml}$

EXPERIMENT NOTES

1. Students need practice in making solutions to a particular concentration. This skill will be required frequently in future experiments. Have students make the necessary calculations for Part I in advance and check their results at the start of lab.
2. Spend some time on a prelab demonstrating how solutions are made correctly with volumetric flasks. Explain that most dilute solutions in this course can be made using a graduated cylinder by assuming that the solvent volume equals the solution volume. The graduated cylinder is also used when diluting a stock solution. When more precise dilutions are required, pipets or burets will be specified.
3. While the general effect of a solute on the boiling point of the solvent is obvious, the experimental results are not overly quantitative. You may wish to review the electrical conductivity tests of molecular and ionic solutions to explain the higher boiling point of the $CaCl_2$ solution. Some students may notice that the boiling points of the solutions do not remain constant.
4. Prepare a set of standard $KMnO_4$ solutions for Part III. Use concentrations of 1.0×10^{-3} M, 8.0×10^{-4} M, 6.0×10^{-4} M, 4.0×10^{-4} M and 2.0×10^{-4} M. Suggested

concentrations for the unknowns are 9.0×10^{-4} M, 7.0×10^{-4} M, 5.0×10^{-4} M, 3.0×10^{-4} M, and 1.0×10^{-4} M.

5. If you have access to a colorimeter or spectrophotometer, consider using it as discussed in the Extension for Part III. (See Section 21.3 of the text.) Have students measure the absorbances of the known solutions at a wavelength of 525 nm. Plot a calibration curve of absorbance values versus molarity. Measure the absorbance of the unknown solution and determine its concentration from the curve.

MATERIALS NEEDED

Apparatus

beakers, 50 ml (2)
beaker, 100 ml
graduated cylinder, 25 ml or 50 ml
stirring rod
iron ring
wire gauze
thermometer
burner
watch glass
test tubes, large (2) or vials, flat bottom (2)
medicine dropper
wash bottle
funnel
crucible tongs

Chemicals

*sucrose	(18 g)
*NaCl	(10 g)
$CaCl_2$, 3.00 M; 441 g hydrate/ℓ	(20 ml)
$KMnO_4$, 1.0×10^{-3} M; 0.16 g/ℓ	(20 ml)
$KMnO_4$ standards (See Note 4)	
$KMnO_4$ unknowns (See Note 4)	

Supplies

filter paper

SAMPLE DATA TABLE

Part I

mass of $C_{12}H_{22}O_{11}$	17.1 g	vol of 2.00 M $CaCl_2$	16.7 ml

Part II

B.P. of H_2O	99.9°C	B.P. of $C_{12}H_{22}O_{11}$ soln	101.8°C
		B.P. of $CaCl_2$ soln	103.8°C

Part III

mass of beaker and watch glass	99.90 g	vol of NaCl soln	10.0 ml
mass of beaker, watch glass and NaCl	103.91 g	mass of NaCl	4.01 g

$KMnO_4$ unknown no.	(5.0×10^{-4} M)	$KMnO_4$ known soln	6.0×10^{-4} M
vol of unknown soln	20.0 ml	vol of soln needed for color matching	15.0 ml
vol of stock soln used in Procedure 2 (b)	12.0 ml		

CALCULATIONS AND QUESTIONS

1. a. $\Delta BP\ (CaCl_2) = 103.8°C - 99.9°C = 3.9°C$

$\Delta BP\ (C_{12}H_{22}O_{11}) = 101.8°C - 99.9°C = 1.9°C$

$$\frac{\Delta BP\ (CaCl_2)}{\Delta BP\ (C_{12}H_{22}O_{11})} = \frac{3.9°C}{1.9°C} = 2.1$$

b. The addition of a solid solute to water always lowers its vapor pressure. The amount of the lowering is directly proportional to the concentration of the solution. For this reason the boiling point of a solution is always higher than that of the pure solvent and increases with concentration. The boiling points of 2M $CaCl_2$ and $C_{12}H_{22}O_{11}$ solutions are not the same because ionic solutes break apart in water into individual ions. In the case of $CaCl_2$, three ions are formed for each unit of $CaCl_2$:

$$CaCl_2(s) \rightarrow Ca^{2+}(aq) + 2\ Cl^-(aq)$$

For this reason the boiling point elevation ratio in 2(a) is expected to be 3:1.

NOTE: The observed ratio is not equal to 3:1 because the solutions are too concentrated to behave as the simple theory predicts. The theoretical boiling point elevations are only approached for dilute solutions (< 0.1 M).

2. a. $$M\ (NaCl) = \frac{4.01\text{ g NaCl}}{0.0100\ \ell\text{ soln}} \times \frac{1\text{ mol NaCl}}{58.5\text{ g NaCl}} = 6.85\text{ mol}/\ell$$

NOTE: The theoretical concentration of a saturated NaCl solution is 6.2 M.

b. $$\text{mass of NaCl soln} = 10.0\text{ cm}^3 \times \frac{1.2\text{ g}}{\text{cm}^3} = 12\text{ g}$$

$$\%\ NaCl = \frac{4.01\text{ g NaCl}}{12\text{ g soln}} \times 100 = 33\%$$

3. $M_2 = \dfrac{V_1}{V_2} \times M_1$

$$M_2 = \frac{15.0\ \text{ml}}{20.0\ \text{ml}} \times 6.0 \times 10^{-4}\ \text{M} = 4.5 \times 10^{-4}\ \text{M}$$

EXTENSIONS

Follow the procedures given in Experiment Note 5. Typically, an absorbance (A) of 1.8 is obtained for a 1.0×10^{-3} M $KMnO_4$ solution at 525 nm.

EXPERIMENT 22

Isomerism in Organic Chemistry

TIMING

The experiment requires about 45 minutes. It should be done early in Chapter 14 to reinforce the text material.

PRELIMINARY STUDY

2. Structural isomers of C_4H_9Cl:

```
    H   H   H   H            H   H   H   H              H          H          H
    |   |   |   |            |   |   |   |              |          |          |
Cl—C — C — C — C — H    H—C — C — C — C — H    H———C———C———C———Cl
    |   |   |   |            |   |   |   |              |          |          |
    H   H   H   H            H   Cl  H   H              H      H—C—H      H
                                                                   |
                                                                   H

          H        Cl        H
          |        |         |
H———C———C———C———H
          |        |         |
          H        |         H
                   |
               H—C—H
                   |
                   H
```

EXPERIMENT NOTES

1. Isomerism is a basic concept in the study of organic chemistry. It is most easily learned when applied to the hydrocarbons. Chlorocarbons are also used in the experiment to show the effect of adding a different atom or group.
2. The possibility of geometric isomerism in compounds which contain double bonds is introduced. A necessary condition for geometric isomerism is that a cis-trans pair of structures exists.
3. Remind students that the molecular models of aromatic compounds do not represent the correct bonding description of the molecules. The alternating single and double bonds are misleading. All of the bonds in aromatic compounds are not identical. (Text-Sec. 14.4)
4. You may wish to extend the experiment by asking students to provide the IUPAC name of each isomer. These are listed in Extension 2.

MATERIALS NEEDED

Apparatus
ball and stick molecular model kit (one per two students)

SAMPLE DATA TABLE

Compound — *Structural Formulas*

a. CH_2Cl_2

```
     H
     |
 H - C - Cl
     |
     Cl
```

b. $C_2H_4Cl_2$

```
     H   H              H   H
     |   |              |   |
 Cl- C - C - H      Cl- C - C - Cl
     |   |              |   |
     Cl  H              H   H
```

c. C_3H_8

```
     H   H   H
     |   |   |
 H - C - C - C - H
     |   |   |
     H   H   H
```

d. C_3H_7Cl

```
     H   H   H              H   Cl  H
     |   |   |              |   |   |
 Cl- C - C - C - H      H - C - C - C - H
     |   |   |              |   |   |
     H   H   H              H   H   H
```

e. C_2H_4

```
 H         H
  \       /
   C  =  C
  /       \
 H         H
```

f. $C_2H_2Cl_2$

```
 H         Cl     Cl         H     Cl         Cl
  \       /         \       /        \       /
   C  =  C           C  =  C          C  =  C
  /       \         /       \        /       \
 H         Cl     H          Cl    H          H
                      (trans)           (cis)
```

g. C_4H_{10}

```
     H   H   H   H                H          H          H
     |   |   |   |                |          |          |
 H - C - C - C - C - H     H ---- C -------- C -------- C ---- H
     |   |   |   |                |          |          |
     H   H   H   H                H      H - C - H      H
                                             |
                                             H
```

h. C_4H_8

```
        H
        |
      H-C-H      H            H         H
         \      /              \       /
          C=C   H               C=C
         /   \  |          H   /   \   H
        H     C-H          |  /     \  |
              |          H-C         C-H
              H            |           |
                           H           H

        (trans)                 (cis)

H       H  H  H          H       H
 \      |  |  |          |       |
  C=C - C - C - H     H- C - C - C - H
 /      |  |             |   ||  |
H       H  H             H   C   H
                            / \
                           H   H
```

(NOTE: Two additional cyclic isomers are not considered.)

i. C_5H_{12}

```
   H  H  H  H  H                         H
   |  |  |  |  |                         |
H- C- C- C- C- C -H                    H-C-H
   |  |  |  |  |                    H    |    H
   H  H  H  H  H                    |    |    |
                                  H-C----C----C-H
   H  H  H  H                       |    |    |
   |  |  |  |                       H    |    H
H- C- C- C- C -H                       H-C-H
   |  |  |  |                            |
   H  H  |  H                            H
         |
       H-C-H
         |
         H
```

j. C_6H_6

```
          H
          |
    H    C    H
     \ //  \ /
      C     C
      |     ||
      C     C
     / \\  / \
    H    C    H
         |
         H
```

k. $C_6H_5(CH_3)$

```
         H       H   H
   H     |        \ /
    \    C         C
     C //  \\ C  /   \
     ||      |        H
     C       C
    /  \   // \
   H     C      H
         |
         H
```

1. $C_6H_4Cl_2$

```
          Cl                        Cl                        Cl
 H        |         Cl     H        |         H      H        |         H
   \    / C \\    /          \    / C \\    /          \    / C \\    /
     C          C              C          C              C          C
     ||         |              ||         |              ||         |
     C          C              C          C              C          C
   /    \ C //    \          /    \ C //    \          /    \ C //    \
 H        |         H      H        |         Cl     H        |         H
          H                         H                         Cl
```

QUESTIONS

1. a. CH_2Cl_2, C_3H_8, C_2H_4, C_6H_6, and $C_6H_5(CH_3)$ have only one isomeric form.
   b. $C_2H_2Cl_2$ and C_4H_8 have geometric isomers.
   c. C_2H_4, $C_2H_2Cl_2$ and C_4H_8 are unsaturated compounds.
   d. C_2H_4, $C_2H_2Cl_2$, C_6H_6 and $C_6H_4Cl_2$ are planar.
2. The C-C-C bond angles are 109° in C_3H_8, 120° in C_4H_8 (3 of the 4 isomers), and 120° in C_6H_6.

EXTENSIONS

1.

```
    H   H
    |   |
H — C — C — H
    |   |
H — C — C — H
    |   |
    H   H
```

Cyclobutane is a nonplanar molecule. (If planar, the C-C-C bond angles would be 90°.) The actual geometry is:

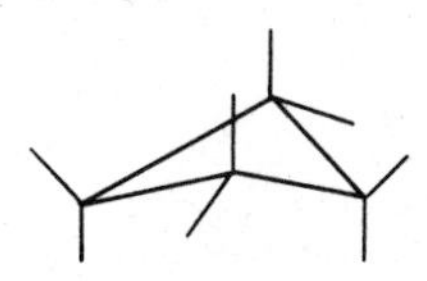

2. The IUPAC names are listed in the order in which the isomers appear in the data table.
   a. dichloromethane
   b. 1, 1-dichloroethane; 1, 2-dichloroethane
   c. propane
   d. 1-chloropropane; 2-chloropropane
   e. ethene
   f. 1, 1-dichlorethene; trans-1, 2-dichloroethene; cis-1, 2-dichloroethene
   g. butane; 2-methylpropane
   h. trans-2-butene; cis-2-butene; 1-butene; 2-methylpropene
   i. pentane; 2, 2-dimethylpropane; 2-methylbutane
   j. benzene
   k. methylbenzene
   l. 1, 2-dichlorobenzene; 1, 3-dichlorobenzene; 1, 4-dichlorobenzene

Properties of Hydrocarbons

TIMING

The experiment requires about 50 minutes when the special stopper assembly (Part I) is provided. The experiment should be done during Chapter 14.

PRELIMINARY STUDY

2. The molecule C_5H_{10} fits the general formula C_nH_{2n}. This means that it could be an alkene or a cycloalkane. (Alkanes, alkynes, and aromatics do not fit the general formula, C_nH_{2n}.) A bromine test would distinguish between an alkene and a cycloalkane. If the compound is an alkene (pentene) it would react with Br_2 in an addition reaction while cyclopentane would not.

EXPERIMENT NOTES

1. The experiment allows the student to perform some hydrocarbon reactions which are simple and safe. Toluene is used as a typical aromatic compound in place of benzene because it is less volatile and much less toxic.
2. Test the cyclohexane and toluene solutions in advance with bromine water. They are easily contaminated and, if so, will give misleading results in the addition tests of Part III.

MATERIALS NEEDED

Apparatus	Chemicals	
mortar and pestle	$NaC_2H_3O_2$	(4 g)
test tube, large	soda lime	(8 g)
test tubes, small (5)	Br_2- H_2O	(2 ml)
stopper with jet tip, right angle tube (to fit large t.t.)	CaC_2	(2 g)
	cyclohexane	(10 ml)
	cyclohexene	(3 ml)
test tube clamp	toluene	(10 ml)
burner	naphthalene	(1 g)
test tube rack	phenolphthalein; 0.05 g/100 ml 50% C_2H_5OH	
beaker, 400 ml		
corks (to fit small t.t.) (3)		
	phenolphthalein soln.	(2 drops)

SAMPLE DATA TABLE

	Procedure	Observations
(I-2)	$NaC_2H_3O_2$ + soda lime	A gas is evolved when the mixture is heated.
	$CH_4 + O_2$	The CH_4 gas burns cleanly when ignited.
(I-3)	$CH_4 + Br_2$	No reaction.
(II-1)	$CaC_2 + H_2O$	A gas is evolved which can be collected by water displacement.
	reaction soln + phenolphthalein	The solution tests basic (pink).
(II-2)	$C_2H_2 + O_2$	The C_2H_2 burns with a smoky flame when ignited.
(II-3)	$C_2H_2 + Br_2$	The Br_2 color disappears.
(III-1)	cyclohexane + Br_2	No reaction.
	cyclohexene + Br_2	The Br_2 color disappears.
	toluene + Br_2	No reaction.
(III-2)	naphthalene in H_2O	Insoluble
	naph. in cyclohexane	Soluble
	naph. in toluene	Soluble
(III-3)	cyclohexane in H_2O	Insoluble (immiscible)
	cyclohexane in toluene	Soluble (miscible)

QUESTIONS AND CALCULATIONS

1. a. C_2H_2 (Part II) and cyclohexene (Part III) reacted with bromine.

```
H         H
 \       /
  C  =  C
 /       \
Br        Br
```

```
            H   H
             \ /
              C
H           /   \           H
 \        /       \        /
  C                 C
 /|                 |\
Br|                 | H
H |                 | H
 \|                 |/
  C                 C
 /  \             /  \
Br    \         /     H
        \     /
          C
         / \
        H   H
```

b. C_2H_2 is an alkyne; cyclohexene is an alkene. Both classes are unsaturated hydrocarbons.

2. a. $2C_2H_2(g) + 5O_2(g) \rightarrow 4CO_2(g) + 2H_2O$

   b. $C_6H_{10}(l) + Br_2(aq) \rightarrow C_6H_{10}Br_2(l)$

3. Acetylene would react with bromine, while methane would not.
   Acetylene burns with a smoky flame while methane burns with a clean flame.

4. A combustible gas (C_2H_2) is produced. The gas reacted with Br_2 showing that it is unsaturated.
   The solution tests basic with phenolphthalein indicating the presence of OH^- ions.
   The solid CaC_2 disappears indicating the possible presence of Ca^{2+} ions.

5. $$\text{mass } CaC_2 = 100\ \ell\ C_2H_2 \times \frac{1 \text{ mol } C_2H_2}{22.4\ \ell\ C_2H_2} \times \frac{1 \text{ mol } CaC_2}{1 \text{ mol } C_2H_2} \times \frac{64.1 \text{ g } CaC_2}{1 \text{ mol } CaC_2} = 286 \text{ g}$$

6. a. Naphthalene was soluble in both toluene and cyclohexane and insoluble in water. In general, molecular compounds are insoluble in water and soluble in molecular liquids. (The strong hydrogen bonding between water molecules cannot be broken apart by most molecular compounds.)
   b. Cyclohexane was soluble in toluene and insoluble in water. The explanation for this is essentially the same as given in 6 (a).
   c. Cyclohexene is expected to be soluble in either toluene or cyclohexane.

EXTENSIONS

Using the laboratory natural gas, follow the procedures of Experiment 10 to determine the molar mass of the gas. If it is assumed that the mixture contains only CH_4 and C_2H_6 then it follows that:

$$X\ (\text{MM of } CH_4) + (1 - X)\ (\text{MM of } C_2H_6) = \text{MM of natural gas}$$

where X = mole fraction of CH_4 and $1 - X$ = mole fraction of C_2H_6.

EXPERIMENT 24

Organic Functional Groups

TIMING

The experiment requires about 50 minutes. It should be done during Chapter 15.

EXPERIMENT NOTES

1. The student should learn to recognize functional groups and classes of organic compounds as a result of this experiment. Consider following the experiment with a quiz in which students must identify molecular models of the different classes.
2. The O_2 molecule cannot be correctly pictured with a Lewis structure. It is satisfactory for our purposes to represent it with the 2-hole balls held together by two springs representing a double bond.
3. The forming of an ester from the condensation of an alcohol and an acid is important to the understanding of polymers in Chapter 16.

MATERIALS NEEDED

Apparatus
ball and stick molecular model kit (one per two students)

SAMPLE DATA TABLE

Class	*Structural Formula*	*Compound Name*	*Geometry of the Functional Group*
alcohol	H H H |	| H–C–C–C–OH |	| H H H	propyl alcohol	bent
	H H H |	| H–C–C–C–H |	| H OH H	isopropyl alcohol (2-propanol)	bent

<table>
<tr><th><u>Class</u></th><th><u>Structural Formula</u></th><th><u>Compound Name</u></th><th><u>Geometry of the Functional Group</u></th></tr>
<tr><td>aldehyde</td><td><pre>
    H   H   O
    | | ‖
H — C — C — C — H
    | |
    H   H
</pre></td><td>propionaldehyde
(propanal)</td><td>triangular</td></tr>
<tr><td>ketone</td><td><pre>
    H   O   H
    | ‖ |
H — C — C — C — H
    | |
    H   H
</pre></td><td>dimethylketone
or acetone
(2-propanone)</td><td>triangular</td></tr>
<tr><td>acid</td><td><pre>
    H   H   O
    | | ‖
H — C — C — C — OH
    | |
    H   H
</pre></td><td>propionic acid
(propanoic acid)</td><td><pre>
            O
            ‖
triangular (— C —)
bent (— OH)
</pre></td></tr>
<tr><td>ether</td><td><pre>
    H   H   H
    | | |
H — C — C — O — C — H
    | | |
    H   H   H
</pre></td><td>methyl ethyl
ether
(methoxyethane)</td><td>bent</td></tr>
<tr><td>ester</td><td><pre>
    H   O   H
    | ‖ |
H — C — C — O — C — H
    | |
    H   H
</pre></td><td>methyl acetate
(methyl ethanoate)</td><td><pre>
                 O
                 ‖
triangular (— C —)
bent   (— O —)
</pre></td></tr>
<tr><td></td><td><pre>
    O   H   H
    ‖ | |
H — C — O — C — C — H
            | |
            H   H
</pre></td><td>ethyl formate
(ethyl methanoate)</td><td><pre>
                 O
                 ‖
triangular (— C —)
bent   (— O —)
</pre></td></tr>
</table>

<table>
<tr><th><u>Reaction</u></th><th><u>Structural Formula of Product</u></th><th><u>Class of Product</u></th><th><u>Name of Product</u></th></tr>
<tr><td>a. oxidation of methyl alcohol</td><td><pre>
    O
    ‖
    C
   / \
  H   H
</pre></td><td>aldehyde</td><td>formaldehyde
(methanal)</td></tr>
<tr><td>b. condensation of ethyl alcohol</td><td><pre>
    H   H   H   H
    | | | |
H — C — C — O — C — C — H
    | | | |
    H   H   H   H
</pre></td><td>ether</td><td>diethyl ether
(ethoxyethane)</td></tr>
</table>

c. oxidation of propionaldehyde

```
     H   H   O
     |   |   ||
 H - C - C - C - OH
     |   |
     H   H
```

acid — propionic acid (propanoic acid)

d. condensation of methyl alcohol and acetic acid

```
     H   O       H
     |   ||      |
 H - C - C - O - C - H
     |           |
     H           H
```

ester — methyl acetate (methyl ethanoate)

QUESTIONS

1. a. Molecules and classes which can form H-bonds:
   propyl alcohol, alcohol
   propionic acid, acid

   Example of H-bond formation:

```
     H   H   H
     |   |   |
 H - C - C - C - O - - - H   H   H   H
     |   |   |   |       |   |   |   |
     H   H   H   H - - - O - C - C - C - H
                             |   |   |
                             H   H   H
```

   b. There are two alcohols with the formula of $C_3H_7 - OH$:

```
     H   H   H                   H   H   H
     |   |   |                   |   |   |
 H - C - C - C - OH    and   H - C - C - C - H
     |   |   |                   |   |   |
     H   H   H                   H   OH  H
```

   There are two esters with the formula of $C_3H_6O_2$:

```
     H   O       H                 O       H   H
     |   ||      |                 ||      |   |
 H - C - C - O - C - H    and  H - C - O - C - C - H
     |           |                         |   |
     H           H                         H   H
```

   c. Propyl alcohol and methyl ethyl ether: C_3H_8O

   Propionaldehyde and dimethylketone: C_3H_6O

   Propionic acid and methyl acetate: $C_3H_6O_2$

2. a. $$H-\overset{H}{\underset{H}{C}}-OH + 1/2\,O_2 \rightarrow \underset{H\quad H}{\overset{\overset{O}{\|}}{C}} + H_2O$$

   b. $$H-\overset{H}{\underset{H}{C}}-\overset{H}{\underset{H}{C}}-OH + HO-\overset{H}{\underset{H}{C}}-\overset{H}{\underset{H}{C}}-H \rightarrow H-\overset{H}{\underset{H}{C}}-\overset{H}{\underset{H}{C}}-O-\overset{H}{\underset{H}{C}}-\overset{H}{\underset{H}{C}}-H + H_2O$$

   c. $$H-\overset{H}{\underset{H}{C}}-\overset{H}{\underset{H}{C}}-\overset{\overset{O}{\|}}{C}-H + 1/2\,O_2 \rightarrow H-\overset{H}{\underset{H}{C}}-\overset{H}{\underset{H}{C}}-\overset{\overset{O}{\|}}{C}-OH$$

   d. $$H-\overset{H}{\underset{H}{C}}-\overset{\overset{O}{\|}}{C}-OH + HO-\overset{H}{\underset{H}{C}}-H \rightarrow H-\overset{H}{\underset{H}{C}}-\overset{\overset{O}{\|}}{C}-O-\overset{H}{\underset{H}{C}}-H + H_2O$$

3. a. HCHO, aldehyde, formaldehyde (methanal)
   b. CH_3COOH, acid, acetic acid (ethanoic acid)
   c. $CH_3COCH_2CH_3$, ketone, methyl ethyl ketone
   d. $CH_3OCH_2CH_2CH_3$, ether, methyl propyl ether
   e. $CH_3CH_2CHOHCH_2CH_3$, alcohol, 3-pentanol
   f. $CH_3COOCH_2CH_3$, ester, ethyl acetate (ethyl ethanoate)

EXTENSIONS

1. $$C_6H_5-\overset{\overset{O}{\|}}{C}-OH + HO-\overset{H}{\underset{H}{C}}-H \rightarrow C_6H_5-\overset{\overset{O}{\|}}{C}-O-\overset{H}{\underset{H}{C}}-H + H_2O$$

EXPERIMENT 25

Organic Syntheses

TIMING

The experiment requires about 40 minutes for Part I (soap) and 30 minutes for Part II (mauve). It should be done during Chapter 15.

PRELIMINARY STUDY

2. A fatty acid is a long-chain carboxylic acid while a fat is a triester of fatty acids and the trihydroxy alcohol, glycerol. The structural formula for the saturated C_{16} fatty acid (palmitic acid) is:

```
    H  H  H  H  H  H  H  H  H  H  H  H  H  H  H     O
    |  |  |  |  |  |  |  |  |  |  |  |  |  |  |    //
H - C -C -C -C -C -C -C -C -C -C -C -C -C -C -C - C
    |  |  |  |  |  |  |  |  |  |  |  |  |  |  |    \
    H  H  H  H  H  H  H  H  H  H  H  H  H  H  H     OH
```

EXPERIMENT NOTES

1. Students make a soap and investigate the properties of soaps and detergents, common household products. The first synthetic dye is made and used in the second part of the experiment.
2. Use a commercial liquid detergent for comparison with the lye soap in Part I.
3. It is necessary to use impure aniline (a mixture of aniline and toluidines) in Part II. Pure aniline does not produce the desired product. To prepare the impure aniline, add 0.5 ml ortho-toluidine and 0.2 g of para-toluidine to 5 ml of aniline and mix well.
4. Aniline and its typical impurities and reaction products are toxic. As these substances are generally fat soluble they are readily absorbed through the skin. For this reason, the instructor should dispense the aniline to the students in Part II and should wear plastic gloves while doing so.
5. If desired, another coal tar dye, fluorescein, can be prepared by the directions given in Extension 2. Its preparation requires about 10 minutes. Fluorescein is a fluorescent dye used to dye the water in sea rescue operations.

6. A simple data table using headings of ''Reactants'' and ''Observations'' is sufficient for this experiment.

MATERIALS NEEDED

Apparatus	Chemicals	
beaker, 250 ml	*cottonseed oil or lard	(10 g)
beakers, 100 ml (2)	ethyl alcohol	(10 ml)
beaker, 50 ml	NaOH, 6 M; 240 g/ℓ	(15 ml)
iron ring	NaCl, satd. soln.; 360 g/ℓ	(50 ml)
wire gauze	*kerosene or light oil	(2 ml)
burner	*detergent, liquid	(5 ml)
stirring rods (2)	$CaCl_2$, 0.1 M; 11 g/ℓ	(1 ml)
test tubes, medium (6)	$MgCl_2$, 0.1 M; 9.5 g/ℓ	(1 ml)
test tube rack	$FeCl_3$, 0.1 M; 16 g/ℓ	(1 ml)
wash bottle	H_2SO_4, 3 M; 167 ml conc./ℓ	(15 ml)
funnel	aniline, impure (See Note 3)	(5 ml)
crucible tongs	$K_2Cr_2O_7$	(0.5 g)
	tannic acid, 10 % soln	(10 ml)
Supplies	phenolphthalein; 0.05 g/100 ml 50% C_2H_5OH	
filter paper		
cotton cloth, small squares	phenolphthalein soln.	(1 ml)

CALCULATIONS AND QUESTIONS

1. R for stearic acid in a fat is $CH_3(CH_2)_{16}$—.
2. Oleic acid, $C_{17}H_{33}COOH$, is unsaturated with one double bond.

a.

$$
\begin{array}{l}
\quad\ \ \mathrm{H} \qquad \mathrm{O} \\
\quad\ \ | \qquad\ \ \| \\
\mathrm{H-C-O-C-C_{17}H_{33}} \\
\quad\ \ | \qquad\ \ \mathrm{O} \\
\quad\ \ | \qquad\ \ \| \\
\mathrm{H-C-O-C-C_{17}H_{33}} \\
\quad\ \ | \qquad\ \ \mathrm{O} \\
\quad\ \ | \qquad\ \ \| \\
\mathrm{H-C-O-C-C_{17}H_{33}} \\
\quad\ \ | \\
\quad\ \ \mathrm{H}
\end{array}
$$

glycerol trioleate

b.

$$
\begin{array}{l}
\text{H}-\overset{\text{H}}{\overset{|}{\text{C}}}-\text{O}-\overset{\text{O}}{\overset{\|}{\text{C}}}-\text{C}_{17}\text{H}_{33}\\
\qquad |\\
\text{H}-\text{C}-\text{O}-\overset{\text{O}}{\overset{\|}{\text{C}}}-\text{C}_{17}\text{H}_{33} \qquad + 3\,\text{Na}^{+}\,(aq) \; + \; 3\,\text{OH}^{-}\,(aq) \rightarrow 3\;[\text{Na}^{+}, {}^{-}\text{O}-\overset{\text{O}}{\overset{\|}{\text{C}}}-\text{C}_{17}\text{H}_{33}]\\
\qquad |\\
\text{H}-\underset{\text{H}}{\underset{|}{\text{C}}}-\text{O}-\overset{\text{O}}{\overset{\|}{\text{C}}}-\text{C}_{17}\text{H}_{33} \qquad\qquad + \; \text{HO}-\underset{\text{H}}{\underset{|}{\overset{\text{H}}{\overset{|}{\text{C}}}}}-\underset{\text{OH}}{\underset{|}{\overset{\text{H}}{\overset{|}{\text{C}}}}}-\underset{\text{H}}{\underset{|}{\overset{\text{H}}{\overset{|}{\text{C}}}}}-\text{OH}
\end{array}
$$

3. a. The soap dissolves slightly in water, becomes sudsy, and has a cleansing action when rubbed on the hands.
   b. The kerosene-water emulsion gradually separates upon standing. Both the soap and detergent stabilize the emulsion. (This demonstrates the hydrocarbon dissolving property of the nonpolar end of the soap molecule and the water dissolving property of its polar end.)
   c. The soap solution forms precipitates with the $CaCl_2$, $MgCl_2$, and $FeCl_3$ solutions. (The precipitates are the metal salts of the fatty acids in the soap. Ca^{2+}, Mg^{2+}, and Fe^{3+} ions are commonly present in hard water.) The detergent does not form precipitates with these solutions.
   d. Both the soap solution and the detergent solution test basic (pink) with phenolphthalein indicator. The soap solution forms a precipitate when acidified. The precipitate consists of the insoluble fatty acids used to make the soap. The detergent solution does not form a precipitate when acidified.

   In general, detergents are slightly better emulsifiers than soap. They do not form precipitates in hard water (curd) or in acidic solution.

4. $\text{moles fat} = \dfrac{10 \text{ g fat}}{900 \text{ g fat/mol}} = 0.01 \quad \text{moles NaOH} = 0.015\ \ell \text{ soln} \times \dfrac{6 \text{ mol NaOH}}{\ell \text{ soln}} = 0.09$

   The reaction (Equation 24.1) requires 3 mol NaOH per mol of fat. For 0.01 mol of fat, 0.03 mol of NaOH is required. As 0.09 mol of NaOH was used, the NaOH was present in excess. In general, the least expensive reactant (NaOH) is used in excess in order to react as much as possible with the more expensive reactant (fat).
5. The molecular formula of quinine, $C_{20}H_{24}N_2O_2$, has many, many isomers. Even if the substance formed had the same formula it would most likely have been one of the isomers of quinine rather than quinine itself. Making a molecule as complex as quinine with Perkin's simple approach was like trying to find a needle in a haystack.
6. A good dye must bind firmly to the cloth. It must not fade out with repeated washings, ironings, and dry cleanings. In addition, it must be unaffected by sunlight. Lastly, it should have a desirable color.

EXTENSIONS

1. Azo dyes have the structure Ar–N=N–Ar′ where Ar and Ar′ represent various aromatic groups. The structures of two azo dyes are shown on the next page.

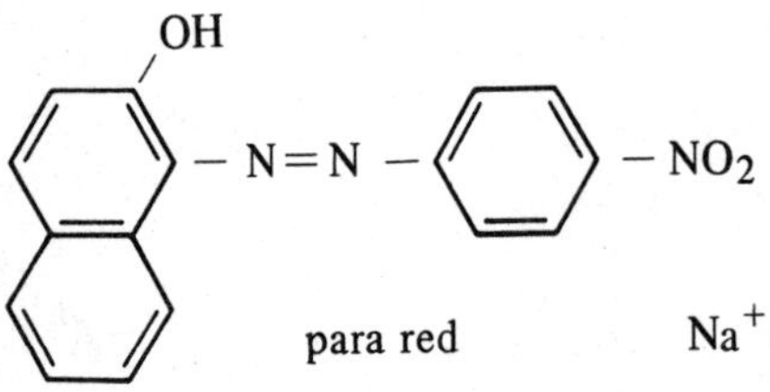

OH

$-N=N-$ $-NO_2$

para red

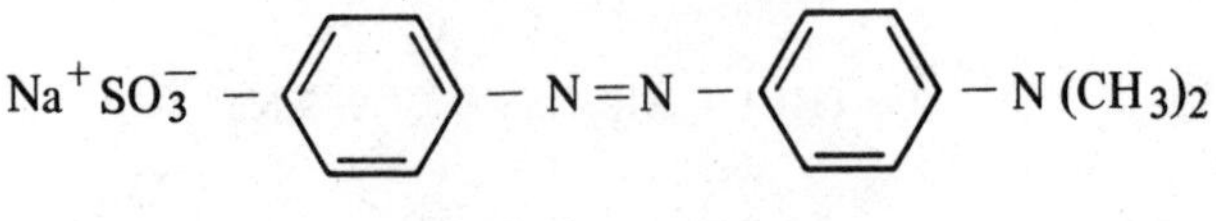

$Na^+SO_3^-$ $-N=N-$ $-N(CH_3)_2$

methyl orange

The structure of mauve according to Noller, 2nd Ed., is:

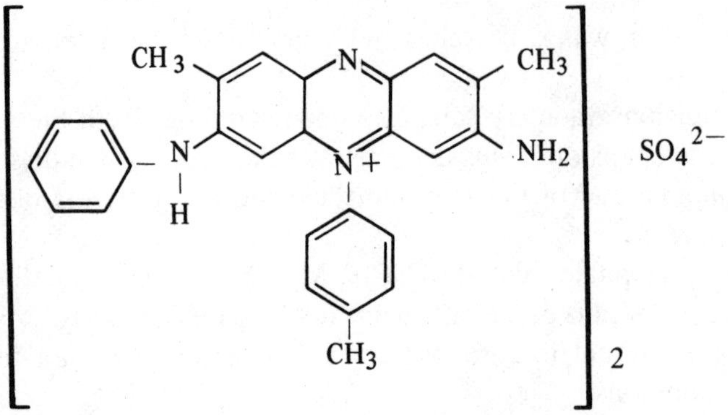

CH_3 N CH_3

N N^+ NH_2 SO_4^{2-}

H

CH_3 2

The earliest dyes used were natural dyes from plants or animals. Two examples are indigo from the indigo plant and royal purple from sea snails.

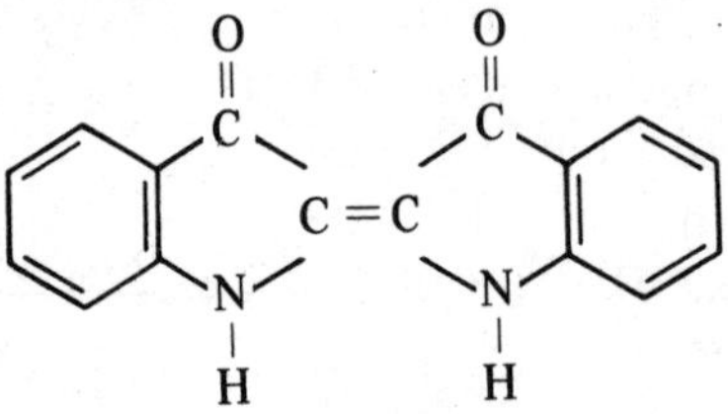

O O

C C

C=C

N N

H H

indigo

O O

C C

C=C

Br N N Br

H H

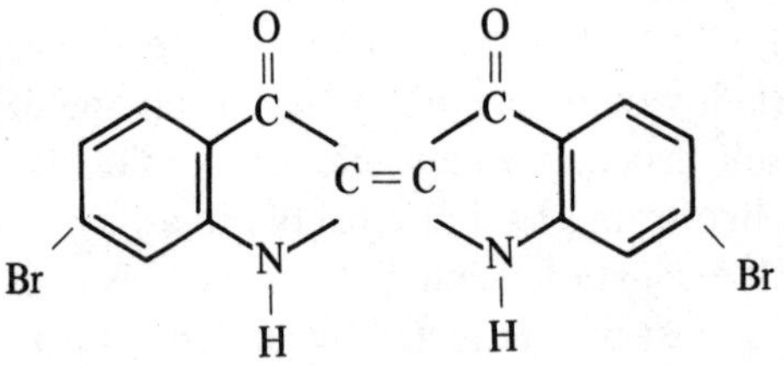

royal purple

All of the dyes are aromatic compounds with various groups containing the nitrogen atom (−N=N−, −N=, −N−).

H

Preparation and Properties of Polymers

TIMING

The experiment requires about 45 minutes. It should be done during Chapter 16 after section 16.3.

PRELIMINARY STUDY

2. a.

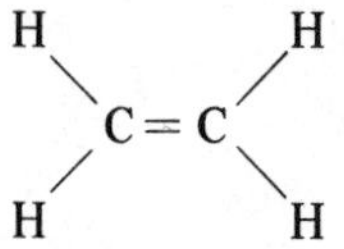

```
H       H
 \     /
  C = C
 /     \
H       H
```

b.

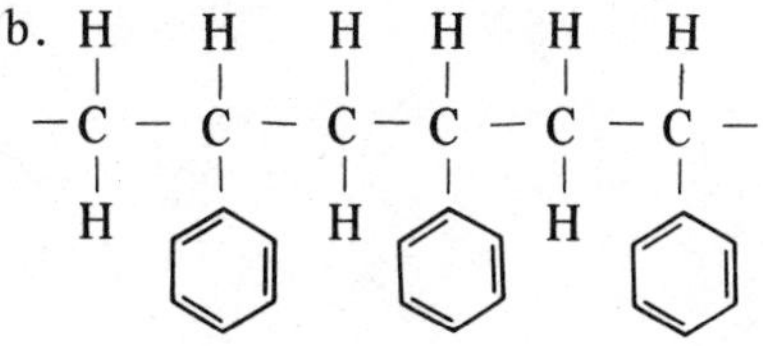

```
  H   H   H   H   H   H
  |   |   |   |   |   |
- C - C - C - C - C - C -
  |   |   |   |   |   |
  H   ⌬   H   ⌬   H   ⌬
```

c. Dacron monomers

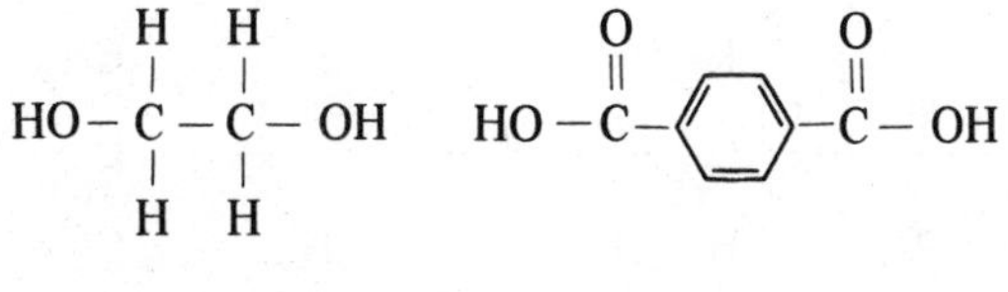

```
      H   H                  O           O
      |   |                  ||          ||
HO -  C - C - OH      HO  -  C -(⌬)-  C - OH
      |   |
      H   H
```

Nylon 6—6 monomers

```
       H   H   H   H   H   H                 O   H   H   H   H   O
       |   |   |   |   |   |                 ||  |   |   |   |   ||
H2N -  C - C - C - C - C - C - NH2     HO -  C - C - C - C - C - C - OH
       |   |   |   |   |   |                     |   |   |   |
       H   H   H   H   H   H                     H   H   H   H
```

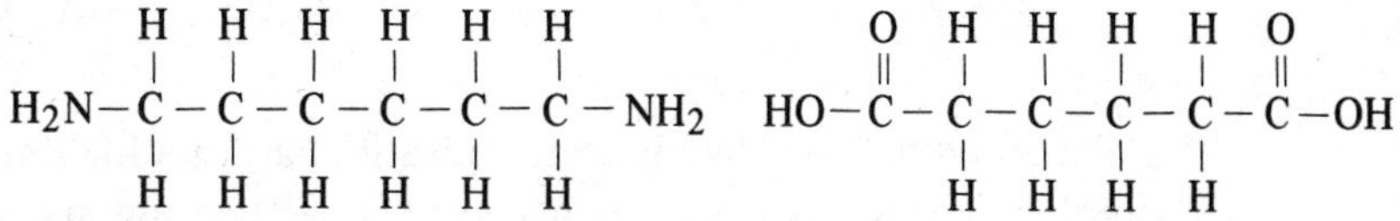

EXPERIMENT NOTES

1. Benzoyl peroxide is used in Part I as a chemical initiator to begin the breaking of double bonds to start the addition polymerization. As benzoyl peroxide is very reactive it should be dispensed by the teacher to the student when needed.
2. The polymethylmethacrylate of Part I may be freed from the test tube by carefully smashing the tube on the desk top.
3. We use a drum made of a coffee can to wind large amounts of the nylon rope in Part III. The drum has an attached handle through its center and is supported at both ends to a wooden platform.

4. A simple data table using headings of "Reactants" and "Observations" is sufficient for this experiment.

MATERIALS NEEDED

Apparatus	Chemicals	
beaker, 400 ml	methyl methacrylate	(5 ml)
beaker, 100 ml	benzoyl peroxide (See Note 1)	(0.1 g)
test tubes, small (3)	glycerol	(1 ml)
stirring rod	phthalic anhydride	(3 g)
test tube holder	$NaC_2H_3O_2$	(0.1 g)
glass plate	methyl ethyl ketone	(5 ml)
spatula	sebacoyl chloride soln.	(20 ml)
tweezers	(10 ml in 500 ml CCl_3CH_3)	
iron ring	hexamethylenediamine soln.	(10 ml)
wire gauze	(8 g NaOH and 10 g	
burner	hexamethylenediamine	
	in 250 ml water)	

Supplies

wood splints

CALCULATIONS AND QUESTIONS

1. a. Polymethylmethacrylate is an addition polymer.

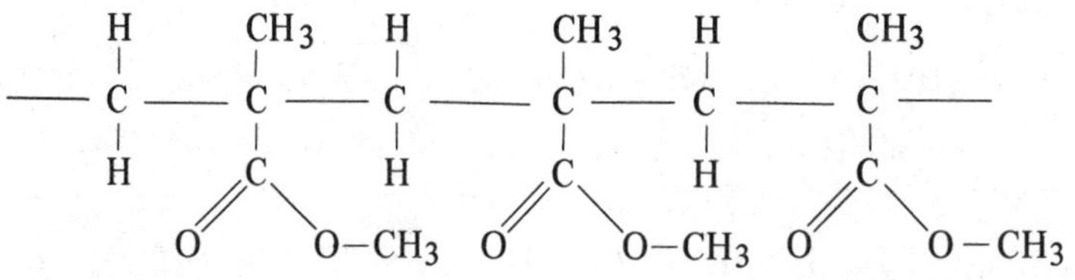

```
     H       CH3    H        CH3     H        CH3
     |       |      |        |       |        |
 --- C ----- C ---- C ------ C ----- C ------ C ---
     |       |      |        |       |        |
     H       C      H        C       H        C
           //  \           //  \            //  \
          O     O-CH3     O     O-CH3      O     O-CH3
```

b. Polymethylmethacrylate is thermoplastic, hard, moldable, and has a transparency similar to glass. It is commonly used in advertising signs and displays, lighting fixture lenses, medallions and plaques, and automotive parts such as tail lights.

2. Glyptal is a condensation polymer and a polyester.

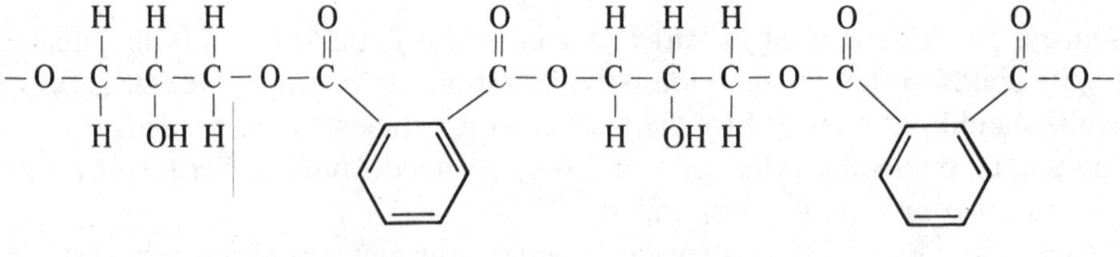

```
    H  H  H       O          O     H  H  H       O            O
    |  |  |       ||         ||    |  |  |       ||           ||
 -O-C -C -C -O -C            C -O -C -C -C -O -C              C -O-
    |  |  |        \        /      |  |  |        \          /
    H  OH H         (C6H4)         H  OH H         (C6H4)
```

3. Nylon is a condensation polymer and a polyamide.

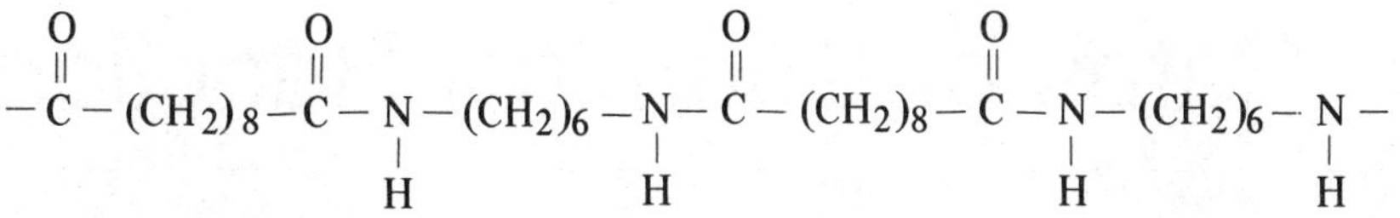

$$-\overset{O}{\overset{\|}{C}}-(CH_2)_8-\overset{O}{\overset{\|}{C}}-\underset{H}{\underset{|}{N}}-(CH_2)_6-\underset{H}{\underset{|}{N}}-\overset{O}{\overset{\|}{C}}-(CH_2)_8-\overset{O}{\overset{\|}{C}}-\underset{H}{\underset{|}{N}}-(CH_2)_6-\underset{H}{\underset{|}{N}}-$$

4. MM methyl methacrylate ($C_5H_8O_2$) = 100

   a. $\text{MM (polymer)} = 10^5 \text{ units} \times \frac{10^2}{\text{units}} = 10^7$

   b. For n monomers there are 2n C–C bonds:

   $$\text{length} = 2 \times 10^5 \text{ bonds} \times \frac{0.15 \text{ nm}}{\text{bond}} \times \frac{10^{-9} \text{ m}}{\text{nm}} = 3 \times 10^{-5}\text{m}$$

EXTENSIONS

1. a. Polystyrene is most soluble in toluene. The structure of toluene is most similar to that of polystyrene.
   c. The polystyrene melts when heated.
   d. The complete combustion of polystyrene produces CO_2 and water vapor. In actuality, considerable amounts of unburned carbon and some monomer is released.

2.

Monomer: $$H-\underset{H}{\underset{|}{N}}-(CH_2)_5-\overset{O}{\overset{\|}{C}}-OH$$

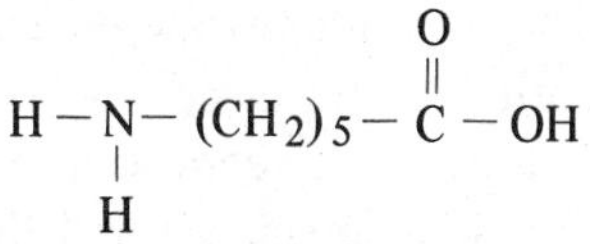

Polymer: $$-\underset{H}{\underset{|}{N}}-(CH_2)_5-\overset{O}{\overset{\|}{C}}-\underset{H}{\underset{|}{N}}-(CH_2)_5-\overset{O}{\overset{\|}{C}}-\underset{H}{\underset{|}{N}}-(CH_2)_5-\overset{O}{\overset{\|}{C}}-$$

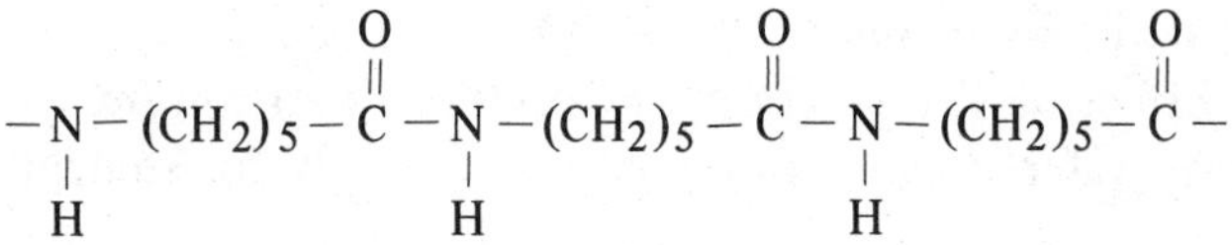

3. Cross-linked Glyptal:

```
      H   H   H        O             O        H   H   H
      |   |   |        ||            ||       |   |   |
 -O - C - C - C - O - C\            /C - O - C - C - C - O -
      |   |   |          \   ___   /          |   |   |
      H   |   H           \ /   \ /           H   |   H
          |                |     |                |
          O                 \___/                 O
          |                                       |
      H   |   H        O             O        H   |   H
      |   |   |        ||            ||       |   |   |
 -O - C - C - C - O - C\            /C - O - C - C - C - O -
      |   |   |          \   ___   /          |   |   |
      H   H   H           \ /   \ /           H   H   H
                           |     |
                            \___/
```

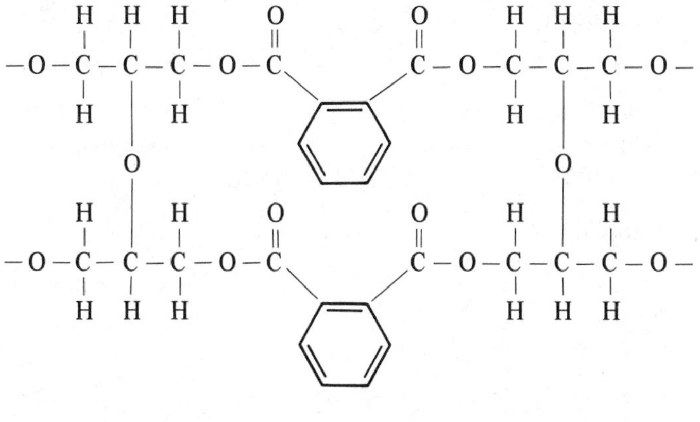

Rates of Chemical Reactions

TIMING

The experiment requires 50 minutes. It should be done early in Chapter 17.

PRELIMINARY STUDY

2. a. moles $S_2O_3^{2-}$ = V × M = 0.0030 ℓ × 0.10 M = 3.0×10^{-4}

$$M = \frac{3.0 \times 10^{-4} \text{ mol}}{0.0100 \ \ell} = 0.030 \text{ mol}/\ell$$

b. $$M(\text{after mixing}) = \frac{3.0 \times 10^{-4} \text{ mol}}{0.0200 \ \ell} = 0.015 \text{ mol}/\ell$$

EXPERIMENT NOTES

1. As this may be the student's first use of a buret, demonstrate its operation and reading. Burets with Teflon stopcocks are most reliable and easiest to use.
2. Rate data is obtained by using a sulfur "clock." The end-point of the reaction, while not as dramatic as the iodine clock, is readily detectable and the results are reproducible. The reaction being studied is simpler than the iodine clock reaction and more easily understood by the students.
3. Part II (The Effect of Temperature on Reaction Rate) can be speeded up by maintaining a warm water supply and a cold water supply for student use.

MATERIALS NEEDED

Apparatus

burets, 50 ml (3)
(may be shared)
test tubes, regular (10)
test tube rack
beaker, 400 ml (2)
thermometer
buret clamps (2)

Chemicals

HCl, 1 M; 83 ml conc./ℓ	(60 ml)
$Na_2S_2O_3$, 0.1 M; 25 g hydrate/ℓ	(40 ml)
$Na_2S_2O_3$, unknowns (0.09 M, 0.07 M, 0.05 M, 0.03 M)	(10 ml)

Supplies

ice

SAMPLE DATA TABLE

Part I

Room Temp 20°C

Tube No.	Time (sec)	Relative Rate 1/Time (1/Sec)	Concentration of $S_2O_3^{2-}$ As Prepared	Concentration of $S_2O_3^{2-}$ Mixed with HCl
1	35	0.029	0.10 M	0.050 M
2	42	0.024	0.080 M	0.040 M
3	53	0.019	0.060 M	0.030 M
4	82	0.012	0.040 M	0.020 M
5	154	0.00649	0.020 M	0.010 M
Unknown	63	0.016		

Part II

Temp	Time (sec)	1/Time (1/sec)
10°C	50	0.020
35°C	20	0.050

CALCULATIONS AND QUESTIONS

1. a. See data table. b. See data table.
2. a.

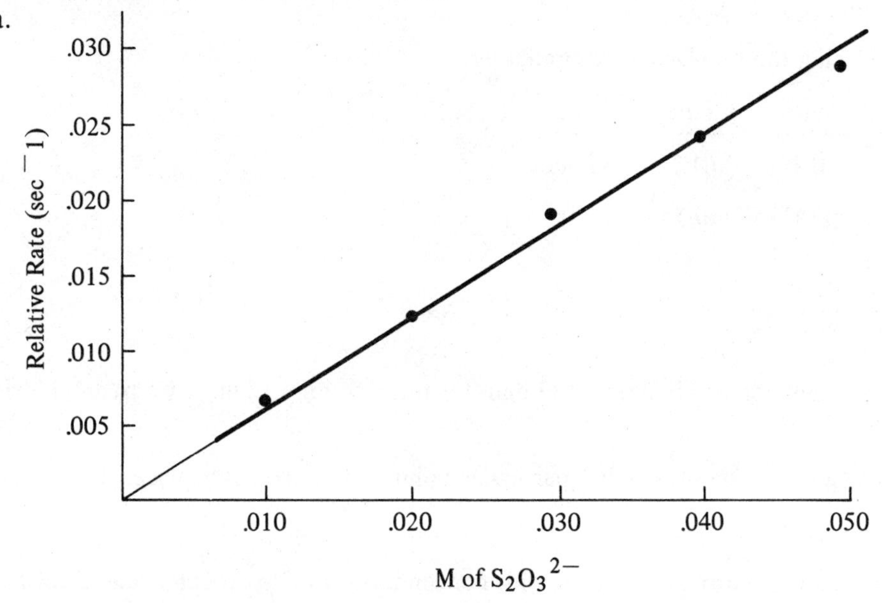

b. The relative rate varies directly with the concentration.
Mathematically: Rate = constant × (conc. $S_2O_3^{2-}$). If the concentration of the $S_2O_3^{2-}$ ion is doubled, the rate is approximately doubled.

c. Using the graph, the concentration (after mixing) is 0.026 M. Before mixing the concentration of the unknown would be 2(0.026 M) or 0.052 M.

3. a. See data table.

b.

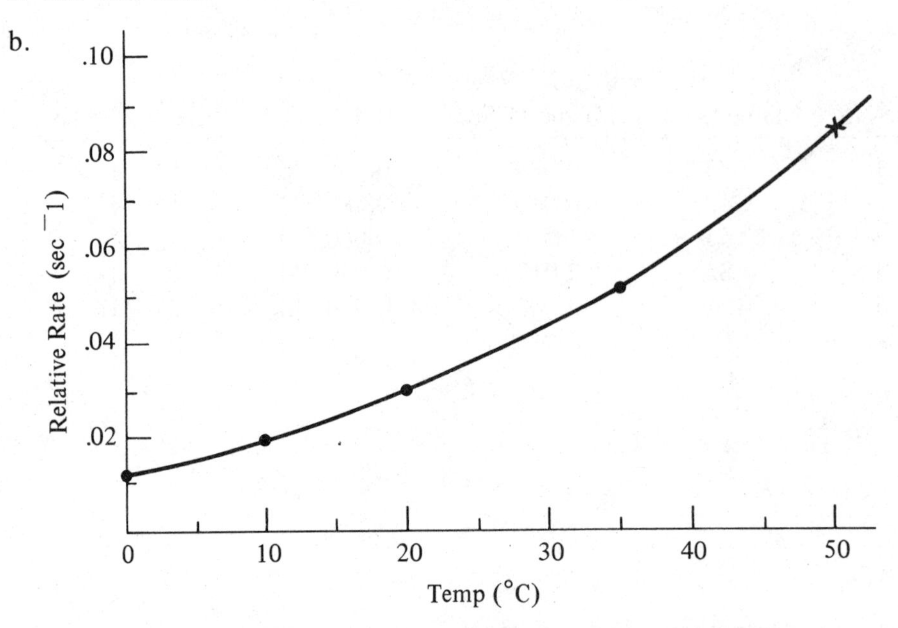

c. The relative rate increases ever more rapidly as the temperature goes up. (While it is difficult to determine with only three data points, the relationship is non-linear.)

d. Using the graph and extrapolating:

Temp	1/Time	Time
0°C	0.012	83 sec
50°C	0.085	12 sec

EXTENSIONS

Possible methods of obtaining rate data for the reaction of zinc with hydrochloric acid are:

a. Measure the rate at which a particular volume of water is displaced by H_2 gas.

b. Measure the change in mass of zinc after a particular reaction time.

The reaction is carried out at varying concentrations of hydrochloric acid (0.01 M to 6 M).

Systems in Chemical Equilibrium

TIMING

The experiment requires about 50 minutes. It should be done during Chapter 18.

PRELIMINARY STUDY

2. $A^+(aq) + B^-(aq) \rightleftharpoons C(aq)$; $\Delta H = -20$ kcal. The amount of the product C present at equilibrium can be maximized by increasing the concentration of the reactant A^+ and by decreasing the system's temperature. (NOTE: A similar reaction is used in Part III of the experiment.)

EXPERIMENT NOTES

1. For the most part, student experiments involving gaseous equilibrium systems are not feasible. For this reason, all reactions of this experiment involve ionic equilibria. The selected reactions are relatively easy to understand and all involve a color change. It may be necessary to make the point once again that the charge must be balanced in ionic equations.
2. Part I of the experiment introduces the student to equilibrium changes in acid-base indicators. In this respect it lays ground work for Experiment 29.

MATERIALS NEEDED

Apparatus	Chemicals	
test tubes, regular (2)	bromthymol blue soln.	(2–3 drops)
test tube rack	HCl, 0.1 M; 8.3 ml conc./ℓ	(5 ml)
beakers, 100 ml (2)	HCl, 1 M; 83 ml conc./ℓ	(5 ml)
stirring rod	HCl, 6 M; 500 ml conc./ℓ	(25 ml)
wire gauze	NaOH, 0.1 M; 4.0 g/ℓ	(5 ml)
iron ring	NaOH, 1 M; 40 g/ℓ	(5 ml)
burner	$K_2Cr_2O_7$, 0.1 M; 29 g/ℓ	(5 ml)
graduated cylinder, 10 ml	$CoCl_2 \cdot 6\ H_2O$	(1 g)
	bromthymol blue; 0.1 g/6 ml 0.01 M NaOH + 94 ml H_2O	

SAMPLE DATA TABLE

Procedure	Observations
Part I:	
1. bromthymol blue + HCl	soln color is light yellow
above mixture + NaOH	soln color changes through green to blue
2. above mixture + HCl	soln color changes back to yellow
Part II:	
1. $K_2Cr_2O_7$ + NaOH	soln color changes from orange to yellow
2. above mixture + HCl	soln color changes back to orange
3. above mixture + NaOH	(until color is intermediate between 1 and 2)
Part III:	
1. a. $CoCl_2 \cdot 6\ H_2O$ + HCl	soln color is blue
b. $CoCl_2 \cdot 6\ H_2O$ + H_2O	soln color is pink
2. a. mixture 1(a) + H_2O	soln color changes from blue to pink
above mixture + heat	soln color changes from pink to blue when heated, and back to pink when cooled.
b. mixture 1(b) + HCl	soln color changes from pink to blue
mixture 1(b) + heat	soln color changes from pink to blue when heated and back to pink when cooled.

QUESTIONS

1. a. The solution color changes from yellow, through green, to blue as NaOH is added.
   b. When the color is green, appreciable amounts of both B^- (blue) and Y (yellow) are present.
   c. As NaOH is added $[H^+]$ decreases, [Y] decreases, and $[B^-]$ increases.
2. Adding HCl to the blue bromthymol blue solution causes the solution to turn yellow.

   $H^+(aq) + B^-(aq) \rightleftharpoons Y(aq)$

   Using Le Chatelier's principle, the system attempts to consume the H^+ ions added, thereby causing $[B^-]$ to decrease and [Y] to increase.

3. a. NaOH. Adding NaOH decreases $[H^+]$ by the forming of water. Using Le Chatelier's principle, the system attempts to replace the H^+ ions which are removed.

   $$2\ H^+(aq) + 2\ CrO_4^{2-}(aq) \rightleftharpoons Cr_2O_7^{2-}(aq) + H_2O$$

   To produce H^+ ions the $[Cr_2O_7^{2-}]$ must decrease and the $[CrO_4^{2-}]$ must increase, causing the solution color to change to yellow.

   b. The CrO_4^{2-} – $Cr_2O_7^{2-}$ system is reversible as either ion can be changed to the other as shown by the color changes when HCl or NaOH is added.

4. a. When water was added the system attempted to consume it by converting the blue $Co(H_2O)_4Cl_2$ to the pink $Co(H_2O)_6^{2+}$ ion.

   $$Co(H_2O)_6^{2+}(aq) + 2\ Cl^-(aq) \rightleftharpoons Co(H_2O)_4Cl_2(aq) + 2\ H_2O.$$

   b. The reaction is reversible as either colored form can be predominant as shown by the color changes when HCl or water is added. In agreement with Le Chatelier, an equilibrium system shifts in favor of the endothermic reaction when heat is added. As the solution turns blue when heated, Reaction 28.4 is endothermic, as written.

   c. The pink solution contains mostly $Co(H_2O)_6^{2+}$ ions. In agreement with Le Chatelier's principle, the solution can be made to turn blue by increasing the temperature or by adding HCl to increase $[Cl^-]$.

EXTENSIONS

1. $BaCrO_4$ is considerably less soluble than $BaCr_2O_7$. A $BaCrO_4$ precipitate can be dissolved by adding an acidic solution. The adding of H^+ ions causes the CrO_4^{2-} ions to be converted to $Cr_2O_7^{2-}$ ions.
2. The reactions that occur here are:

   $$Cu^{2+}(aq) + 4\ NH_3(aq) \rightleftharpoons Cu(NH_3)_4^{2+}(aq)$$

   $$Cu(NH_3)_4^{2+} + 4\ H^+(aq) \rightleftharpoons Cu^{2+}(aq) + 4\ NH_4^+(aq)$$

Measurement of pH with Acid-Base Indicators

TIMING

The experiment requires about 45 minutes. It should be done during Chapter 19.

PRELIMINARY STUDY

2. For 0.001 M HNO_3 (strong acid), $[H^+] = 0.001$ M or 10^{-3} M.

   $pH = -\log [H^+] = -\log (10^{-3}) = 3$

3. $K_a = \dfrac{[H^+][F^-]}{[HF]}$

4. From Equation 29.9: $\dfrac{[B^-]}{[Y]} = \dfrac{K_a}{[H^+]}$

   If $K_a = 100\,[H^+]$, then $\dfrac{[B^-]}{[Y]} = \dfrac{100\,[H^+]}{[H^+]} = 100$

Since $[B^-] = 100\,[Y]$, the solution will be blue.

EXPERIMENT NOTES

1. This experiment helps the student to relate the pH of a solution to $[H^+]$ through the use of indicators. You may also wish to demonstrate the use of pH paper or a pH meter. In addition, the experiment provides a simple method for the determination of an equilibrium constant, K_a.
2. Colors and ranges for some indicators are given below. Appropriate indicator substitutions may be made if desired.

Indicator	Acid Form	Base Form	pH Range
Thymol blue	Red	Yellow	1.2–2.8
Orange IV	Red	Yellow	1.3–3.0
Methyl Orange	Red	Yellow	3.1–4.4
Congo Red	Blue	Red	3.0–5.0
Bromcresol green	Yellow	Blue	3.8–5.4
Methyl red	Red	Yellow	4.3–6.2

MATERIALS NEEDED

Apparatus	Chemicals	
test tubes, regular (21)	HCl, 0.10 M; 8.3 ml conc./ℓ	(35 ml)
test tube racks (2)	thymol blue soln; 0.5 g/100 ml	(1 ml)
graduated cylinder, 50 ml	methyl orange soln; 0.1 g/100 ml	(1 ml)
stirring rod	methyl red soln; 0.1 g/100 ml	(1 ml)
beaker, 100 ml	HCl unknown, 10^{-1} to 10^{-5} M	(30 ml)
	weak acid unknowns, 1.0 M	(30 ml)

Formula	K_a
acetic acid; 5.7 ml conc./ℓ	1.8×10^{-5}
propionic acid; 74 ml/ℓ	1.5×10^{-5}
lactic acid; 90 g/ℓ	1.4×10^{-4}
glycolic acid; 76 g/ℓ	1.5×10^{-4}
benzoic acid; 122 g/ℓ	6.3×10^{-5}

SAMPLE DATA TABLE

			Indicator Color		
Test Tube	$[H^+]$	pH	Thymol Blue	Methyl Orange	Methyl Red
1	1×10^{-1} M	1	red	red	red
2	1×10^{-2} M	2	red-orange	red	red
3	1×10^{-3} M	3	yellow	red-orange	red
4	1×10^{-4} M	4	yellow	orange	red
5	1×10^{-5} M	5	yellow	orange-yellow	orange
HCl unknown	1×10^{-4} M (estimated)		yellow	orange	red
weak acid	5×10^{-3} M (estimated)		orange	red	red

CALCULATIONS AND QUESTIONS

1. a. See data table for pH values.
   Each dilution by a factor of 10 increases the pH of the HCl solution by one unit.

   b.

Indicator	pH Range	Low pH	High pH	K_a
thymol blue	1–3	red	yellow	10^{-2}
methyl orange	3–5	red	yellow	10^{-4}
methyl red	4–6	red	yellow	10^{-5}

   c. The pH of the unknown HCl solution was about 4 since its color matched that of the 1×10^{-4} HCl standard.

2. a. $[H^+] = 5 \times 10^{-3}$ M

   b. $[A^-] = [H^+] = 5 \times 10^{-3}$ M

   $[HA] \cong M_{HA} = 1.0$ M

   c. $$K_a = \frac{[H^+]\,[A^-]}{[HA]}$$

   $$K_a = \frac{(5 \times 10^{-3})\,(5 \times 10^{-3})}{1.0} = 2 \times 10^{-5}$$

   d. $HA(aq) \rightleftharpoons H^+(aq) + A^-(aq)$

   If some NaA is added to the HA solution, the $[A^-]$ will increase. According to Le Chatelier's principle, the system will attempt to reduce the $[A^-]$ by converting the A^- ion to HA. This will cause a decrease in $[H^+]$ and a resulting increase in pH.

EXTENSIONS

1. The equilibrium expression for water is:

   $K = [H^+] \times [OH^-] = 1.0 \times 10^{-14}$. Therefore,

   the $[H^+]$ in pure water is 1.0×10^{-7} M.

   When acid solutions are diluted past 10^{-5} M, the hydrogen ions obtained from water must also be considered. It is not possible to dilute the solution beyond pH 7, since that is the pH of pure water.
2. Red cabbage contains a number of different pH sensitive dyes. If the odor doesn't bother you, a wide range of colors can be obtained.

Properties of Acids and Bases

TIMING

The experiment requires about 50 minutes. It is recommended for Chapter 19.

PRELIMINARY STUDY

1. acidic: HNO_3 and NH_4Br
   basic: NaCN   neutral: KCl
2. HNO_3 and NaOH: $H^+(aq) + OH^-(aq) \rightarrow H_2O$

   HCl and NaF: $H^+(aq) + F^-(aq) \rightarrow HF(aq)$

   HBr and NH_3: $H^+(aq) + NH_3(aq) \rightarrow NH_4^+(aq)$

EXPERIMENT NOTES

1. The important properties of acids and bases are introduced by carrying out the reactions of this experiment. Students should be able to predict the acidity or basicity of an ionic solution and to write equations for acid-base reactions.
2. The idea of a visual completion to a neutralization reaction (an "end-point") is introduced in Procedure 2. Quantitative acid-base titrations will be done in Experiments 32 and 33.
3. There are many instances in this experiment where rate and equilibrium principles can be reemphasized: the effect of concentration, the addition or removal of a reactant, and the relationship between K_a and the extent of ionization.

MATERIALS NEEDED

Apparatus	Chemicals	
test tubes, small (10)	NaCl	(0.1 g)
test tube rack	Na_2CO_3	(0.1 g)
beaker, 50 or 100 ml	NaOH	(0.1 g)
stirring rod	NH_4Cl	(0.1 g)
	$FeCl_3$	(0.1 g)
	marble chips ($CaCO_3$)	(2 g)

Supplies

red and blue litmus paper
corks

$NaOH$, 1 M; 40 g/ℓ	(20 ml)
HCl, 1 M; 83 ml conc./ℓ	(25 ml)
HCl, 6 M; 500 ml conc./ℓ	(3 ml)
CH_3COOH, 1 M; 57 ml conc./ℓ	(2 ml)
CH_3COOH, 6 M; 342 ml conc./ℓ	(3 ml)
NH_3, 1 M; 67 ml conc./ℓ	(2 ml)
$NaCH_3COO$, 1 M; 82 g/ℓ	(2 ml)
NH_4Cl, 1 M; 53 g/ℓ	(2 ml)
$FeCl_3$, 0.1 M; 16 g/ℓ	(2 ml)
phenolphthalein soln; 0.05 g/100 ml 50% C_2H_5OH	

SAMPLE DATA TABLE

Procedure 1

Solution	Litmus Test	Solution	Litmus Test
NaCl	neutral	NH_4Cl	acidic (red)
Na_2CO_3	basic (blue)	$FeCl_3$	acidic (red)
NaOH	basic (blue)		

Procedures 2–5	Observations
2. HCl + NaOH + indicator	Soln turns pink after about an equal amount of NaOH is added. Soln returns to colorless after more HCl is added.
3. a. CH_3COOH + NaOH	The CH_3COOH has a vinegar-like odor. The odor is gone after the NaOH is added.
b. $NaCH_3COO$ + HCl	The $NaCH_3COO$ soln has no odor. After HCl is added the mixture has a vinegar odor.
4. a. NH_3 + HCl	The NH_3 soln has a sharp, pungent odor. After HCl is added the NH_3 odor is gone.
b. NH_4Cl + NaOH	The NH_4Cl soln has no odor. After NaOH is added the odor of NH_3 is present.
5. a. $CaCO_3$ + 6 M HCl	A gas (CO_2) is produced at a rapid rate.
$CaCO_3$ + 6 M CH_3COOH	A gas is given off very slowly.
b. $FeCl_3$ + NaOH	A brown ppt. of $Fe(OH)_3$ forms.
$Fe(OH)_3$ + HCl	The ppt. is dissolved and the soln regains its light yellow color.

QUESTIONS

1. $CO_3^{2-}(aq) + H_2O \rightleftharpoons HCO_3^-(aq) + OH^-(aq)$

   $NaOH(s) \rightarrow Na^+(aq) + OH^-(aq)$

   $NH_4^+(aq) \rightleftharpoons NH_3(aq) + H^+(aq)$

   $Fe^{3+}(aq) + H_2O \rightleftharpoons Fe(OH)^{2+}(aq) + H^+(aq)$

2. The NaCl solution is neutral because the Na^+ ion is derived from a strong base (NaOH) and the Cl^- ion is derived from a strong acid (HCl). Because of this, neither ion interacts with water. KNO_3 is an example of another salt whose solution would be neutral.

3. Proc. 2: HCl and NaOH $\underline{H^+}$, Cl^- and Na^+, $\underline{OH^-}$

   $H^+(aq) + OH^-(aq) \rightarrow H_2O$

   Proc. 3(a): CH_3COOH and NaOH $\underline{CH_3COOH}$ and Na^+, $\underline{OH^-}$

   $CH_3COOH(aq) + OH^-(aq) \rightarrow CH_3COO^-(aq) + H_2O$

   Proc. 3(b): $NaCH_3COO$ and HCl Na^+, $\underline{CH_3COO^-}$ and $\underline{H^+}$, Cl^-

   $CH_3COO^-(aq) + H^+(aq) \rightarrow CH_3COOH(aq)$

   Proc. 4(a): NH_3 and HCl $\underline{NH_3}$ and $\underline{H^+}$, Cl^-

   $NH_3(aq) + H^+(aq) \rightarrow NH_4^+(aq)$

   Proc. 4(b): NH_4Cl and NaOH $\underline{NH_4^+}$, Cl^- and Na^+, $\underline{OH^-}$

   $NH_4^+(aq) + OH^-(aq) \rightarrow NH_3(aq) + H_2O$

4. a. HCl is a strong acid and is completely ionized while CH_3COOH is a weak acid ($K_a = 1.8 \times 10^{-5}$) and is only partially ionized. The higher concentration of hydrogen ions in HCl causes its reaction rate to be much faster than that of CH_3COOH.

   b. $CaCO_3(s) + 2\ H^+(aq) \rightarrow CO_2(g) + Ca^{2+}(aq) + H_2O$

   c. $Fe(OH)_3(s) + 3\ H^+(aq) \rightarrow Fe^{3+}(aq) + 3\ H_2O$

EXTENSIONS

The HSO_4^- ion can act as an acid by the reaction:

$HSO_4^-(aq) \rightleftharpoons H^+(aq) + SO_4^{2-}(aq)$

It cannot act as a base by reacting with water, since the product would be H_2SO_4, a strong acid. $NaHSO_4$ in solution would be acidic.

The SO_4^{2-} ion can only act as a base as it is derived from a weak acid:

$$SO_4^{2-}(aq) + H_2O \rightleftharpoons HSO_4^-(aq) + OH^-(aq)$$

A solution of Na_2SO_4 would be slightly basic.

Ionic Precipitation Reactions

TIMING

The experiment requires about 30 minutes. A second set of solutions can be completed within a 50 minute period. The experiment is recommended for early in Chapter 20.

PRELIMINARY STUDY

2. a. K^+, CrO_4^{2-} b. Fe^{3+}, Cl^- c. Na^+, OH^-

   d. Al^{3+}, NO_3^-

3. Soluble pairs: Ba^{2+}, Cl^-; Cu^{2+}, SO_4^{2-}; K^+, Cl^-; Cu^{2+}, NO_3^-; Cu^{2+}, Cl^-; K^+, NO_3^-

   Insoluble pairs: Ba^{2+}, SO_4^{2-}. Precipitate must be $BaSO_4$.

EXPERIMENT NOTES

1. Solution sets may be distributed in 30 ml dropping bottles. The sets listed under ''Chemicals'' are graded according to their level of difficulty. The student data tables are easier to check if each bottle is labeled A-1, A-2, A-3, etc., in addition to the solute's formula.

2. Many other solution sets can be used. If you devise your own set attempt to pick combinations in which the student can reason the formula of the precipitate. The student should not have to resort to solubility rules.

MATERIALS NEEDED

Apparatus	Chemicals
spot plate	solution set (see Note 1)
	solution unknown

NOTE: The amount of solute needed to make 100 ml of solution of the indicated concentration is given.

Set A (Easy)

1. 0.1 M Na_2SO_4 (1.4 g anhydrous)
2. 0.07 M $Al_2(SO_4)_3$ (4.7 g hydrate)
3. 0.1 M $Sr(NO_3)_2$ (2.1 g)
4. 0.1 M $BaCl_2$ (2.4 g hydrate)
5. 0.1 M $Ba(NO_3)_2$ (2.6 g)
6. 0.07 M $AlCl_3$ (1.7 g hydrate)

Set A gives:

2 - $SrSO_4$ ppts
4 - $BaSO_4$ ppts

Set B (easy)

1. 0.1 M $FeCl_3$ (2.7 g hydrate)
2. 0.1 M $Co(NO_3)_2$ (2.9 g hydrate)
3. 0.1 M $CoCl_2$ (2.4 g hydrate)
4. 0.2 M NaOH (0.8 g)
5. 0.2 M KOH (1.1 g)
6. 0.2 M $NaNO_3$ (1.7 g)

Set B gives:

2 - $Fe(OH)_3$ ppts
4 - $Co(OH)_2$ ppts

Set C (Moderate)

1. 0.2 M $Sr(NO_3)_2$ (4.2 g)
2. 0.07 M $Al_2(SO_4)_3$ (4.7 g hydrate)
3. 0.1 M KNO_3 (1.0 g)
4. 0.1 M Na_2CrO_4 (1.6 g)
5. 0.1 M K_2CrO_4 (1.9 g)
6. 0.1 M $BaCl_2$ (2.4 g hydrate)

Set C gives:

1 - $BaSO_4$ ppt
1 - $SrSO_4$ ppt
2 - $SrCrO_4$ ppts
2 - $BaCrO_4$ ppts
2 - $Al_2(CrO_4)_3$ ppts

Set D (Moderate)

1. 0.1 M Na_2CrO_4 (1.6 g)
2. 0.1 M K_2CrO_4 (1.9 g)
3. 0.1 M $FeCl_3$ (2.7 g hydrate)
4. 0.2 M KOH (1.1 g)
5. 0.1 M $MgSO_4$ (2.5 g hydrate)
6. 0.1 M $BaCl_2$ (2.4 g hydrate)

Set D gives:

1 - $Mg(OH)_2$ ppt
1 - $BaSO_4$ ppt
1 - $Fe(OH)_3$ ppt
2 - $BaCrO_4$ ppt
2 - $Fe_2(CrO_4)_3$ ppts

Set E (Difficult)

1. 0.1 M $NiCl_2$ (2.4 g hydrate)
2. 0.1 M $MgCl_2$ (2.0 g hydrate)
3. 0.1 M Na_2SO_4 (1.4 g anhydrous)
4. 0.2 M NaOH (0.8 g)
5. 0.1 M $Ba(OH)_2$ (1.7 g hydrate)
6. 0.1 M $MgSO_4$ (2.5 g hydrate)

Set E gives:

2 - $Ni(OH)_2$ ppts
2* - $BaSO_4$ ppts
4* - $Mg(OH)_2$ ppt
*1 double ppt

Set F (Difficult)

1. 0.1 M $AgNO_3$ (1.7 g)
2. 0.1 M $MgCl_2$ (2.0 g hydrate)
3. 0.1 M Na_2SO_4 (1.4 g anhydrous)
4. 0.2 M NaOH (0.8 g)
5. 0.1 M $Ba(OH)_2$ (1.7 g hydrate)
6. 0.1 M $MgSO_4$ (2.5 g hydrate)

Set F gives:

1 - AgCl ppt
2 - Ag_2O ppts
2* - $BaSO_4$ ppts
4* - $Mg(OH)_2$ ppts
*1 double ppt

SAMPLE DATA TABLE

Set D	1.	2.	3.	4.	5.	6.
Solutions	Na_2CrO_4	K_2CrO_4	$FeCl_3$	KOH	$MgSO_4$	$BaCl_2$
1. Na_2CrO_4	X	NR	ppt.	NR	NR	yellow ppt.
2. K_2CrO_4	X	X	ppt.	NR	NR	yellow ppt.
3. $FeCl_3$	X	X	X	ppt.	NR	NR
4. KOH	X	X	X	X	wht ppt.	NR
5. $MgSO_4$	X	X	X	X	X	white ppt.
6. $BaCl_2$	X	X	X	X	X	X

QUESTIONS (Answers are for Set D)

1. a.

Solutions Mixed	Possible Precipitates
Na_2CrO_4 and $FeCl_3$	NaCl and $Fe_2(CrO_4)_3$
Na_2CrO_4 and $BaCl_2$	NaCl and $BaCrO_4$
K_2CrO_4 and $FeCl_3$	KCl and $Fe_2(CrO_4)_3$
K_2CrO_4 and $BaCl_2$	KCl and $BaCrO_4$
$FeCl_3$ and KOH	KCl and $Fe(OH)_3$
KOH and $MgSO_4$	K_2SO_4 and $Mg(OH)_2$
$MgSO_4$ and $BaCl_2$	$MgCl_2$ and $BaSO_4$

b. No Precipitates

Na_2CrO_4, K_2CrO_4, NaOH, Na_2SO_4, $MgCrO_4$, KOH, K_2SO_4, $MgCl_2$, $Fe_2(SO_4)_3$, $FeCl_3$, $BaCl_2$, KCl, $Ba(OH)_2$, $MgSO_4$

NOTE: KCl, K_2SO_4, and $MgCl_2$ are eliminated as possible precipitates. NaCl is indirectly eliminated as $Fe_2(CrO_4)_3$ and $BaCrO_4$ were shown to be precipitates when KCl was eliminated.

2. $2\ Fe^{3+}(aq) + 3\ CrO_4{}^{2-}(aq) \rightarrow Fe_2(CrO_4)_3(s)$

iron (III) chromate

$Ba^{2+}(aq) + CrO_4{}^{2-}(aq) \rightarrow BaCrO_4(s)$

barium chromate

$Fe^{3+}(aq) + 3\ OH^-(aq) \rightarrow Fe(OH)_3(s)$

iron (III) hydroxide

$Mg^{2+}(aq) + 2\ OH^-(aq) \rightarrow Mg(OH)_2(s)$

magnesium hydroxide

$Ba^{2+}(aq) + SO_4^{2-}(aq) \rightarrow BaSO_4(s)$

barium sulfate

3. The unknown results will match the behavior of one of the solutions in the set of known solutions.

EXTENSIONS

2. The theoretical K_{sp} of $CaSO_4$ is 3×10^{-5}.

Sample calculations for 0.90 g $CaSO_4 \cdot 2\ H_2O$ dissolved:

$$\text{mols } CaSO_4 = 0.90 \text{ g } CaSO_4 \cdot 2\ H_2O \times \frac{136 \text{ g } CaSO_4}{172 \text{ g } CaSO_4 \cdot 2\ H_2O} = 0.71 \text{ g}$$

$$\text{solubility } (CaSO_4) = \frac{0.71 \text{ g}}{1\ \ell} \times \frac{1 \text{ mol}}{136 \text{ g } CaSO_4} = 5.2 \times 10^{-3} \text{ mol}/\ell$$

$$K_{sp} = [Ca^{2+}]\,[SO_4^{2-}] = (\text{solubility})^2$$

$$K_{sp} = (5.2 \times 10^{-3})^2 = 2.7 \times 10^{-5}$$

Acid-Base Titrations

TIMING

The experiment requires about 50 minutes. While it is designed for Chapter 21, it can also be done during Chapter 19 if desired.

PRELIMINARY STUDY

1. a. moles NaOH = $V_{NaOH} \times M_{NaOH}$

$$\text{moles NaOH} = 0.0200\ \ell \times \frac{1.50\ \text{mol}}{\ell}\ \text{NaOH} = 3.00 \times 10^{-2}$$

b. $M_{HCl} \times V_{HCl}$ = moles NaOH

$$M_{HCl} = \frac{\text{moles NaOH}}{V_{HCl}} = \frac{3.00 \times 10^{-2}\ \text{mol}}{0.0165\ \ell} = 1.82$$

2. a. $NH_3(aq) + H^+(aq) \rightarrow NH_4^+(aq)$

b. The solution will be acidic since the NH_4^+ ion is acidic.

$$NH_4^+(aq) \rightleftharpoons NH_3(aq) + H^+(aq)$$

EXPERIMENT NOTES

1. As this is the first volumetric analysis, take some time in a prelab period to demonstrate the proper use of the buret in a titration. The titration technique will be used again in Experiments 33 and 36.

2. If a shortage of burets exists, use 10 ml pipets in Part II instead of a second buret. A pipet-filling bulb should be used.

3. While it is your choice, we recommend that all students standardize the same base in Part I. An NaOH concentration of 0.1 to 0.2 M will require a volume of 25 to 50 ml to neutralize 1 g of KHP. Then provide a variety of acid unknowns for Part II with concentrations varying from 0.2 M to 0.6 M.

MATERIALS NEEDED

Apparatus	Chemicals	
beakers, 150 or 250 ml (2)	$KHC_8H_4O_4$	(2 g)
burets, 50 ml (2) (See Note 2)	phenolphthalein soln; 0.05 g/100 ml 50% C_2H_5OH	
graduated cylinder, 50 ml	NaOH, 0.1 – 0.2 M; 4–8 g/ℓ	(100 ml)
Erlenmeyer flasks, 250 ml (2)	HCl, 0.2 – 0.6 M; 17–50 ml conc./ℓ	(20 ml)
funnel	(See Note 3)	
buret clamp		
wash bottle		

Supplies

filter paper

SAMPLE DATA TABLE

Part I: Run no. 1

mass of paper	1.14 g	vol. of NaOH (initial)	2.00 ml
mass of paper + $KHC_8H_4O_4$	2.14 g	vol. of NaOH (final)	43.57 ml
mass of $KHC_8H_4O_4$	1.00 g	vol. of NaOH used	41.57 ml

Part II: Run no. 1

vol. of HCl (initial)	0.00 ml	vol. of NaOH (initial)	0.00 ml
vol. of HCl (final)	10.00 ml	vol. of NaOH (final)	21.39 ml
vol. of HCl used	10.00 ml	vol. of NaOH used	21.39 ml

CALCULATIONS AND QUESTIONS

1. a. $$\text{moles KHP} = 1.00\ \text{g KHP} \times \frac{1\ \text{mol}}{204\ \text{g KHP}} = 4.90 \times 10^{-3}$$

   b. $M_{NaOH} \times V_{NaOH} = \text{moles KHP}$

$$M_{NaOH} = \frac{\text{moles KHP}}{V_{NaOH}} = \frac{4.90 \times 10^{-3}\ \text{mols}}{0.04157\ \ell} = 0.118\ \text{mol}/\ell$$

2. a. $\text{moles NaOH} = M_{NaOH} \times V_{NaOH}$

$$\text{moles NaOH} = 0.118\ \frac{\text{mol NaOH}}{\ell} \times 0.02139\ \ell = 2.52 \times 10^{-3}$$

b. $M_{HCl} \times V_{HCl}$ = moles NaOH

$$M_{HCl} = \frac{\text{moles NaOH}}{V_{HCl}} = \frac{2.52 \times 10^{-3}\ \text{mol}}{0.01000\ \ell} = 0.252\ \text{mol}/\ell$$

3. The unknown hydrochloric acid solution contains 0.252 mol HCl in every liter.

EXTENSIONS

1. The student should understand from Experiment 29 why a different indicator is required for the titration of NH_3. An end point with methyl orange is more difficult to distinguish than with phenolphthalein so the student may need to make a practice run to find the approximate equivalence point.

Testing Consumer Products

TIMING

Any two of the three parts may be completed in 50 minutes. The experiment should be done during Chapter 21.

PRELIMINARY STUDY

2. MM CH_3COOH = 60.0

$$\% \; CH_3COOH \; (1 \; M) = \frac{60.0 \text{ g } CH_3COOH}{1.0 \times 10^3 \text{ g soln}} \times 100 = 6.0$$

EXPERIMENT NOTES

1. Added interest in the experiment is gained by having the students bring vinegar, ammonia and antacid samples from home. Students become quite involved in the relative merits of different antacids. If possible, test different brands.
2. You may wish to have half the class analyze vinegar while the other half analyzes ammonia. If pipets are not used, the vinegar and ammonia may be loaded into separate burets which are shared by the class.
3. The technique of back-titration used in Part III should be discussed in a prelab session.
4. If possible, use the NaOH and HCl solutions which were standardized in Experiment 32. This emphasizes why solutions are standardized.

MATERIALS NEEDED

Apparatus	Chemicals
buret, 50 ml (See Exp. Note 2)	*vinegar (10 ml)
buret clamp	*household ammonia (10 ml)
pipet and bulb, 5 ml	methyl orange soln; 0.1 g/100 ml
	phenolphthalein soln; 0.05 g/100 ml 50% C_2H_5OH
Erlenmeyer flasks, 250 ml (2)	NaOH (See Exp. Note 4)
graduated cylinder, 50 ml	HCl (See Exp. Note 4)
mortar and pestle	*antacid (1 tablet)
wash bottle	

SAMPLE DATA TABLE

Analysis of Vinegar

vol. of vinegar 5.00 ml
M of NaOH 0.150
vol. of NaOH (initial) 0.00 ml
vol. of NaOH (final) 25.00 ml
vol. of NaOH used 25.00 ml

Analysis of Ammonia

vol. of NH_3 soln. 5.0 ml
M of HCl, 0.971
vol. of HCl (initial) 0.00 ml
vol. of HCl (final) 12.92 ml
vol. of HCl (used) 12.92 ml

Analysis of Antacid

brand of antacid Rolaid
mass of tablet 1.44 g
mass of paper 0.96 g
mass of paper & antacid 1.71 g
mass of antacid used 0.75 g

M of HCl 0.100
vol. of HCl used 50.00 ml
vol. of NaOH (initial) 0.00 ml
vol. of NaOH (final) 4.79 ml
vol. of NaOH used 4.79 ml
M of NaOH 0.150

CALCULATIONS AND QUESTIONS

1. (a) moles NaOH $= 0.02500\ \ell \times \dfrac{0.150 \text{ mol}}{\ell} = 3.75 \times 10^{-3}$

(b) moles CH_3COOH = moles NaOH $= 3.75 \times 10^{-3}$

(c) MM CH_3COOH = 60.05 g/mol

(d) mass $CH_3COOH = 3.75 \times 10^{-3} \text{ mol} \times \dfrac{60.05 \text{ g}}{\text{mol}} = 0.225 \text{ g}$

(e) mass vinegar $= 5.00 \text{ ml} \times \dfrac{1.00 \text{ g}}{\text{ml}} = 5.00 \text{ g}$

(f) $\% \ CH_3COOH = \dfrac{0.225 \text{ g } CH_3COOH}{5.00 \text{ g vinegar}} \times 100 = 4.50$

2. (a) moles HCl $= 0.01292\ \ell \times \dfrac{0.971 \text{ mol}}{\ell} = 1.25 \times 10^{-2}$

(b) moles NH_3 = moles HCl $= 1.25 \times 10^{-2}$

(c) MM NH_3 = 17.03 g/mol

(d) mass $NH_3 = 1.25 \times 10^{-2} \text{ mol} \times \dfrac{17.03 \text{ g}}{\text{mol}} = 0.213 \text{ g}$

(e) mass NH_3 soln. $= 5.00 \text{ ml} \times \dfrac{1.00 \text{ g}}{\text{ml}} = 5.00 \text{ g}$

(f) $\% \ NH_3 = \dfrac{0.213 \text{ g } NH_3}{5.00 \text{ g } NH_3 \text{ soln.}} \times 100 = 4.26$

3. (a) moles HCl $= 0.05000 \ \ell \times \dfrac{0.100 \text{ mol}}{\ell} = 5.00 \times 10^{-3}$

(b) moles NaOH $= 0.00479 \ \ell \times \dfrac{0.150 \text{ mol}}{\ell} = 7.18 \times 10^{-4}$

(c) moles base (antacid) $= 5.00 \times 10^{-3} - 7.18 \times 10^{-4} = 4.28 \times 10^{-3}$

(d) moles HCl neutralized by antacid = moles base (antacid)

$= 4.28 \times 10^{-3}$

(e) moles HCl $= 1.00 \text{ g antacid} \times \dfrac{4.28 \times 10^{-3} \text{ mol HCl}}{0.75 \text{ g antacid}} = 5.7 \times 10^{-3}$

EXTENSIONS

1. (a) vol. HCl neutralized $= 5.7 \times 10^{-3} \text{ mol HCl} \times \dfrac{1000 \text{ ml}}{0.10 \text{ mol HCl}} = 57 \text{ ml}$

mass HCl solution neutralized $= 57 \text{ ml} \times \dfrac{1.00 \text{ g HCl solution}}{\text{ml}} = 57 \text{ g}$

The Rolaid antacid in this experiment neutralized 57 times its mass of stomach acid (0.1 M HCl).

Developing a Qualitative Analysis Scheme

TIMING

Procedures 1–3 can be easily completed in a single period. Procedures 4 and 5 generally require part of a second period. The experiment should be done early in Chapter 22.

EXPERIMENT NOTES

1. Students gain experience in developing a workable scheme for the separation and identification of ions. The procedures and techniques practiced here prepare the student for Experiment 35, The Qualitative Analysis of the Group I Ions.

2. In Procedure 5, a single ion unknown can be identified by performing the four solubility tests on separate portions of the unknown. The unknown ion can be identified by matching results with those obtained for the knowns. If a multiple ion unknown is used, the pH of the solution must be considered. In particular, the hydroxides and oxalates will not precipitate from an acidic solution. Students will need to make solutions neutral to litmus before testing for precipitation.

MATERIALS NEEDED

Apparatus	Chemicals	
test tubes, small (8)	H_2SO_4, 1 M; 56 ml conc./ℓ	(10 ml)
test tube rack	NaOH, 1 M; 40 g/ℓ	(10 ml)
burner	$(NH_4)_2C_2O_4$, 0.25 M; 35.5 g hydrate/ℓ	(10 ml)
flame tester, Nichrome wire	K_2CrO_4, 1 M; 194 g/ℓ	(10 ml)
iron ring	CH_3COOH, 1 M; 57 ml conc./ℓ	(10 ml)
wire gauze	$Mg(NO_3)_2$, 0.1 M; 15 g/ℓ	(10 ml)
beaker, 50 ml	$Ca(NO_3)_2$, 0.1 M; 16 g/ℓ	(10 ml)
centrifuge (optional)	$Sr(NO_3)_2$, 0.1 M; 21 g/ℓ	(10 ml)
	$Ba(NO_3)_2$, 0.1 M; 26 g/ℓ	(10 ml)
	HNO_3, 6 M; 375 ml conc./ℓ	(10 ml)

SAMPLE DATA TABLE

1. Solubility Tests:

	H_2SO_4	NaOH	$(NH_4)_2C_2O_4$	K_2CrO_4 CH_3COOH
$Mg(NO_3)_2$	NR	wht. ppt.	NR	NR
$Ca(NO_3)_2$	NR	wht. ppt.	wht. ppt.	NR
$Sr(NO_3)_2$	wht. ppt.	NR	wht. ppt.	NR
$Ba(NO_3)_2$	wht. ppt.	NR	wht. ppt.	yell. ppt.

2. Flame Tests:

Distilled H_2O	$Mg(NO_3)_2$	$Ca(NO_3)_2$	$Sr(NO_3)_2$	$Ba(NO_3)_2$
colorless	colorless	red	red	green

NOTE: The Ca and Sr colors can be distinguished if they are observed carefully.

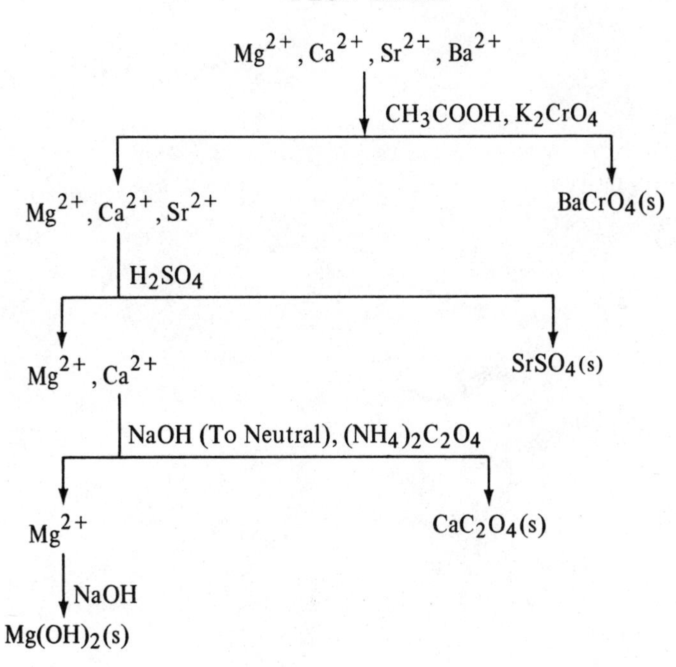

CALCULATIONS AND QUESTIONS

1. $Ba^{2+}(aq) + CrO_4^{2-}(aq) \rightarrow BaCrO_4(s)$
   $Sr^{2+}(aq) + SO_4^{2-}(aq) \rightarrow SrSO_4(s)$
   $Ca^{2+}(aq) + C_2O_4^{2-}(aq) \rightarrow CaC_2O_4(s)$
   $Mg^{2+}(aq) + 2\ OH^-(aq) \rightarrow Mg(OH)_2(s)$

2. a. The Li^+ ion flame test may conflict with either the Ca^{2+} ion or the Sr^{2+} ion as all give a reddish color flame.

   b. The green Cu^{2+} flame conflicts with the Ba^{2+} ion flame test which is also green.

3. (a) Mg^{2+}, Ca^{2+} : $(NH_4)_2C_2O_4$ (c) Ca^{2+}, Sr^{2+} : H_2SO_4 or NaOH

   (b) Mg^{2+}, Ba^{2+} : any of the 4 (d) Ca^{2+}, Ba^{2+} : all except $(NH_4)_2C_2O_4$

EXPERIMENT 35

Qualitative Analysis of the Group I Ions: Ag^+, Pb^{2+}, Hg_2^{2+}

TIMING

Part I of the experiment requires about 45 minutes. Part II, Analysis of an Unknown, requires another 5–15 minutes. The experiment should be done during Chapter 22.

EXPERIMENT NOTES

1. The scheme used is the standard scheme for the analysis of the Group I ions. Preliminary tests are studied first so as to become familiar with the reactions.

2. Point out the importance of complete precipitations and clean separations in the analysis of a multiple ion unknown. The lack of either can lead to improper conclusions on succeeding tests. The proper control of pH is also crucial in Procedure 5.

MATERIALS NEEDED

Apparatus	Chemicals	
test tubes, small (6)	$AgNO_3$, 0.1 M; 17 g/ℓ	(1 ml)
test tube rack	HCl, 6 M; 500 ml conc./ℓ	(2 ml)
stirring rod	NH_3, 6 M; 400 ml conc./ℓ	(5 ml)
burner	HNO_3, 6 M; 375 ml conc./ℓ	(5 ml)
iron ring	$Pb(NO_3)_2$, 0.2 M; 66 g/ℓ	(1 ml)
wire gauze	K_2CrO_4, 1 M; 194 g/ℓ	(1 ml)
beaker, 250 ml	$Hg_2(NO_3)_2$, 0.1 M; 52 g/ℓ	(1 ml)
centrifuge	Group I unknowns*	(10 ml)
wash bottle		

*Ion concentrations should be the same as the knowns.

Supplies

litmus paper, red and blue

SAMPLE DATA TABLE

Part I: Preliminary Tests

Procedure	Observations
1. $AgNO_3(aq) + HCl(aq)$	wht. ppt. ($AgCl$)
$AgCl(s) + NH_3(aq)$	ppt. dissolves ($Ag(NH_3)_2^+$)
$Ag(NH_3)_2^+(aq) + HNO_3(aq)$	wht. ppt. ($AgCl$)
2. $Pb(NO_3)_2(aq) + HCl(aq)$	wht. ppt. ($PbCl_2$)
$PbCl_2(s)$ + hot water	ppt. dissolves (Pb^{2+})
$Pb^{2+}(aq) + K_2CrO_4(aq)$	yell. ppt. ($PbCrO_4$)
3. $Hg_2(NO_3)_2(aq) + HCl(aq)$	wht. ppt. (Hg_2Cl_2)
$Hg_2Cl_2(s) + NH_3(aq)$	black ppt. ($HgNH_2Cl$ and Hg)

CALCULATIONS AND QUESTIONS

1. Part I Procedures:

   a. $Ag^+(aq) + Cl^-(aq) \rightarrow AgCl(s)$

   $AgCl(s) + 2\ NH_3(aq) \rightarrow Ag(NH_3)_2^+(aq) + Cl^-(aq)$

   $Ag(NH_3)_2^+(aq) + Cl^-(aq) + 2\ H^+(aq) \rightarrow AgCl(s) + 2\ NH_4^+(aq)$

   b. $Pb^{2+}(aq) + 2\ Cl^-(aq) \rightarrow PbCl_2(s)$

   $PbCl_2(s) \xrightarrow{\text{hot } H_2O} Pb^{2+}(aq) + 2\ Cl^-(aq)$

   $Pb^{2+}(aq) + CrO_4^{2-}(aq) \rightarrow PbCrO_4(s)$

   c. $Hg_2^{2+}(aq) + 2\ Cl^-(aq) \rightarrow Hg_2Cl_2(s)$

   $Hg_2Cl_2(s) + 2\ NH_3(aq) \rightarrow HgNH_2Cl(s) + Hg(l) + NH_4^+(aq) + Cl^-(aq)$

2. Procedure 1: precipitation, complex ion formation, complex ion decomposition

   Procedure 2: precipitation, solution, precipitation

   Procedure 3: precipitation, oxidation-reduction

3. a. Hot water in Procedure 1 would dissolve some $PbCl_2$ in the Group I precipitate. This might cause the Pb^{2+} ion to be missed in Procedure 3.

   b. Cold water in Procedure 2 would not dissolve much $PbCl_2$. The Pb^{2+} ion would probably be missed in Procedure 3. The undissolved $PbCl_2$ might confuse the Procedure 4 results.

   c. Insufficient HNO_3 in Procedure 5 would leave NH_3 in the solution. The Ag^+ might remain as the $Ag(NH_3)_2{}^+$ complex ion, rather than precipitate as AgCl. If no AgCl formed, the test for Ag^+ would be missed.

4. a. AgCl, Hg_2Cl_2 : NH_3

   b. AgCl, $PbCl_2$: NH_3 or hot water

   c. $PbCl_2$, Hg_2Cl_2 : hot water

EXTENSIONS

1. Identification can be accomplished with any of the reagents listed below; the colors of precipitates obtained are as indicated, using excess reagent:

	Ag^+	Pb^{2+}	$Hg_2{}^{2+}$
6 M NH_3	soluble	white	black
1 M K_2CrO_4	dk red	yellow	orange
6 M NaOH	dk tan	soluble	black

Analysis by an Oxidation-Reduction Titration

TIMING

The experiment requires about 50 minutes. It is designed to be done during Chapter 23.

PRELIMINARY STUDY

2. a. moles $S_2O_3^{2-} = V_{Na_2S_2O_3} \times M_{Na_2S_2O_3}$

$$\text{moles } S_2O_3^{2-} = 0.0240\ \ell \times \frac{0.550 \text{ mol}}{\ell} = 1.32 \times 10^{-2}$$

b. moles $OCl^- = \frac{1}{2}$ moles $S_2O_3^{2-}$ (Equation 36.4)

moles $OCl^- = \frac{1}{2}(1.32 \times 10^{-2}) = 6.60 \times 10^{-3}$

c. moles NaOCl = moles $OCl^- = 6.60 \times 10^{-3}$

$$\text{mass NaOCl} = 6.60 \times 10^{-3} \text{ mol NaOCl} \times \frac{74.4 \text{ g}}{1 \text{ mol NaOCl}} = 0.491 \text{ g}$$

Assume a bleach density of 1.00 g/cm^3

$$\% \text{ NaOCl} = \frac{0.491 \text{ g NaOCl}}{10.0 \text{ g bleach}} \times 100 = 4.91\%$$

EXPERIMENT NOTES

1. The analytical technique involves an indirect titration. An excess of KI is added so that all of the OCl^- ions are converted to Cl^- ions. The I_2 produced in this reaction is then reduced quantitatively back to the I^- ion. The end-point is detected by the loss of the dark blue color caused by a starch indicator.

MATERIALS NEEDED

Apparatus	Chemicals	
Erlenmeyer flasks, 250 ml (2)	*household bleach	(20 ml)
pipet and bulb, 10 ml	KI, 1 M; 166 g/ℓ	(40 ml)
graduated cylinder, 25 or 50 ml	HCl, 6 M; 500 ml conc./ℓ	(20 ml)

buret, 50 ml
buret clamp

$Na_2S_2O_3$, 1.00 M; 248.2 g hydrate/ℓ
starch solution, 3% (make fresh) (30 ml)

SAMPLE DATA TABLE

molarity of $Na_2S_2O_3$ 1.00 M
vol. of $Na_2S_2O_3$ (initial) 0.00 ml
vol. of $Na_2S_2O_3$ (final) 13.97 ml
vol. of $Na_2S_2O_3$ used 13.97 ml

vol. of bleach 10.00 ml

CALCULATIONS AND QUESTIONS

1. a. moles $Na_2S_2O_3 = 0.01397\ \ell \times \frac{1.00\ \text{mol}}{\ell} = 0.0140$

   b. moles $NaOCl = \frac{1}{2}(0.0140) = 7.00 \times 10^{-3}$

   c. mass $NaOCl = 7.00 \times 10^{-3}\ \text{mol NaOCl} \times \frac{74.5\ \text{g}}{1\ \text{mol NaOCl}} = 0.522\ \text{g}$

   d. $\%\ NaOCl = \frac{0.522\ \text{g NaOCl}}{10.0\ \text{g bleach}} \times 100 = 5.22\%$

2. Equation 36.2:

   $2I^-(aq) \rightarrow I_2(aq) + 2\ e^-$

   $OCl^-(aq) + 2\ H^+ + 2\ e^- \rightarrow Cl^-(aq) + H_2O$

   Equation 36.3:

   $2\ S_2O_3^{2-}(aq) \rightarrow S_4O_6^{2-}(aq) + 2\ e^-$

   $I_2(aq) + 2\ e^- \rightarrow 2\ I^-(aq)$

   Each oxidation half-equation produces two electrons.

EXTENSIONS

1. Check the label on the cleanser to make certain that it contains sodium hypochlorite. The NaOCl content varies considerably from brand to brand.

EXPERIMENT 37

Voltaic Cells

TIMING

The experiment requires about 50 minutes. It should be done during Chapter 24.

PRELIMINARY STUDY

2. E^{o}_{ox} $(Zn \rightarrow Zn^{2+})$ $= +\ 0.76$ V

$$E^{o}_{red}\ (Cu^{2+} \rightarrow Cu) = +\ 0.34\ \text{V}$$
$$E^{o} = +\ 1.10\ \text{V}$$

EXPERIMENT NOTES

1. The experiment specifies porous bottom crucibles which fit into a 50 ml beaker. Alternatively, 4 cm × 9 cm porous cups are available which fit into a 250 ml beaker. Cells can also be made using salt bridges. A glass U-tube is filled with an electrolyte (1 M KNO_3) and plugged at both ends with cotton. The half-cells are in separate 150 ml beakers and are connected with the salt bridge. Salt bridge cells have relatively low electrical resistance.
2. Vacuum tube volt-meters or pH meters with a voltage scale should be used for the voltage measurements to obtain accurate voltages. Check with the school physics teacher, if necessary, for the loan of this item. If such meters are not available use salt bridges to make the cells and ordinary volt-ohmmeter to measure voltage.

MATERIALS NEEDED

Apparatus	Chemicals	
beakers, 50 ml (2)	$Zn(NO_3)_2$, 0.1 M; 19 g/ℓ	(40 ml)
crucible, porous bottom (See Note 1)	$Cu(NO_3)_2$, 0.1 M; 19 g/ℓ	(40 ml)
	$Pb(NO_3)_2$, 0.1 M; 33 g/ℓ	(20 ml)
stirring rod	NH_4Cl	(5 g)
voltmeter (See Note 2)	MnO_2	(5 g)
	NH_3, 6 M; 400 ml conc./ℓ	(5 ml)
	NaOH, 6 M; 240 g/ℓ	(5 ml)

Chemicals

graphite rod
zinc strip
copper strip
lead strip

SAMPLE DATA TABLE

Part I: Voltage of dry cell 1.30 volts

	Cell components	Cell voltage	Negative electrode
Part II:	Zn/Zn^{2+}, Cu/Cu^{2+}	1.05 V	Zn
	Zn/Zn^{2+}, Pb/Pb^{2+}	0.60 V	Zn
	Pb/Pb^{2+}, Cu/Cu^{2+}	0.47 V	Pb
Part III:	NH_3 added to Cu^{2+}	0.83 V	Zn
	NaOH added to Zn^{2+}	1.23 V	Zn

CALCULATIONS AND QUESTIONS

1. A dry cell can be packaged into a small size and used in any position with no worry of leakage. It also maintains a more constant voltage in use due to its smaller cell resistance.

2. a. $Zn(s) + Cu^{2+}(aq) \rightarrow Zn^{2+}(aq) + Cu(s)$

   $Zn(s) + Pb^{2+}(aq) \rightarrow Zn^{2+}(aq) + Pb(s)$

   $Pb(s) + Cu^{2+}(aq) \rightarrow Pb^{2+}(aq) + Cu(s)$

   b. $Zn(s) \rightarrow Zn^{2+}(aq) + 2\ e^-$ $Cu^{2+}(aq) + 2\ e^- \rightarrow Cu(s)$

   $Zn(s) \rightarrow Zn^{2+}(aq) + 2\ e^-$ $Pb^{2+}(aq) + 2\ e^- \rightarrow Pb(s)$

   $Pb(s) \rightarrow Pb^{2+}(aq) + 2\ e^-$ $Cu^{2+}(aq) + 2\ e^- \rightarrow Cu(s)$

3. a. $E^o_{ox}\ (Zn \rightarrow Zn^{2+}) + E^o_{red}\ (Cu^{2+} \rightarrow Cu) = E^o_{cell}$

   $E^o_{red}\ (Cu^{2+} \rightarrow Cu) = 1.05\ V - 0.76\ V = 0.29\ V$

   $E^o_{ox}\ (Zn \rightarrow Zn^{2+}) + E^o_{red}\ (Pb^{2+} \rightarrow Pb) = E^o_{cell}$

   $E^o_{red}\ (Pb^{2+} \rightarrow Pb) = 0.60\ V - 0.76\ V = -\ 0.16\ V$

   b. $E^o_{cell} = E^o_{ox}\ (Pb \rightarrow Pb^{2+}) + E^o_{red}\ (Cu^{2+} \rightarrow Cu)$

   $E^o_{cell} = 0.16\ V + 0.29\ V = 0.45\ V$

   The measured voltage of the lead-copper cell was 0.47 V. This is in close agreement with the predicted voltage of 0.45 V.

4. When the zinc electrode was placed in the copper nitrate solution, the cell was shorted out and the voltage dropped. Copper metal was deposited on the surface of the zinc electrode:

   $Zn(s) + Cu^{2+}(aq) \rightarrow Zn^{2+}(aq) + Cu(s)$

5. Adding NH_3 to the $Cu(NO_3)_2$ solution converted the Cu^{2+} ions to the complex ion, $Cu(NH_3)_4{}^{2+}$. In agreement with Le Chatelier's Principle, the decreased concentration of the Cu^{2+} ion decreases the force which drives the reaction to the right. This is accompanied by a decrease in the cell voltage. Adding NaOH to the $Zn(NO_3)_2$ solution converted the Zn^{2+} ions to the $Zn(OH)_4{}^{2-}$ complex ion. This causes the reaction to proceed more in the forward direction in an attempt to replace the Zn^{2+} ions. This is accompanied by an increase in the cell voltage.

EXTENSIONS

1. a. $\Delta V = 1.05\ V - 0.83\ V = 0.22\ V$

   $$\frac{0.22\ V}{0.03\ V} = 7$$

   This means that the concentration of the Cu^{2+} decreased by a factor of 10, seven times, or $(1/10)^7$. Its new concentration is then 10^{-7} times its original concentration.

$[Cu^{2+}] = 10^{-7}\ (0.1\ M) = 10^{-8}\ M$

   b. $Cu^{2+}(aq) + 4\ NH_3(aq) \rightleftarrows Cu(NH_3)_4{}^{2+}(aq)$

   $$K = \frac{[Cu(NH_3)_4^{2+}]}{[Cu^{2+}]\ [NH_3]^4} = \frac{0.1}{10^{-8}(6)^4} = 8 \times 10^3$$

Winning a Metal From Its Ore

TIMING

The experiment requires about 50 minutes. While it has been designed for Chapter 25, it can also be used during Chapter 23, Oxidation-Reduction Reactions.

PRELIMINARY STUDY

2. $\text{mass Al} = 1 \text{ kg ore} \times \frac{61 \text{ kg } Al_2O_3}{100 \text{ kg ore}} \times \frac{54.0 \text{ kg Al}}{102 \text{ kg } Al_2O_3} = 0.32 \text{ kg}$

   $\text{mass Al recovered} = 0.81\ (0.32 \text{ kg}) = 0.26 \text{ kg}$

EXPERIMENT NOTES

1. Use artificial ores in Part I to simulate a rich crushed ore. While galena, PbS, is the more common commercial lead ore, white lead, $PbCO_3$, is used in Part II instead because it is easier to roast and reduce.

MATERIALS NEEDED

Apparatus

- test tube, large
- crucible and cover
- clay triangle
- iron ring
- burner
- stirring rod
- mortar
- heat resistant pad
- beaker, 100 ml
- crucible tongs
- wash bottle

Chemicals

- red lead ore (2 g) (50% Pb_3O_4 and 50% SiO_2)
- galena ore (2 g) (50% PbS and 50% SiO_2)
- *mineral oil (20 ml)
- $PbCO_3$ (10 g)
- charcoal (1 g)

Supplies

- filter paper

SAMPLE DATA TABLE

Process	Observations
Flotation	The lead ore and sand both settle to the bottom when shaken with water. After mineral oil is added, shaking tends to disperse the lead ores with the oil droplets in the top layer. The sand settles to the bottom with the water layer.
Roasting	A yellow product (PbO) is obtained when the $PbCO_3$ is heated.
Reduction	When heated with charcoal, the yellow solid is converted to liquid lead metal.

mass of $PbCO_3$ 10.00 g mass of Pb 6.80 g

CALCULATIONS AND QUESTIONS

1. a. $$\text{mass Pb} = 10.00 \text{ g } PbCO_3 \times \frac{207.2 \text{ g Pb}}{267.2 \text{ g } PbCO_3} = 7.754 \text{ g}$$

   b. $$\% \text{ yield} = \frac{6.80 \text{ g Pb (recovered)}}{7.754 \text{ g Pb}} \times 100 = 87.7\%$$

2. a. $PbCO_3(s) \rightarrow PbO(s) + CO_2(g)$

   b. $PbO(s) + CO(g) \rightarrow Pb(l) + CO_2(g)$

   oxidizing agent (PbO) reducing agent (CO)

3. $2\ PbS(s) + 3\ O_2(g) \rightarrow 2\ PbO(s) + 2\ SO_2(g)$

   $PbO(s) + CO(g) \rightarrow Pb(l) + CO_2(g)$

EXTENSIONS

1. $CuCO_3(s) \rightarrow CuO(s) + CO_2(g)$

   $CuO(s) + CO(g) \rightarrow Cu(l) + CO_2(g)$

   1 mol $CuCO_3$ $\triangleq$ 1 mol Cu

   123.5 g $CuCO_3$ $\triangleq$ 63.5 g Cu

2. The copper in a $CuSO_4$ solution can be recovered by:

   a. a reaction with iron (nails or scrap iron).

   b. electrolysis of the solution.

Trace Analysis by Colorimetry

TIMING

The experiment requires about 40 minutes. While the experiment is designed for Chapter 26, it also fits in with Chapter 21, Quantitative Analysis.

EXPERIMENT NOTES

1. The technique of adding a reagent to a colorless solution to form a colored product is commonly used in quantitative analysis. The eyeball method used here for detection is amazingly good. If a colorimeter is available, consider using that instrument as described in Extension 2.

2. Have unknowns ready for students who wish to do Extension 1. The procedure to be used is described in Experiment 21.

MATERIALS NEEDED

Apparatus	Chemicals	
beaker, 100 ml	$Fe(NO_3)_3$, 1.10×10^{-3} M; 0.266 g/ℓ	(30 ml)
graduated cylinder, 10 ml	$K_4Fe(CN)_6$, 0.02 M; 7.4 g/ℓ	(9 ml)
graduated cylinder, 100 ml	$Cu(NO_3)_2$, 1.10×10^{-3} M; 0.206 g/ℓ	(30 ml)
Erlenmeyer flask, 250 ml	$HgCl_2$, 1.10×10^{-3} M; 0.299 g/ℓ	(30 ml)
test tubes, med. (10) (or flat-bottom vials)		
test tube rack		

SAMPLE DATA TABLE

Cation tested	Solution color	Lowest concentration
Fe^{3+}	blue to green	10^{-5} M–10^{-6} M
Cu^{2+}	red-brown to rust	10^{-4} M–10^{-5} M
Hg^{2+}	green to yellow	10^{-4} M–10^{-5} M

NOTE: Detection limits may vary due to natural eye differences.

CALCULATIONS AND QUESTIONS

1. a. 10^{-5} M Fe^{3+}

   b. $$\text{mass Fe} = \frac{10^{-5}\text{ mol Fe}}{1\ \ell\text{ soln}} \times \frac{55.85\text{ g Fe}}{1\text{ mol Fe}} = 6 \times 10^{-4}\text{ g}$$

   c. $$\text{mass Fe} = 10^{6}\text{ g soln} \times \frac{6 \times 10^{-4}\text{ g Fe}}{10^{3}\text{ g soln}} = 0.6\text{ g or }0.6\text{ ppm}$$

2. a. 10^{-4} M Cu^{2+}

   $$\text{mass Cu} = \frac{10^{-4}\text{ mol Cu}}{1\ \ell\text{ soln}} \times \frac{63.54\text{ g Cu}}{1\text{ mol Cu}} = 6 \times 10^{-3}\text{ g}$$

   $$\text{mass Cu} = 10^{6}\text{ g soln} \times \frac{6 \times 10^{-3}\text{ g Cu}}{10^{3}\text{ g soln}} = 6\text{ g or }6\text{ ppm}$$

   b. 10^{-4} M Hg^{2+}

   $$\text{mass Hg} = \frac{10^{-4}\text{ mol Hg}}{1\ \ell\text{ soln}} \times \frac{200.6\text{ g Hg}}{1\text{ mol Hg}} = 2 \times 10^{-2}\text{ g}$$

   $$\text{mass Hg} = 10^{6}\text{ g soln} \times \frac{2 \times 10^{-2}\text{ g Hg}}{10^{3}\text{ g soln}} = 20\text{ g or }20\text{ ppm}$$

3. The method could be made more sensitive by:
   a. concentrating the solution by evaporation before adding the $K_4Fe(CN)_6$.
   b. using an instrumental method of detection. (See Extension 2.)
4. Some possible disadvantages of the test method are that:
   a. the original solution must be colorless.
   b. the solution must contain no other ions (than the one being tested for) which react to give a colored product.
5. The detection limit for Hg was found to be 20 ppm (Question 2). The method could not detect Hg between the maximum safe limit of 0.1 ppm and 20 ppm.

EXTENSIONS

1. Provide unknowns whose ion concentrations are between the detectable limits of the experiment.
2. Have students measure the absorbances of the solutions at varying concentrations to determine the detectable limit. Use the wave-length setting prescribed. If desired, a plot of absorbance vs concentration may be drawn and used to determine the concentration of an unknown. With most simple spectrophotometers the minimum detectable concentration is about equal to that with the visual method.

MATERIALS NEEDED

Glassware and Plasticware

Apparatus

Item	Used in Experiment
Buret, 50 ml	27, 32, 33, 36
Beaker, 50 ml	frequent use
Beaker, 100 ml	frequent use
Beaker, 150 ml	8
Beaker, 250 ml	frequent use
Beaker, 400 ml	frequent use
Beaker, 600 ml	11
Bottle, wide-mouth	10, 12, 14
Crucible and cover	6, 7, 16, 38
Crucible, porous bottom	37
Drying tube	10
Erlenmeyer flask, 250 ml	32, 33, 36, 39
Eudiometer, 50 ml	11
Evaporating dish	2, 14, 19
Florence flask, 300 ml	10
Funnel	2, 7, 21, 25, 32
Glass plate	2, 9, 12, 26
Glass tubing	7, 10, 12, 23
Graduated cylinder, 10 ml	frequent use
Graduated cylinder, 50 ml	frequent use
Graduated cylinder, 100 ml	39
Graduated cylinder, 250 ml	10
Graduated cylinder, 500 ml (or hydrometer jar)	11
Medicine dropper	3, 21
Mortar and pestle	23, 33, 38
Pipet, capillary	5
Pipet, 5 ml	33
Pipet, 10 ml	36
Spot plate	31
Stirring rod	frequent use
Test tube, small (13 × 100 mm or 15 × 125 mm)	frequent use
Test tube, regular (16 × 150 mm or 18 × 150 mm)	frequent use

Test tube, large (25 × 150 mm or 25 × 200 mm)	frequent use
Thistle tube	10
U-tube	37
Vial, flat bottom	21, 39
Wash bottle, 500 ml	frequent use
Watch glass	8, 21

Hardware

Item	Used in Experiment
Heat resistant pad	2, 6, 16, 38
Beaker tongs	18
Buret clamp	27, 32, 33, 36
Burner	frequent use
Can, 12 oz	9
Can, no. 3 tall	9
Clay triangle	6, 7, 16, 38
Corks (variety)	frequent use
Crucible tongs	frequent use
Dart	15
Deflagrating spoon	12, 14
Flame tester, Nichrome wire	34
Iron ring	frequent use
Metric ruler, 15 cm	1, 5, 11
Pneumatic trough	12
Ring stand	frequent use
Rubber tubing, 3/16″ or 1/4″	10, 12
Safety goggles	frequent use
Spatula	frequent use
Stoppers (variety)	frequent use
Suction bulb	33, 36
Target board, 2 ft. × 2 ft.	15
Test tube brush	frequent use
Test tube clamp	frequent use
Test tube holder	frequent use

Test tube rack	frequent use
Thermometer, $-10°C$ to $110°C$	frequent use
Tray, 18 in. × 24 in.	5
Tweezers	26
Wire gauze	frequent use

Glass Equipment

Item	Used in Experiment
Balance, centigram	frequent use
Barometer	10, 11
Centrifuge	34, 35
Conductivity tester	19
Hot plate, electric	8
Molecular model kit	17, 22, 24
Voltmeter	37

Inorganic Reagents

Chemicals

Name	Used in Experiment
Aluminum chloride $AlCl_3$	31
Aluminum sulfate $Al_2(SO_4)_3$	31
Ammonia (concentrated)	frequent use
Ammonium chloride NH_4Cl	19, 30, 37
Ammonium nitrate NH_4NO_3	9
Ammonium oxalate $(NH_4)_2 C_2O_4$	34
Barium chloride (hydrate) $BaCl_2 \cdot 2H_2O$	16, 31
Barium nitrate $Ba(NO_3)_2$	31, 34
Bromine water	23
Calcium	14
Calcium carbide CaC_2	23
Calcium carbonate (marble chips) $CaCO_3$	10, 30
Calcium chloride (anhydrous) $CaCl_2$	9, 10, 19
Calcium chloride (hydrate) $CaCl_2 \cdot 2H_2O$	21, 25
Calcium hydroxide $Ca(OH)_2$	7
Calcium nitrate $Ca(NO_3)_2$	34
Calcium oxide CaO	37
Calcium sulfate (hydrate) $CaSO_4 \cdot 2H_2O$	2, 16
Charcoal (lump)	12
Charcoal (powdered)	7, 38
Chlorine water	14

Cobalt chloride (hydrate) $CoCl_2 \cdot 6H_2O$	28, 31
Cobalt nitrate $Co(NO_3)_2$	31
Copper (strips)	37
Copper (wire)	11
Copper (wool)	7
Copper carbonate $CuCO_3$	7
Copper nitrate $Cu(NO_3)_2$	37, 39
Copper sulfate (hydrate) $CuSO_4 \cdot 5H_2O$	7, 16
Graphite (rods)	37
Hydrochloric acid (concentrated)	frequent use
Iron (turnings)	19
Iron (III) chloride $FeCl_3$	25, 30, 31
Iron (III) nitrate $Fe\ (NO_3)_3$	39
Lead (strips)	37
Lead carbonate $PbCO_3$	38
Lead nitrate $Pb(NO_3)_2$	frequent use
Lead oxide (red lead) Pb_3O_4	38
Lead sulfide (galena) PbS	38
Lithium	14
Magnesium (ribbon)	6, 11
Magnesium chloride $MgCl_2$	25
Magnesium nitrate $Mg(NO_3)_2$	34
Magnesium sulfate $MgSO_4$	31
Manganese dioxide MnO_2	12, 14, 37
Mercury (II) chloride $HgCl_2$	39
Mercury (I) nitrate $Hg_2(NO_3)_2$	35
Nitric acid (concentrated)	frequent use
Potassium	14
Potassium carbonate (hydrate) $K_2CO_3 \cdot 3/2H_2O$	2, 16
Potassium chlorate $KClO_3$	12
Potassium chromate K_2CrO_4	2, 31, 34, 35
Potassium dichromate $K_2Cr_2O_7$	20, 25, 28
Potassium ferrocyanide $K_4Fe(CN)_6$	39
Potassium hydrogen phthalate $KHC_8H_4O_4$	32
Potassium hydroxide KOH	31
Potassium iodide KI	36
Potassium nitrate KNO_3	9, 31
Potassium permanganate $KMnO_4$	21
Silicon	19
Silicon carbide SiC	19

Silicon dioxide (silica) SiO_2	19, 38
Silver nitrate $AgNO_3$	35
Soda lime	23
Sodium	14
Sodium acetate $NaCH_3COO$	frequent use
Sodium bromide NaBr	14
Sodium carbonate (hydrate) $Na_2CO_3 \cdot H_2O$	9, 16, 30
Sodium chloride NaCl	frequent use
Sodium chromate $Na_2CrO_4 \cdot 4H_2O$	16, 31
Sodium hydroxide NaOH	frequent use
Sodium iodide NaI	14
Sodium nitrate $NaNO_3$	9, 19, 31
Sodium sulfate Na_2SO_4	31
Sodium thiosulfate $Na_2S_2O_3$	27, 36
Strontium nitrate $Sr(NO_3)_2$	31, 34
Sulfur	5, 14
Sulfuric acid	frequent use
Tin (foil)	1
Zinc (strips)	37
Zinc nitrate $Zn(NO_3)_2$	37

Acid-Base Indicators

Name	Used in Experiment
Bromthymol blue	28
Litmus paper (red and blue)	frequent use
Methyl orange	29, 33
Methyl red	29
Phenolphthalein	frequent use
Thymol blue	29

Organic Reagents

Name	Used in Experiment
Acetic acid (glacial)	29, 30, 34
Acetone	2
Aniline	25
Benzoic acid	19, 21
Benzoyl peroxide	26
Butyl alcohol	3
Cyclohexane	5, 23
Cyclohexene	23
Ethyl alcohol	3, 25
Ethylene glycol	8
Glycerol	frequent use
Hexamethylenediamine	26

Hexane	3
Lactic acid	29
Methyl alcohol	2, 3
Methyl ethyl ketone	26
Methyl methacrylate	26
Naphthalene	18, 19, 23
Propanol	8
Propionic acid	29
Ortho-toluidine	25
Para-dichlorobenzene	18, 19
Para-toluidine	25
Phthalic anhydride	26
Salicylic acid	2, 19
Sebacoyl chloride	26
Starch (soluble)	36
Stearic acid	2, 5
Tannic acid	25
Thymol	18
Toluene	3, 23
Trichloroethane CCl_3CH_3	3, 14, 19, 26
Urea	19
Urethane	18

Household Chemicals

Name	Used in Experiment
Ammonia (household)	33
Aluminum (foil)	1
Antacid (Rolaid)	33
Beans (five types)	4
Bleach (household)	36
Candles (parrafin)	9
Cottonseed oil	2, 3, 25
Copper coin (penny)	14
Detergent (liquid)	25
Iron nails	7
Isopropyl alcohol (rubbing)	8
Kerosene	25
Linseed oil	2
Mineral oil	38

Sucrose (table sugar)	19, 21
Steel wool	12
Turpentine	2
Vinegar (white)	33

Consumable Non-Chemicals

Item	Used in Experiment
Cellophane tape	14
Cotton	12, 37
Cotton cloth (white)	25
Emery cloth (fine)	14, 37
Filter paper	frequent use
Glass wool	12
Graph paper	3, 13, 20, 27
Labels	frequent use
Masking tape	15
Matches	frequent use
Paper cups (7 oz)	4
Styrofoam cups (10 oz)	8, 9, 18
Weighing paper	frequent use
Wood splints	2, 12, 26

Notes to the Teacher

There are two diagrams that follow these pages. The first diagram is a blank Periodic Table. This blank table can be cut out of the page by the teacher and then photocopied or made into duplicating masters. It is needed by each student when performing Experiment 13. See page 257 of the Teacher's Guide.

The second diagram is a target. The two halves of the target can be cut out by the teacher and taped together. We suggest photocopying several targets to have on hand for your classes when performing Experiment 15. See page 264 of the Teacher's Guide.

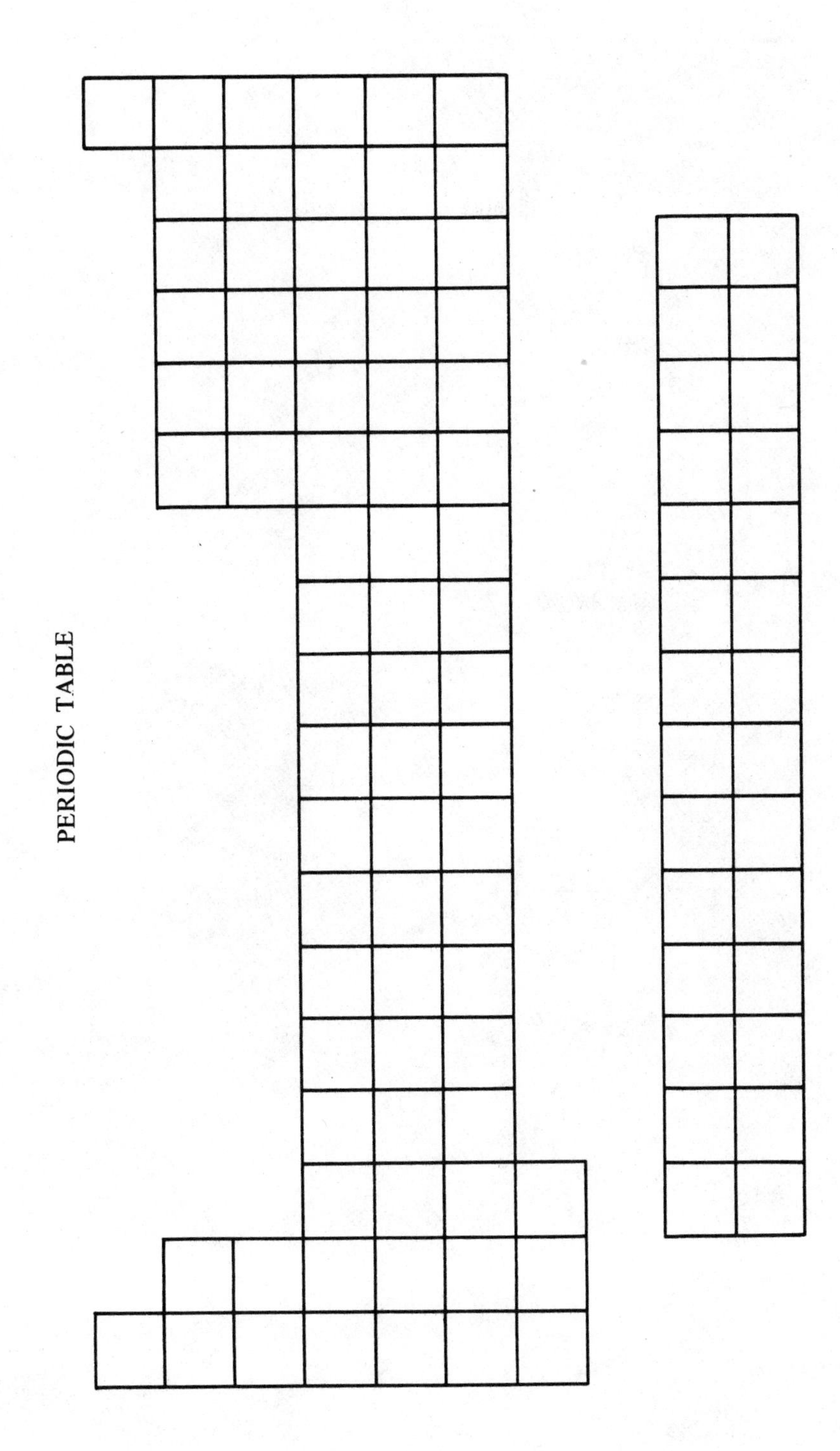
PERIODIC TABLE

HYDROGEN ATOM TARGET